Oxford Revision Guides

AS & A Level

PHYSICS

Through Diagrams

Stephen Pople
Carol Tear

OXFORD
UNIVERSITY PRESS

Great Clarendon Street, Oxford OX2 6DP

Oxford University Press is a department of the University of Oxford.
It furthers the University's objective of excellence in research, scholarship,
and education by publishing worldwide in

Oxford New York

Auckland Bangkok Buenos Aires Cape Town Chennai
Dar es Salaam Delhi Hong Kong Istanbul Karachi Kolkata
Kuala Lumpur Madrid Melbourne Mexico City Mumbai Nairobi
São Paulo Shanghai Taipei Tokyo Toronto

First published 2009

British Library Cataloguing in Publication Data

Data available

ISBN 978-0-19-918095-0

10 9 8

Typeseting, artwork and design by Steve Evans Design and Illustration

Printed in Great Britain by Bell and Bain Ltd, Glasgow

CONTENTS

How to use this book

- If you are studying for an AS or A level in physics, start here! (If you are not aiming for one of these qualifications, you can use this book as a general reference for physics up to advanced level: there is an index to help you find the topic(s) you require.)
- Obtain a copy of the specification you are going to be examined on. Specifications are available from the exam boards' websites: www.aqa.org.uk; www.edexcel.org.uk; www.ocr.org.uk.
- With the table below as a starting point, make your own summary of the content of the specification you will be following.
- Use the pathways on pages 6 and 7 to help match the material in this book with that required by your specification.
- Find out the requirements for any coursework and the dates of your exams and plan your revision accordingly. Page 8 has some helpful advice.
- Begin revising! The self-assessment questions on pages 145–153 will help you to check your progress.

Note:

- This book covers AS and A2 material for all the main specifications and therefore contains some sections that you will not require.
- The material in this book is not divided up into AS and A2 because the level required may vary from one specification to another.
- If your specification is not listed, most of the material you need will still be included in this book, but you will have to construct your own route through the book.

Specification structures

This table summarizes the five main AS and A level specifications. Satisfactory assessment in units 1–3 corresponds to an AS level pass. Satisfactory assessment in the AS units 1–3 and the A2 units 4–6 corresponds to an A level pass. In each column are listed the unit names and main subdivisions as given in the specification. The method of assessment in each unit is listed, together with the percentage of marks assigned to the entire AS or A level. Do check your specification for the latest information.

The Edexcel Physics specification can be taught traditionally, or using the Salters-Horners context-led approach. The Salters-Horners module titles are listed below, but you do not need to learn the contexts to answer the exam questions.

		AQA Physics A	AQA Physics B	Edexcel Physics
AS units	Unit 1	**Particles, quantum phenomena, and electricity** *1h15m written exam (short structured questions)* *AS 40% A 20%*	**Harmony and structure in the Universe** Module 1: The World of music Module 2: From quarks to quasars *1h15m written exam (very short questions & longer questions)* *AS 40% A 20%*	**Physics on the go** Mechanics, materials SH modules: Higher, faster, stronger Good enough to eat Spare part surgery *1h20m written exam (multiple-choice, short & long questions)* *AS 40% A 20%*
	Unit 2	**Mechanics, materials, and waves** *1h15m written exam (short structured questions)* *AS 40% A 20%*	**Physics keeps us going** Module 1: Moving people, people moving Module 2: Energy and the environment *1h15m written exam (very short questions & longer questions)* *AS 40% A 20%*	**Physics at work** Waves, electricity, light SH modules: Sound of music Technology in space Digging up the past *1h20m written exam (multiple-choice, short & long questions)* *AS 40% A 20%*
	Unit 3	**Investigative and practical skills in AS Physics** *Centre-marked practical skills assessment and investigative skills assignment OR Practical skills verification and externally marked practical assignment* *AS 20% A 10%*	**Investigative and practical skills in AS Physics** *Centre-marked practical skills assessment and investigative skills assignment OR Practical skills verification and externally marked practical assignment* *AS 20% A 10%*	**Exploring physics** *Internal assessment (an experiment based on a physics-based visit OR A case study of an application of physics)* *AS 20% A 10%*
A2 units	Unit 4	**Fields and further mechanics** *1h45m written exam (multiple-choice and structured questions)* *A 20%*	**Physics inside and out** Module 1: Experiences out of this world Module 2: What goes around comes around Module 3: Imaging the invisible *1h45m written exam (long questions)* *A 20%*	**Physics on the move** Further mechanics, fields, particles SH modules: Transport on track Medium is the message Probing the heart of matter *1h35m written exam (multiple-choice, short & long questions)* *A 20%*
	Unit 5	**One of Units 5A, 5B, 5C, 5D** *1h45m written exam (structured questions)* Section A: Nuclear and thermal physics Section B: one of: A Astrophysics B Medical physics C Applied physics D Turning points in physics *A 20% (Section A 10%, Section B 10%)*	**Energy under the microscope** Module 1: Matter under the microscope Module 2: Breaking matter down Module 3: Energy from the nucleus *1h45m written exam (long questions)* *A 20%*	**Physics from creation to collapse** Thermal energy, nuclear decay, oscillations, astrophysics, cosmology SH modules: Build or bust Reach for the stars *1h35m written exam (multiple-choice, short & long questions)* *A 20%*

		AQA Physics A	AQA Physics B	Edexcel Physics
A2 units	Unit 6	**Investigative and practical skills in A2 Physics** *Centre-marked practical skills assessment and investigative skills assessment OR Practical skills verification and externally marked practical assignment* *A 10%*	**Investigative and practical skills in A2 Physics** *Centre-marked practical skills assessment and investigative skills assessment OR Practical skills verification and externally marked practical assignment* *A 10%*	**Experimental physics** *2h Planning an experiment* *Carrying out either own or set experiment and analysing results* *A 10%*

		OCR Physics A	OCR Physics B (Advancing Physics)
AS units	Unit 1	**Mechanics** Motion, forces, work, and energy *1h written exam* *AS 30% A 15%*	**Physics in action** Communication Designer materials *1h written exam (short & structured questions)* *AS 30% A 15%*
	Unit 2	**Electrons, waves, and photons** Electric current, resistance, DC circuits, waves, quantum physics *1h45m written exam* *AS 50% A 25%*	**Understanding processes, experimentation, and data handling** Waves and quantum behaviour Space, time, and motion *1h45m written exam* *AS 50% A 25%*
	Unit 3	**Practical skills in physics 1** *Three externally set tasks, internally marked using set mark scheme* *AS 20% A 10%*	**Physics in practice** *Measurement task* *Presentation of researched topic* *AS 20% A 10%*
A2 units	Unit 4	**The newtonian world** Newton's laws and momentum, circular motion and oscillations, thermal physics *1h written exam (synoptic questions)* *A 15%*	**Rise and fall of the clockwork Universe** Models and rules Matter in extremes *1h5m written exam (synoptic short & structured questions)* *A 15%*
	Unit 5	**Field, particles, and frontiers of physics** EM fields, capacitors, nuclear physics, medical imaging, the Universe *1h45m (synoptic questions)* *A 50%*	**Fields and particle pictures** Fields Fundamental particles *2h written exam (synoptic short & structured questions)* *A 25%*
	Unit 6	**Practical skills in physics 2** *Three externally set tasks, internally marked using set mark scheme* *A 10%*	**Researching physics** Practical investigation *Research briefing* *A 10%*

What are...

...short-answer questions?
These questions will require just a few words or sentences as answers.

...structured questions?
This type of question is broken up into smaller parts. Some parts will ask you to define or show you understand a given term; explain a phenomenon or describe an experiment; plot sketch graphs or obtain information from given graphs; draw labelled diagrams or indicate particular features on a given diagram. Other parts will lead you to the solution of a complex problem by asking you to solve it in stages.

...comprehension questions?
In these questions you will be given a passage (short or extended) on a topic and then tested on your understanding of the topic and the scientific concepts in it.

...data-analysis questions?
In this type of question you will be given data in a variety of forms: graphs, tables, in text, as a list. You will then be asked to analyse the data to derive new results or information and may be asked to link the results with explanations of the scientific principles involved.

...synoptic questions?
When answering these you will have to apply physics principles or skills in contexts that are likely to be unfamiliar to you. Some questions will require you to show that you understand how different aspects of physics relate to one another or are used to explain different aspects of a particular application. Questions of this type will require you to draw on the knowledge, understanding, and skills developed during your study of the whole course. 20% of the A level marks are allocated to synoptic questions.

Pathways

The following pathways identify the main sections in the book that relate to the topics required by each specification.
Note:

- You will not necessarily need all the material that is given in any section.
- There may be material in other sections (e.g. applications) that you need to know.
- You should identify the relevant material by referring to the specification you are following.
- If this is your own copy of the book, highlight all the relevant topics throughout the book.

AQA Physics A

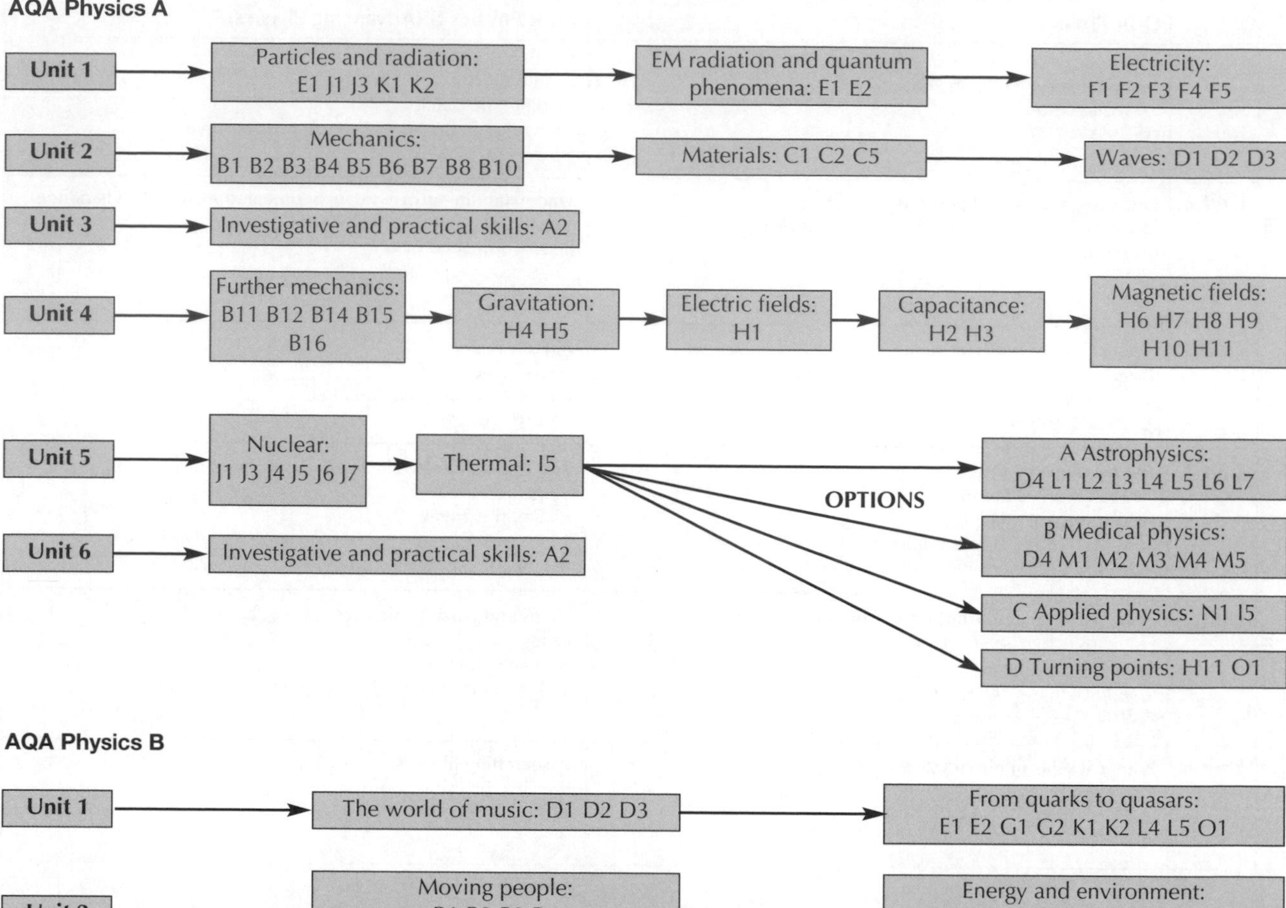

AQA Physics B

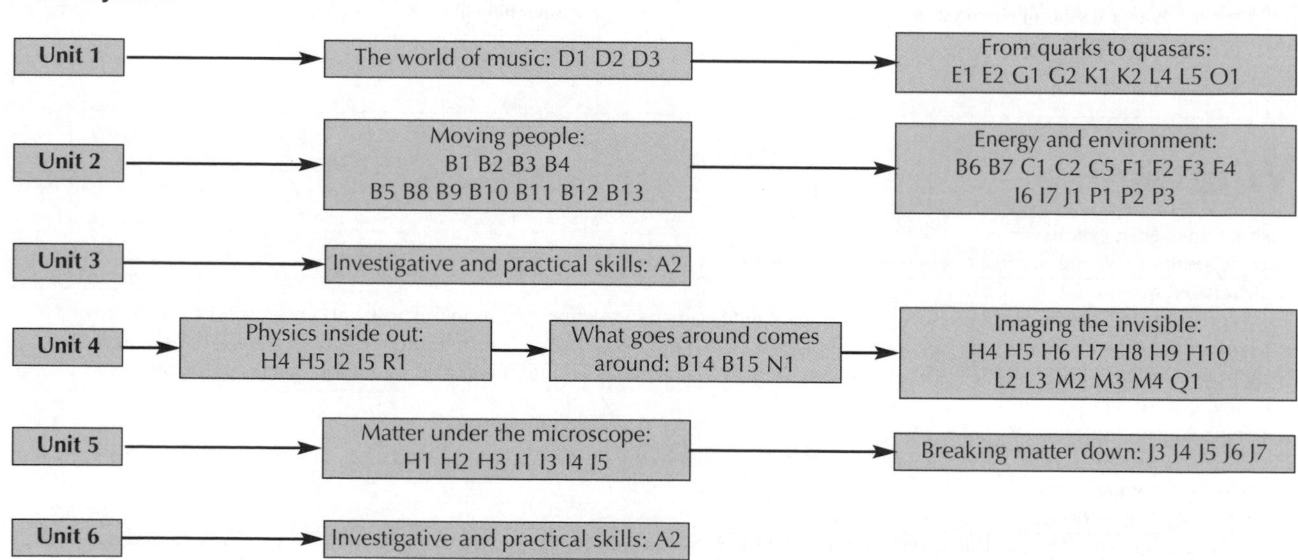

Edexcel Specification A

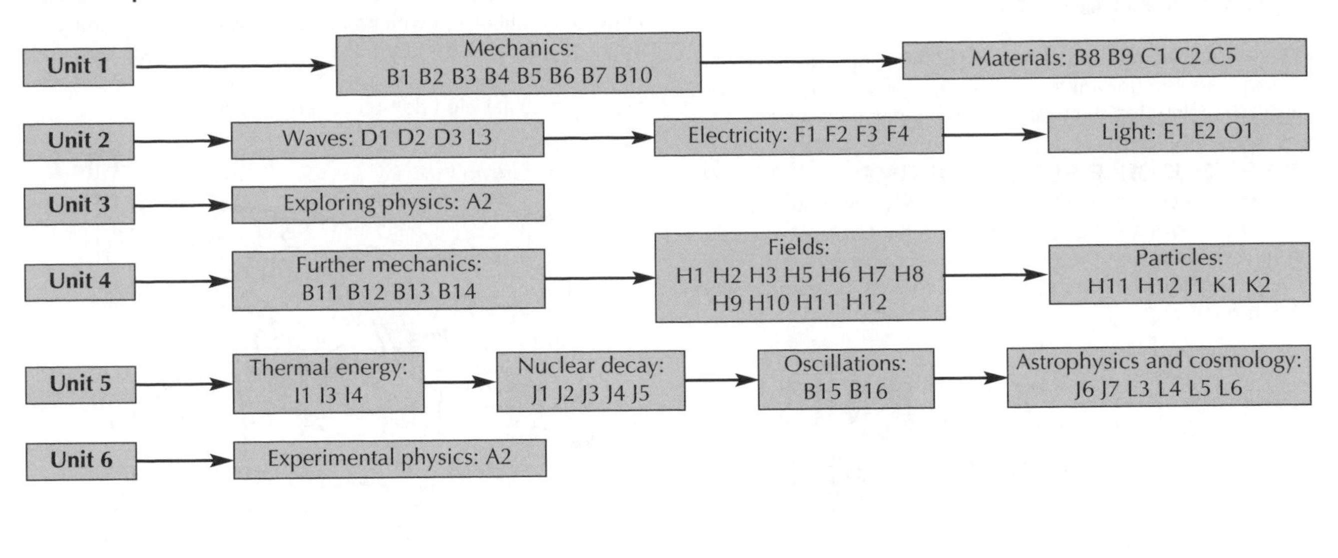

OCR Physics A

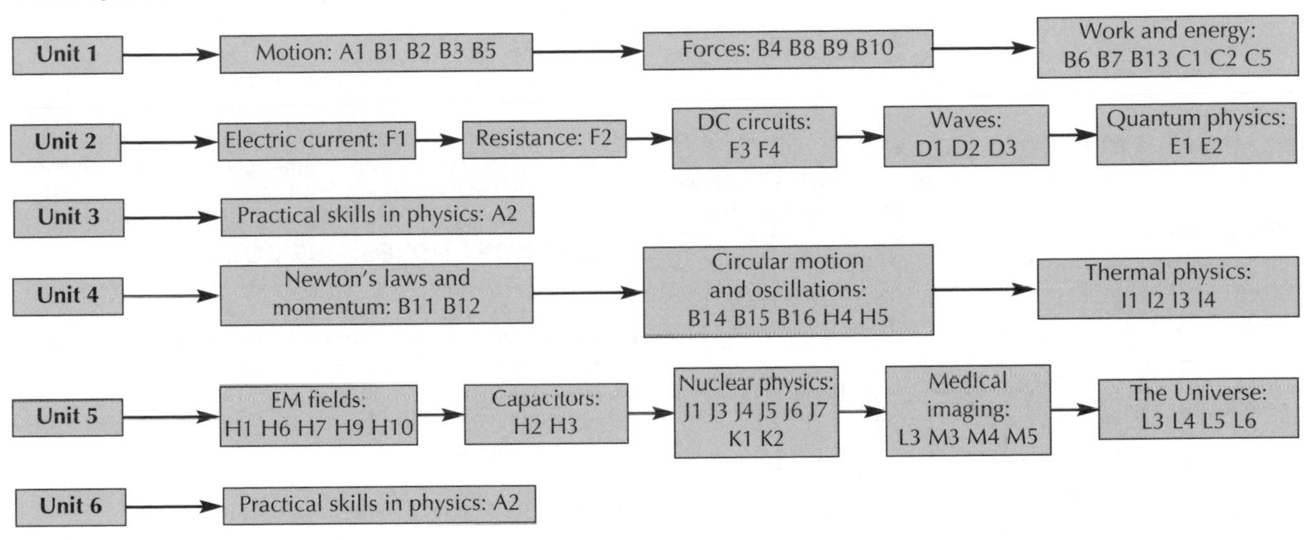

OCR Physics B

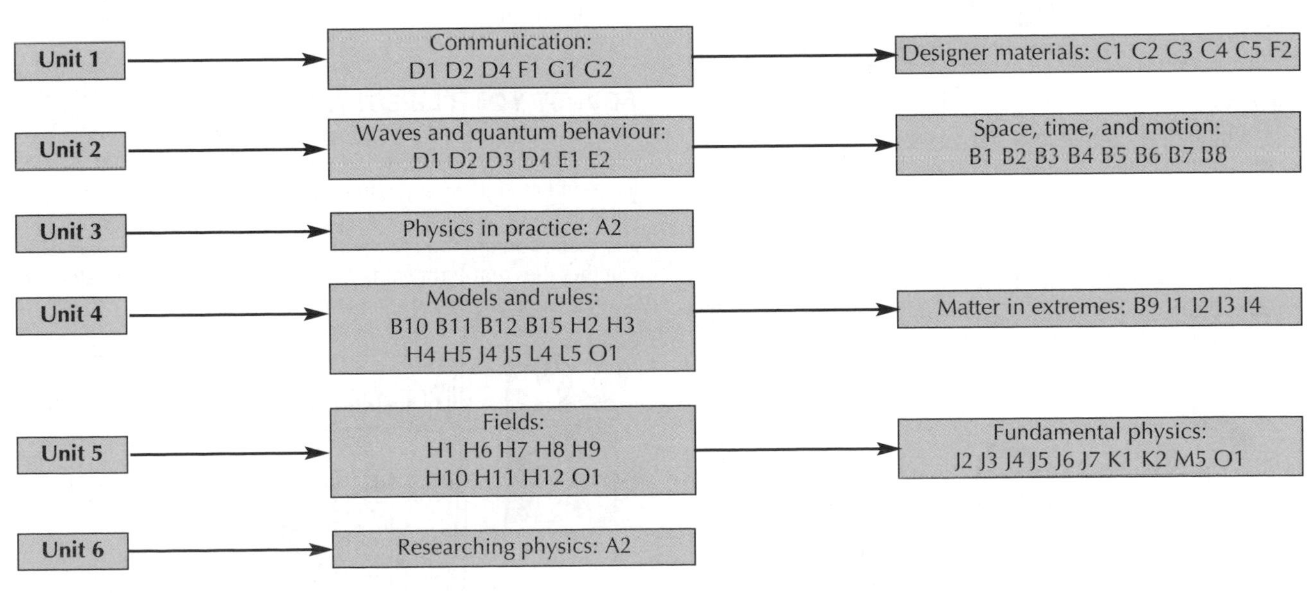

How to revise

There is no one method of revising which works for everyone. It is therefore important to discover the approach that suits you best. The following rules may serve as general guidelines.

GIVE YOURSELF PLENTY OF TIME

Leaving everything until the last minute reduces your chances of success. Work will become more stressful, which will reduce your concentration. There are very few people who can revise everything 'the night before' and still do well in an examination the next day.

PLAN YOUR REVISION TIMETABLE

You need to plan you revision timetable some weeks before the examination and make sure that your time is shared suitably between all your subjects.
Once you have done this, follow it – don't be side-tracked. Stick your timetable somewhere prominent where you will keep seeing it – or better still put several around your home!

RELAX

Concentrated revision is very hard work. It is as important to give yourself time to relax as it is to work. Build some leisure time into your revision timetable.

GIVE YOURSELF A BREAK

When you are working, work for about an hour and then take a short tea or coffee break for 15 to 20 minutes. Then go back to another productive revision period.

FIND A QUIET CORNER

Find the conditions in which you can revise most efficiently. Many people think they can revise in a noisy busy atmosphere – most cannot! Any distraction lowers concentration. Revising in front of a television doesn't generally work!

KEEP TRACK

Use checklists and the relevant examination board specification to keep track of your progress. The Pathways and Specification Outlines in the previous section will help. Mark off topics you have revised and feel confident with. Concentrate your revision on things you are less happy with.

MAKE SHORT NOTES, USE COLOURS

Revision is often more effective when you do something active rather than simply reading material. As you read through your notes and textbooks make brief notes on key ideas. If this book is your own property you could highlight the parts of pages that are relevant to the specification you are following.
Concentrate on understanding the ideas rather than just memorizing the facts.

PRACTISE ANSWERING QUESTIONS

As you finish each topic, try answering some questions. There are some in this book to help you (see pages 145–153). You should also use questions from past papers. At first you may need to refer to notes or textbooks. As you gain confidence you will be able to attempt questions unaided, just as you will in the exam.

ADJUST YOUR LIFESTYLE

Make sure that any paid employment and leisure activities allow you adequate time to revise. There is often a great temptation to increase the time spent in paid employment when it is available. This can interfere with a revision timetable and make you too tired to revise thoroughly. Consider carefully whether the short-term gains of paid employment are preferable to the long-term rewards of examination success.

Success in examinations

EXAMINATION TECHNIQUE

The following are some points to note when taking an examination.

- Read the question carefully. Make sure you understand exactly what is required.
- If you find that you are unable to do a part of a question, do not give up. The next part may be easier and may provide a clue to what you might have done in the part you found difficult.
- Note the number of marks per question as a guide to the depth of response needed (see below).
- Underline or note the key words that tell you what is required (see opposite).
- Underline or note data as you read the question.
- Structure your answers carefully.
- Show all steps in calculations. Include equations you use and show the substitution of data. Remember to work in SI units.
- Make sure your answers are to suitable significant figures (usually 2 or 3) and include a unit.
- Consider whether the magnitude of a numerical answer is reasonable for the context. If it is not, check your working.
- Draw diagrams and graphs carefully.
- Read data from graphs carefully; note scales and prefixes on axes.
- Keep your eye on the clock but don't panic.
- If you have time at the end, use it. Check that your descriptions and explanations make sense. Consider whether there is anything you could add to an explanation or description. Repeat calculations to ensure that you have not made a mistake.

DEPTH OF RESPONSE

Look at the **marks allocated to the question**.
This is usually a good guide to the depth of the answer required. It also gives you an idea how long to spend on the question. If there are 60 marks available in a 90 minute exam, your 1 mark should be earned in 1.5 minutes.

Explanations and descriptions

If a *4 mark* question requires an explanation or description, you will need to make *four* distinct relevant points.
You should note, however, that simply mentioning the four points will not necessarily earn full marks. The points need to be made in a coherent way that makes sense and fits the context and demands of the questions.

Calculations

In calculation questions marks will be awarded for method and the final answer.

In a *3 mark* calculation question you *may* obtain all three marks if the final answer is correct, even if you show no working.
However, you should *always show your working* because

- sometimes the working is a requirement for full marks
- if you make an error in the calculation you cannot gain any method marks unless you have shown your working.

In general in a *3 mark* calculation you earn

1 mark for quoting a relevant equation or using a suitable method

1 mark for correct substitution of data or some progress toward the final answer

1 mark for a correct final answer given to suitable significant figures with a correct unit.

Errors carried forward

If you make a mistake in a calculation and need to use this incorrect answer in a subsequent part of the question, you can still gain full marks. Do not give up if you think you have gone wrong. Press on using the data you have.

KEY WORDS

How you respond to a question can be helped by studying the following, which are the more common key words used in examination questions.

Name: The answer is usually a technical term consisting of one or two words.

List: You need to write down a number of points (often a single word) with no elaboration.

Define: The answer is a formal meaning of a particular term.

What is meant by...? This is often used instead of 'define'.

State: The answer is a concise word or phrase with no elaboration.

Describe: The answer is a description of an effect, experiment, or (e.g.) graph shape. No explanations are required.

Suggest: In your answer you will need to use your knowledge and understanding of topics in the specification to deduce or explain an effect that may be in a novel context. There may be no single correct answer to the question.

Calculate: A numerical answer is to be obtained, usually from data given in the question. Remember to give your answer to a suitable number of significant figures and give a unit.

Determine: Often used instead of 'calculate'. You may need to obtain data from graphs, tables, or measurements.

Explain: The answer will be extended prose. You will need to use your knowledge and understanding of scientific phenomena or theories to elaborate on a statement that has been made in the question or earlier in your answer. A question often asks you to 'state and explain...'.

Justify: Similar to 'explain'. You will have made a statement and now have to provide a reason for giving that statement.

Draw: Simply draw a diagram. If labelling or a scale drawing is needed, you will usually be asked for this, but it is sensible to provide labelling even if it is not asked for.

Sketch: This usually relates to a graph. You need to draw the general shape of the graph on labelled axes. You should include enough quantitative detail to show relevant intercepts and/or whether the graph is exponential or some inverse function, for example.

Plot: The answer will be an accurate plot of a graph on graph paper. Often it is followed by a question asking you to 'determine some quantity from the graph' or to 'explain its shape'.

Estimate: You may need to use your knowledge and/or your experience to deduce the magnitude of some quantities to arrive at the order of magnitude for some other quantity defined in the question.

Discuss: This will require an extended response in which you demonstrate your knowledge and understanding of a given topic.

Show that: You will have been given either a set of data and a final value (that may be approximate) or an algebraic equation. You need to show clearly all basic equations that you use and all the steps that lead to the final answer. You should give the answer to one greater significant figure than the value given in the question.

REVISION NOTE

In your revision remember to

- learn the formulae that are not on your formula sheet
- make sure that you know what is represented by all the symbols in equations on your formula sheet.

Practical assessment

Your practical skills will be assessed at both AS and A level. Make sure you know how your practical skills are going to be assessed.
You may be assessed by

- **Internal assessment** where the experiment(s) are set and marked by your school or college and a selection are marked by the exam board.
- **External assessment** where the experiment(s) are set and marked by the exam board.
- **Mixed internal/external** where the exam board sets a task to be marked by your school or college, for example.

PRACTISING THE SKILLS

Whichever assessment type is used, you need to learn and practise the skills during your course.

Specific skills

You will learn specific skills associated with particular topics as a natural part of your learning during the course. Make sure that you have hands-on experience of all the apparatus that is used. You need to have a good theoretical background of the topics on your course so that you can

- devise a sensible hypothesis
- identify all variables in an experiment
- control variables
- choose suitable magnitudes for variables
- select and use apparatus correctly and safely
- tackle analysis confidently
- make judgements about the outcome.

PRACTICAL EXAMINATION

The form of the examination varies from one examination board to another, so make sure you know what your board requires you to do. Questions generally fall into three types which fit broadly into the following categories:
You may be required to

- examine a novel situation, create a hypothesis, consider variables, and design an experiment to test the hypothesis
- examine a situation, analyse data that may be given to you, and evaluate the experiment that led to the data
- obtain and analyse data in an experiment which has been devised by the examination board.

In any experiment you may be required to determine uncertainties in raw data, derived data, and the final result.

Designing experiments and making hypotheses

Remember that you can only gain marks for what you write, so take nothing for granted. Be thorough. A description that is too long is better than one that leaves out important detail.

Remember to

- use your knowledge of AS and A level physics to support your reasoning
- give quantitative reasoning wherever possible
- draw clear labelled diagrams of apparatus
- provide full details of measurements made, equipment used, and experimental procedures
- be prepared to state the obvious.

A good test of a sufficiently detailed account is to ask yourself whether it would be possible to do the experiment you describe without needing any further infomation.

PRACTICAL SKILLS

There are four basic skill areas:

Planning
Implementing
Analysing
Evaluating

The same skills are assessed in both practical examinations and coursework.

GENERAL ASSESSMENT CRITERIA

You will be assessed on your ability to

- identify what is to be investigated
- devise a hypothesis or theory of the expected outcome
- devise a suitable experiment, use appropriate resources, and plan the procedure
- carry out the experiment or research
- describe precisely what you have done
- present your data or information in an appropriate way
- draw conclusions from your results or other data
- evaluate the uncertainties in your experiment
- evaluate the success or otherwise of the experiment and suggest how it might have been improved.

GENERAL SKILLS

The general skills you need to practise are

- the accurate reporting of experimental procedures
- presentation of data in tables (possibly using spreadsheets)
- graph drawing (possibly using IT software)
- analysis of graphical and other data
- critical evaluation of experiments

Carrying out experiments

When **making observations** and **tabulating data** remember to

- consider carefully the range and intervals at which you make your observations
- consider the accuracy to which it is reasonable to quote your observations (how many significant figures are reasonable)
- repeat all readings and remember to average
- be consistent when quoting data
- tabulate all data (including repeats and averages) remembering to give units for all columns
- make sure figures are not ambiguous.

When **deriving data** remember to

- work out an appropriate unit
- make sure that the precision is consistent with your raw data.

When **drawing graphs** remember to

- choose a suitable scale that uses the graph paper fully
- label the axes with quantity and unit
- mark plotted points carefully with a cross using a sharp pencil
- draw the best straight line or curve through the points so that the points are scattered evenly about the line.

When **analysing data** remember to

- use a large gradient triangle in graph analysis to improve accuracy
- set out your working so that it can be followed easily
- ensure that any quantitative result is quoted to an accuracy that is consisted with your data and analysis methods
- include a unit for any result you obtain.

Carrying out investigations

Record

- all your measurements
- any problems you have met
- details of your procedures
- any decisions you have made about apparatus or procedures including those considered and discarded
- relevant things you have read or thoughts you have about the problem.

Define the problem

Write down the aim of your experiment or investigation. Note the variables in the experiment. Define those that you will keep constant and those that will vary.

Suggest a hypothesis

You should be able to suggest the expected outcome of the investigation on the basis of your knowledge and understanding of science. Try to make this as quantitative as you can, justifying your suggestion with equations wherever possible.

Do rough trials

Before commencing the investigation in detail do some rough tests to help you decide on

- suitable apparatus
- suitable procedures
- the range and intervals at which you will take measurements
- consider carefully how you will conduct the experiment in a way that will ensure safety to persons and to equipment.

Remember to consider alternative apparatus and procedures and justify your final decision.

Carry out the experiment

Remember all the skills you have learnt during your course:

- note all readings that you make
- take repeats and average whenever possible
- use instruments that provide suitably accurate data
- consider the accuracy of the measurements you are making
- analyse data as you go along so that you can modify the approach or check doubtful data.

Presentation of data

Tabulate all your observations, remembering to

- include the quantity, any prefix, and the unit for the quantity at the head of each column
- include any derived quantities that are suggested by your hypothesis
- quote measurements and derived data to an accuracy/significant figures consistent with your measuring instruments and techniques, and be consistent
- make sure figures are not ambiguous.

Graph drawing

Remember to

- label your axes with quantity and unit
- use a scale that is easy to use and fills the graph paper effectively
- plot points clearly (you may wish to include 'error bars')
- draw the best line through your plotted points
- consider whether the gradient and area under your graph have significance.

Analysing data

This may include

- the calculation of a result
- drawing of a graph
- statistical analysis of data
- analysis of uncertainties in the original readings, derived quantities, and results.

Make sure that the stages in the processing of your data are clearly set out.

Evaluation of the investigation

The evaluation should include the following points:

- draw conclusions from the experiment
- identify any systematic errors in the experiment
- comment on your analysis of the uncertainties in the investigation
- review the strengths and weaknesses in the way the experiment was conducted
- suggest alternative approaches that might have improved the experiment in the light of experience.

Use of information technology (IT)

You may have used data capture techniques when making measurements or used IT in your analysis of data. In your analysis you should consider how well this has performed. You might include answers to the following questions.

- What advantages were gained by the use of IT?
- Did the data capture equipment perform better than you could have achieved by a non-IT approach?
- How well has the data analysis software performed in representing your data graphically, for example?

THE REPORT

Remember that your report will be read by an assessor who will not have watched you doing the experiment. For the most part the assessor will only know what you did by what you write, so do not leave out important information.

If you write a good report, it should be possible for the reader to repeat what you have done should they wish to check your work.

A **word-processed report** is worth considering. This makes the report much easier to revise if you discover some aspect you have omitted. It will also make it easier for the assessor to read.

Note.
The report may be used as portfolio evidence for assessment of Application of Number, Communication, and IT Key Skills.

Use subheadings

These help break up the report and make it more readable. As a guide, the subheadings could be the main sections of the investigation: aims, diagram of apparatus, procedure, etc.

Key Skills

What are Key Skills?

These are skills that are not specific to any subject but are general skills that enable you to operate competently and flexibly in your chosen career. Visit the Key Skills website (www.keyskillssupport.net) or phone the Key Skills help line to obtain full, up-to-date information.
While studying your AS or A level courses you should be able to gather evidence to demonstrate that you have achieved competence in the Key Skills areas of

- ***Communication***
- ***Application of Number***
- ***Information Technology.***

You may also be able to prove competence in three other key skills areas:

- ***Working with Others***
- ***Improving your own Learning***
- ***Problem Solving.***

Only the first three will be considered here and only an outline of what you must do is included. You should obtain details of what you need to know and be able to do. You should be able to obtain these from your examination centre.

Communication

You must be able to

- create opportunities for others to contribute to group discussions about complex subjects
- make a presentation using a range of techniques to engage the audience
- read and synthesize information from extended documents about a complex subject
- organize information coherently, selecting a form and style of writing appropriate to complex subject matter.

Application of Number

You must be able to plan and carry through a substantial and complex activity that requires you to

- plan your approach to obtaining and using information, choose appropriate methods for obtaining the results you need and justify your choice
- carry out multistage calculations including use of a large data set (over 50 items) and re-arrangement of formulae
- justify the choice of presentation methods and explain the results of your calculations.

Information Technology

You must be able to plan and carry through a substantial activity that requires you to

- plan and use different sources and appropriate techniques to search for and select information based on judgement of relevance and quality
- automated routines to enter and bring together information, and create and use appropriate methods to explore, develop, and exchange information
- develop the structure and content of your presentation, using others' views to guide refinements, and information from difference sources.

A complex subject is one in which there are a number of ideas, some of which may be abstract and very detailed. Lines of reasoning may not be immediately clear. There is a requirement to come to terms with specialized vocabulary.

A substantial activity is one that includes a number of related tasks. The result of one task will affect the carrying out of others. You will need to obtain and interpret information and use this to perform calculations and draw conclusions.

What standard should you aim for?

Key Skills are awarded at four levels (1–4). In your A level courses you will have opportunities to show that you have reached level 3, but you could produce evidence that demonstrates that you are competent at a higher level.
You may achieve a different level in each Key Skill area.

What do you have to do?

You need to show that you have the necessary underpinning knowledge in the Key Skills area and produce evidence that you are able to apply this in your day-to-day work.
You do this by producing a portfolio that contains

- evidence in the form of reports when it is possible to provide written evidence
- evidence in the form of assessments made by your teacher when evidence is gained by observation of your performance in the classroom or laboratory.

The evidence may come from only one subject that you are studying, but it is more likely that you will use evidence from all of your subjects.
It is up to you to produce the best evidence that you can.

The specifications you are working with in your AS or A level studies will include some ideas about the activities that form part of your course and can be used to provide this evidence. Some general ideas are summarized below, but refer to the specification for more detail.

Communication: in science you could achieve this by

- undertaking a long practical or research investigation on a complex topic (e.g. use of nuclear radiation in medicine)
- writing a report based on your experimentation or research using a variety of sources (books, magazines, CD-ROMs, Internet, newspapers)
- making a presentation to your fellow students
- using a presentation style that promotes discussion or criticism of your findings, enabling others to contribute to a discussion that you lead.

Application of Number: in science you could achieve this by

- undertaking a long investigation or research project that requires detailed planning of methodology
- considering alternative approaches to the work and justifying the chosen approach
- gathering sufficient data to enable analysis by statistical and graphical methods
- explaining why you analysed the data as you did
- drawing the conclusions reached as a result of your investigation.

Information Technology: in science you could achieve this by

- using CD-ROMs and the Internet to research a topic
- identifying those sources which are relevant
- identifying where there is contradictory information and identifying which is most probably correct
- using a word processor to present your report, drawing in relevant quotes from the information you have gathered
- using a spreadsheet to analyse data that you have collected
- using data capture techniques to gather information and mathematics software to analyse the data.

Answering the question

This section contains some examples of types of questions with model answers showing how the marks are obtained. You may like to try the questions and then compare your answers with the model answers given.

MARKS FOR QUALITY OF WRITTEN COMMUNICATION

In questions that require long descriptive answers or explanations, marks may be reserved for the quality of language used in your answers.

2 marks if your answer
- uses scientific terms correctly
- is written fluently and/or is well argued
- contains only a few spelling or grammatical errors.

1 mark if your answer
- generally uses scientific terms correctly
- generally makes sense but lacks coherence
- contains poor spelling and grammar.

An answer that is scientifically inaccurate, is disjointed, and contains many spelling and grammatical errors loses both these marks.

The message is:
do not let your communication skills let you down.

ALWAYS SHOW YOUR WORKING

In calculation questions one examination board might expect to see the working for all marks to be gained. Another might sometimes give both marks if you give the correct final answer. It is wise always to show your working. If you make a mistake in processing the data you could still gain the earlier marks for the method you use.

Question 1
Description and explanation question

(a) Describe the nuclear model of an atom that was proposed by Rutherford following observations made in Geiger and Marsden's alpha-particle scattering experiment. *(4 marks)*

(b) Explain why when gold foil is bombarded by alpha particles

(i) some of the alpha particles are deviated through large angles that are greater than 90°; *(3 marks)*

(ii) most of the alpha particles pass through without deviation and lose little energy while passing through the foil. *(2 marks)*

Answer

(a) The atom consists of a small nucleus (✓) which contains most of the mass (✓) of the atom. The nucleus is positive (✓). Electrons orbit the nucleus (✓).

(b) **(i)** A few alpha particles pass close to a nucleus (✓). There is a repelling force between the alpha particle and the gold nucleus because they are both positively charged (✓). This causes deflection of the alpha particle. Because the alpha particle is much less massive than the gold nucleus it may deviate through a large angle (✓).

(ii) Few alpha particles collide with a nucleus since most of matter is empty space occupied only by electrons (✓). The alpha particles deviate only a little and lose very little energy because an electron has a very small mass compared to that of an alpha particle (✓).

Note: In explanations or descriptive questions there are often alternative relevant statements that would earn marks. For example in part **(a)** you could earn credit for stating that electrons have small mass or negative charge.

Question 2
Graph interpretation and graph sketching

The diagram shows how the pressure p varies with the volume V for a fixed mass of gas.

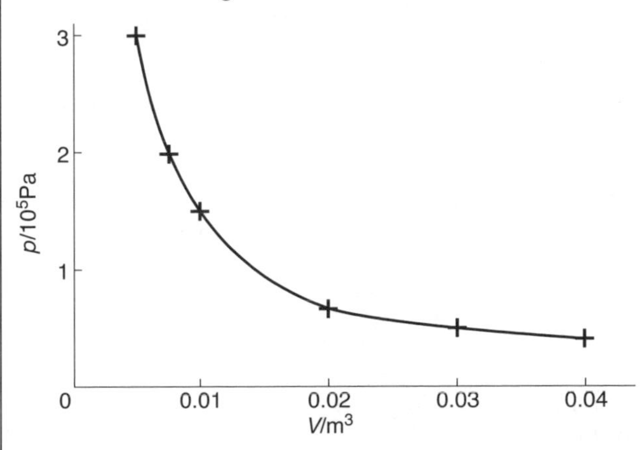

(a) Use data from the graph to show that the changes take place at constant temperature. *(3 marks)*

(b) Sketch a graph to show how the pressure varies with $1/V$ for this gas. *(2 marks)*

Answer

(a) For a change at constant temperature, pV = constant (✓). Use co ordinates from three points A, B, and C on the graph (✓) (NB using only two would lose this mark).
e.g. units (m^3,10^5 Pa). A (0.005, 3) B (0.01,1.5) C (0.03, 0.5).
Product in each case is $0.015 \times 10^5 m^3 Pa$.
The product pV is constant within limits of experimental uncertainties, so the changes take place at constant temperature (✓).

(b) Straight line through the origin (✓).
pV for the line is consistent with data in given graph (✓).

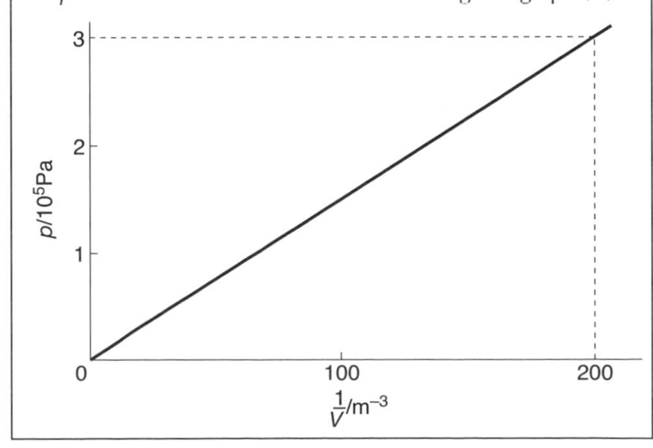

Question 3
Calculation question
The supply in the following circuit has an EMF of 12.0 V and negligible internal resistance.

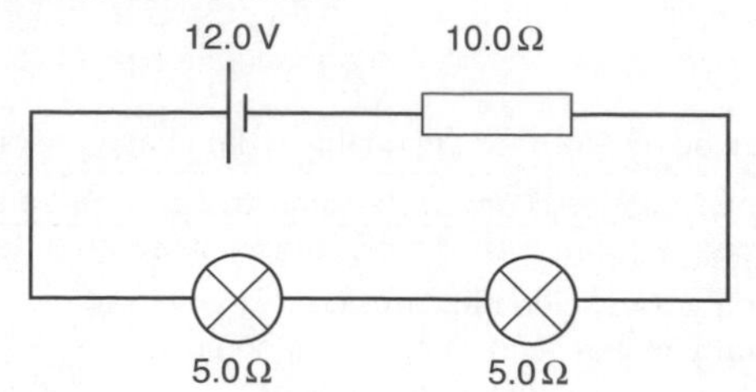

(a) Calculate
 (i) the current through each lamp; *(2 marks)*
 (ii) the power dissipated in each lamp; *(2 marks)*
 (iii) the potential difference across the 10.0 Ω resistor. *(1 mark)*

(b) A student wants to produce the same potential difference across the 10.0 Ω resistor using two similar resistors in parallel.
 (i) Sketch the circuit the student uses. *(1 mark)*
 (ii) Determine the value of each of the series resistors used. Show your reasoning. *(3 marks)*

Answer

(a) (i) Current in circuit = EMF/total resistance (✓)
=12.0/20.0
Current in circuit = 0.60 A (✓)

(ii) Power = I^2R (✓)
= $0.60^2 \times 5.0$
Power = 1.8 W (✓)

(iii) PD = IR = 0.60 × 10.0 = 6.0 V (✓)

(b) (i)

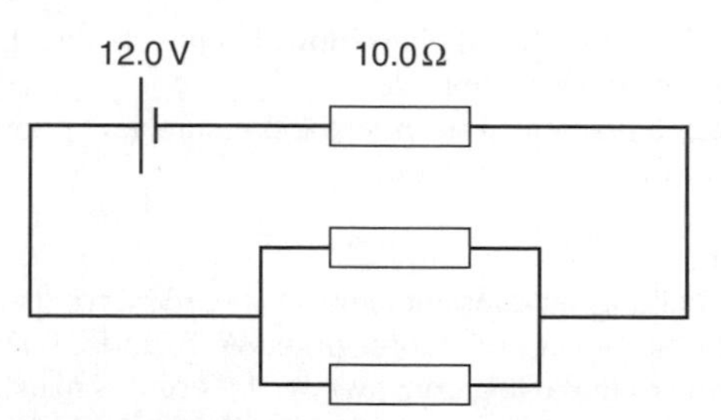

Correct circuit as above. (✓)

(ii) Parallel combination must be 10.0 Ω (✓)
Two similar parallel resistors have total resistance equal to half that of one resistor. (✓)
(*or* $\frac{1}{10} = \frac{1}{R} + \frac{1}{R}$)
Each resistor = 20 Ω (✓)

Question 4
Experiment description
The fundamental frequency f of a stretched string is given by the equation $f = \frac{1}{2l} + \frac{T}{\mu}$, where T is the tension and μ is the mass per unit length of the string.

(a) Sketch the apparatus you would use to test the relationship between f and T. *(2 marks)*
(b) State the quantities that are kept constant in the experiment. *(2 marks)*
(c) Describe how you obtain data using the apparatus you have drawn and how you would use the data to test the relationship. *(7 marks)*

Answer

(a)

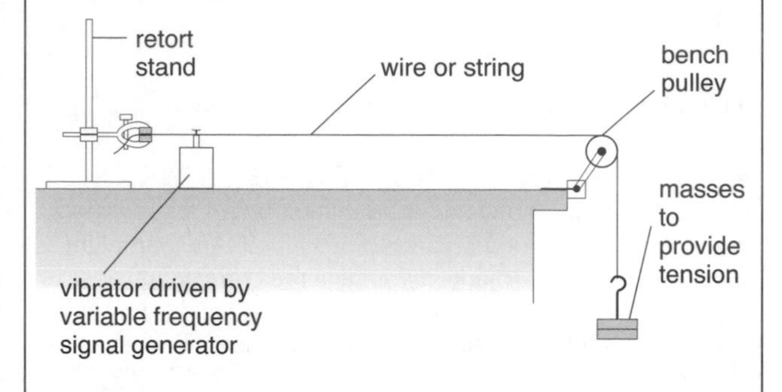

Means of determining frequency. (✓)
Sensible arrangement with means of changing tension. (✓)

(b) The constant quantities are:
- The mass per unit length of the wire. The material and the diameter must not be changed. (✓)
- The length of the wire used. (✓)

(c) A suitable tension is produced by adding masses at the end of the wire. The tension is noted (✓). When the mass used to tension the wire is m the tension is mg (✓). The oscillator frequency drives the vibrator which causes the wire to vibrate (✓). The oscillator frequency is adjusted until the wire vibrates at its fundamental frequency (i.e. a single loop is observed) (✓). The output frequency of the oscillator is noted (✓). The tension is changed and the new frequency at which the wire vibrates with one loop is determined (✓). A graph is plotted of frequency f against the square root of the tension, $\sqrt{T}$ (✓). If $f \propto \sqrt{T}$ the graph should be a straight line through the origin (✓).

Question 5
'Show that' question
A length of wire has diameter 0.5 mm and length 50 cm. The resistance is 2.8 Ω. Show that the resistivity of the wire is about 1×10^{-6} Ωm.

Answer

$\rho = RA/l$ (✓) $A = \pi (0.25 \times 10^{-3})^2$ (✓)

$\rho = 2.8 \times 1.96 \times 10^{-7}/0.5$ Ωm

$\rho = 1.1 \times 10^{-6}$ Ωm (✓)

How Science Works

There are 12 aspects to 'How science works' that are included in all the AS and A level specifications. Some of the specifications have incorporated these ideas into different modules, and some have rewritten them so that the wording is different.
The 12 aspects are listed here with the original wording, together with some guidance on interpretation, and the type of questions that could be asked.

Use theories, models, and ideas to develop and modify scientific explanations

There are many historical examples of scientists making some observations and then using creative thinking and imagination to interpret the data and develop an explanation. The first step is to come up with an **idea** – an initial thought about the reasons for the observations. This is then extended and worked into a model, maybe combining several ideas.

One definition of a **model** is:

A representation of a system that allows for investigation of the properties of the system and, in some cases, prediction of future outcomes.

The model is then tested and, if it works, can be set out as a **theory**.

One definition of a scientific theory is:

A set of statements or principles that explain observations, especially a set that has been repeatedly tested or is widely accepted and can be used to make predictions about natural phenomena.

You may be asked to give an example. Here are some:

Galileo Galilei timed objects rolling down an inclined plane and concluded that falling objects accelerate. Isaac Newton suggested a model for gravity and showed that freely-falling objects have the same acceleration. He developed the theory of gravity (see B5 and H3).

Scientists often use microscopic models, such as that of particle behaviour, to explain macroscopic behaviour. For example, the kinetic theory of gases explains the gas laws (see I4).

Other examples of using models to develop theories include:

Energy transfers for a rollercoaster (B7)
The electron and quantization of charge (E2, O1)
Young's slits (D3)
The photoelectric effect (E1)
Newton's laws (B9, B10)
The Rutherford model of the atom (J1)
Hubble's law (L5)

At this stage in your study of science you are unlikely to be thinking up new theories, but you may be doing experiments to see if your observations fit with a model. For example, plotting the square of the period, T, of a simple pendulum against its length, l, to see if this gives a straight line through the origin. If it does, it shows that T2 ∝ l, which confirms that the oscillations are an example of simple harmonic motion (see B15).

Use knowledge and understanding to pose scientific questions, define scientific problems, present scientific arguments and scientific ideas

As part of your course you use scientific theories to answer scientific questions or address scientific problems. In addition, you are expected to identify scientific questions or problems (within a given context). You may be presented with a hypothesis (an untested theory based on observations) or be asked to suggest one. The hypothesis needs to be tested by experiment, and if a reliable experiment does not support a hypothesis it must be changed.

When presenting arguments and ideas you should be able to distinguish between questions that science can address, and those that science cannot address. For example, whether a view is beautiful is not a question science can answer.

A historical example is the photoelectric effect.

The question was 'For a metal that shows the photoelectric effect, why is there a threshold frequency? (Why does high intensity red light cause no emission of photoelectrons, but low intensity UV radiation does cause emission?)' Albert Einstein suggested a hypothesis based on Max Planck's ideas: that the radiation was quantized and arrived in packets called photons, with energy $E = hf$.

Communicate information and ideas in appropriate ways using appropriate terminology

Using the correct scientific terminology avoids confusion. You should be able to write explanations using correct scientific terms, and support your arguments with equations, diagrams and clear sketch graphs. This applies to exam papers, practical work, and investigations.

Evaluate methodology, evidence and data, and resolve conflicting evidence

Communicate information and ideas in appropriate ways using appropriate terminology

Using the correct scientific terminology avoids confusion. You should be able to write explanations using correct scientific terms, and support your arguments with equations, diagrams and clear sketch graphs. This applies to exam papers, practical work, and investigations.

Use appropriate methodology, including ICT, to answer scientific questions and solve scientific problems

This includes how you conduct experimental work (see pages 10 and 11), for example:

- planning, or following a plan, of an investigation
- identifying the dependent and independent and control variables
- selecting appropriate apparatus and methods (including ICT) to carry out reliable experiments
- choosing instruments with appropriate sensitivity and precision (see below)
- justifying the methods used during experiments (including the use of ICT) to collect valid and reliable data and produce scientific theories
- using ICT (spreadsheets, for example) to develop scientific models or plot graphs, and dataloggers to monitor physical changes.

Sensitivity and precision

Take the example of measuring mass. The more ***sensitive*** a balance is, the smaller variation in mass the balance can detect and measure. A mass smaller than the sensitivity of a balance is not detectable using the balance.

If the mass of an object is measured many times, the ***precision*** is indicated by the spread of the results. If the measurements are all very close, the precision of the instrument is greater.

Accuracy and precision

A measurment with great precision is not the same as one with great accuracy, as illustrated by the diagram:

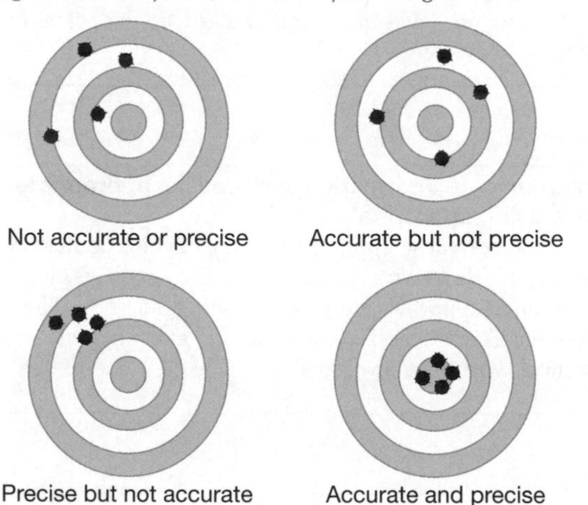

Analyse and interpret data to provide evidence, recognizing correlations and causal relationships

This refers to the methods you use in your experimental work, and what you do with the data you collect (see pages 10 and 11).

You will be expected to:

- record data in tables, and sometimes use equations to calculate values that you add to the table (for example, if you measure the period of a pendulum you might add T^2 to your table)
- plot and use graphs to establish or verify relationships between variables
- calculate the gradient and find the intercepts of straight-line graphs
- analyse data, including graphs, to identify patterns and relationships (correlation and cause, for example).

Correlation and cause

Analyse graphs of datasets that show different correlations.

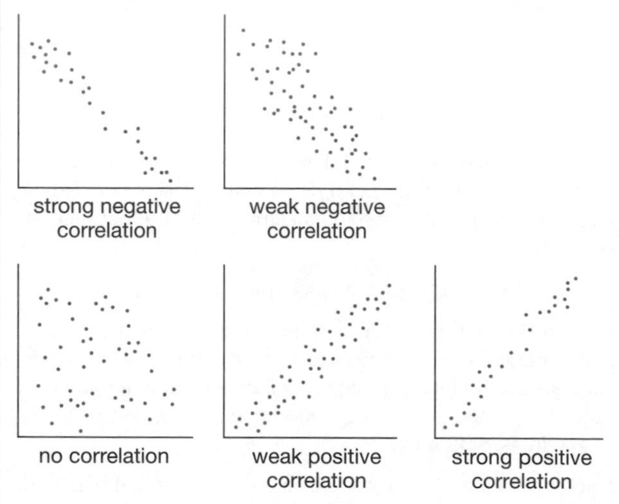

Remember that a correlation does not necessarily show that one thing causes the other. Children with larger feet are better at spelling (but this is because older children have larger feet and are better at spelling then younger children, not because larger feet *cause* better spelling).

Some of the more complex analysis included here is the use of log graphs to analyse nuclear decay or the discharge of a capacitor (see H3 and J4).

Carry out experimental and investigative activities, including appropriate risk management, in a range of contexts

See pages 10 and 11 and A2.
You should be able to show that you can:

- follow experimental procedure in a sensible order
- use appropriate apparatus and methods to make accurate and reliable measurements (see above)
- identify and minimize significant sources of experimental error (see A2)
- identify and take account of risks in carrying out practical work
- produce a risk assessment before carrying out a range of practical work.

An example would be to recognize that you should use eye protection when stretching wires or springs, and ensure that heavy weights cannot cause damage to people or objects if they fall.

Evaluate methodology, evidence and data, and resolve conflicting evidence

(See A2)
You should be able to:

- recognize, and distinguish between, systematic and random errors (see A2)
- estimate the errors in measurements
- use data, graphs, and other experimental evidence to draw conclusions
- use the (estimated) most significant error to assess the reliability of your conclusion
- evaluate the validity of conclusions in the light of the experimental methods used
- recognize conflicting evidence.

These skills can be applied to any experiments, such as using an airtrack to do experiments involving the acceleration and the change of momentum in collisions of airtrack gliders.

Appreciate the tentative nature of scientific knowledge

In everyday speech, 'tentative' usually means hesitant or unsure, but it also means 'not fully worked out', or 'a work in progress'. This is the nature of scientific knowledge.

Once scientists find a theory that works well, they accept and use it for as long as it works, but they recognize and accept that if new observations in the future conflict with the theory, and a better explanation is offered, then the accepted theory will change. This is unlike some other types of knowledge or belief.

A good example is our model of the atom (see J1), which has evolved over many years with contributions from many scientists. During experiments, when events occur that cannot be explained using current scientific theories (such as the alpha particle back-scattering observed by Hans Geiger and Ernest Marsden), scientists modify existing theories, or produce new theories to explain their findings (such as Ernest Rutherford's nuclear model of the atom).

Another example is the way in which Newton's laws of motion, which work perfectly well for everyday speeds, do not correctly describe behaviour at speeds approaching light speed.

The modification of our theory of motion to take into account Einstein's theory of relativity was necessary (see H12, O1).

Einstein's theory was originally theoretical, but it predicted distortion of starlight during an eclipse, which was observed in 1919. In 1971 accurate atomic clocks were available that were sensitive enough to detect the difference in the time taken by clocks flown in different directions around the world for 3 days. When you use a satellite navigation system, the processing takes into account relativity in determining your position.

Other examples of the development of theories include:

- theories of motion – Galileo and Aristotle (B5)
- the photoelectric effect (E1)
- the binding energy of the nucleus (J6)
- Olber's paradox and other cosmological questions (e.g. What is dark matter?) (L5)
- the search for the Higgs boson – a theoretical particle the existence of which may explain why particles have mass (K2)
- the prediction of antiparticles (see K1) in 1930 before the discovery of the positron in 1932.

Consider applications and implications of science and appreciate their associated benefits and risks

You may be asked to discuss the risk associated with an activity from almost any physics topic, in terms of the actual level of the risk and its potential consequences. People's perception of risk depends on factors such as how familiar the risk is. For example, people overestimate the risk of an aircraft crash, and underestimate the risk of a car crash.

There are some physics topics with more obvious benefits and risks than others.

Fossil fuels, electricity generation, and global warming
Since the Industrial Revolution, the application of physics has included many activities that involve burning fossil fuels, which release carbon dioxide. Most scientists now think that global warming is at least partly due to human activities (see P1). You should be aware of the impact this has had on the environment and how scientists are using their current findings to inform decision-makers of the consequence of global warming and advise them how to minimize its effects (see P2).

Medical treatment and imaging
(See M3, M4, M5)
Some medical imaging techniques, for example X-rays, positron emission tomography (PET) scans, and the use of radioactive tracers, involve risks to staff and patients from ionizing radiation. Magnetic resonance imaging (MRI) scanners use very large magnetic fields from superconducting magnets, and there is a risk because any metal will be strongly attracted to these magnets. The benefits to the patient may be a diagnosis without surgery.

Cancer treatment using radioisotopes also involves a risk from ionizing radiation, but the benefit of a cure outweighs the risk.

Using nuclear reactors to provide electrical power
The benefit of nuclear reactors is a large amount of electrical energy for a small amount of nuclear fuel. The risks are associated with ionizing radiation. There is concern about contamination of the environment during operation, and also from the nuclear waste, which has long half-lives (See J7, M5.)

Ultraviolet radiation
Radiation reaching us from the Sun includes UVA, which has an aging effect, and UVB, which can damage the cornea of the eye and cause skin cancer. Sunlight is needed by the body to produce vitamin D, which is required for bone growth in children and has a role in reducing the risk of other cancers. Ultraviolet radiation with shorter wavelengths, UVC, is more dangerous. Sunscreen filters out the ultraviolet radiation so that it does not reach the skin.

Other topics include:

- car safety (see B13), including the global positioning system (GPS), which has benefits we use everyday in satellite navigation systems, but the system can also be used to target air strikes accurately
- material properties (C2)
- geostationary satellites (H5)
- resonance (B16)
- mass spectrometry (H12)
- use of radioactive isotopes (J5)
- capacitors used for flash photos, lasers for fusion research, and back-up power supplies for computers (H2).

Consider ethical issues in the treatment of humans, other organisms, and the environment

You should be able to identify ethical issues arising from the application of science as it impacts on humans and the environment, and discuss scientific solutions from a range of ethical viewpoints.

Scientific research is funded by society, either through public funding or through private companies that obtain their income from commercial activities. Scientists have a duty to consider ethical issues associated with their findings. They set up groups to decide what should be permitted, and also contribute to groups set up by society to make decisions about what should be permitted.

Individual scientists have ethical codes that are often based on humanistic, moral, and religious beliefs.

Science has provided solutions to problems, but it is up to society as a whole (including scientists) to judge whether the solution is acceptable in view of the moral issues that result. Issues such as effects on the planet, and the economic and physical well-being of the living things on it, should be considered. Secure transmission of data is important if people are to be confident that personal data cannot be intercepted in transmission.

When a country is at war there may be difficult decisions for scientists to make. In the Second World War, scientists on both sides were in a race to build the first atom bomb.

Music can now be stored and reproduced to a high standard, but as a result infringement of copyright has become simple. Steps have been taken to make it more difficult to download music illegally – an example of science and technology providing solutions to the problems it creates.

Appreciate the role of the scientific community in validating new knowledge and ensuring integrity

It is important that new data, and new interpretations of data, should be critically evaluated. This is true whether they support established scientific theories or propose new theories.

Scientists communicate their findings to other scientists through journals and conferences. By sharing the findings of their research, scientists provide the scientific community with opportunities to replicate and further test their work. This can result in either confirming new explanations or refuting them.

Peer review

Some scientific journals state in the journal (and on their websites) that they are peer reviewed.

This means that the papers submitted by scientists for publication are sent for peer review before being accepted for publication. A peer is 'a person who is of equal standing with another in a group.' In this case, it is another scientist, or scientists, working in the same, or similar, field of research. Other scientists know that everything in the journal has been considered by another qualified, independent, scientist.

Note that some scientific magazines are not peer reviewed, but may use peer-reviewed articles as a source of information, as well as accepting other articles. A scientist reporting new research would usually publish in a peer-reviewed journal first.

Funding

The interests of the organizations that fund scientific research can influence the direction of that research. In some cases, the validity of the resulting claims may also be influenced.

The UK Government's leading funding agency for research and training in engineering and the physical sciences is the Engineering and Physical Science Research Council (EPSRC). Scientists submit proposals detailing the research they want to do and the equipment and staff they need.

The proposals are peer reviewed, and then all the proposals and the reviewer's comments are considered by a committee of scientists.

Almost all scientists work with the common aim of progressing scientific knowledge and understanding in a valid way and believe that accurate reporting of findings should take precedence over recognition of success of an individual. However, a disadvantage of the system could be that less ethical reviewers have the opportunity to benefit from other scientists' ideas and results, and to prevent publication and funding of their work. To prevent this, the reviewers' comments are sent to the author/researcher for comment. The system could work to exclude scientists with unusual research ideas or theories.

Cold fusion

In 1989, Martin Fleischmann and Stanley Pons reported a nuclear fusion reaction that took place during the electrolysis of heavy water (water containing the hydrogen isotope deuterium, ${}^{2}_{1}H$) using palladium electrodes. Heat was produced, which they claimed was due to nuclear fusion of deuterium. There was great interest all over the world because of the possibility of a cheap and abundant source of power. (It was called 'cold fusion' because nuclear fusion research concentrates on producing the high temperatures necessary to bring nuclei close enough to fuse.) The discovery was rejected after other scientists were unable to reproduce the discovery.

Examples of different parts of the scientific community working together include the following.

- The experimental discovery of electron diffraction confirmed the dual nature of matter particles, first put forward by de Broglie as a hypothesis several years earlier (see E2).
- In the search for a unifying theory, scientists make new discoveries based on theoretical predications, and continue to work to confirm the discovery of others, such as the Higgs boson (see K2).

Appreciate the ways in which society uses science to inform decision-making

Science influences decisions on an individual, local, national, and international level.

Scientific findings lead to new technologies, which enable advances to be made that have potential benefit for humans. However, these have to be balanced against the risks.

In practice, the scientific evidence available to decision-makers may be incomplete – scientific evidence should be considered as a whole. Decision-makers, who include government-appointed science advisers, are influenced by many things. These include their prior beliefs, their vested interests, special interest groups, public opinion, and the media, as well as by expert scientific evidence. The media and pressure groups often select parts of scientific evidence that support a particular viewpoint. This can influence public opinion, which in turn may influence decision-makers. Consequently, decision-makers may make socially and politically acceptable decisions based on incomplete evidence. The following are examples of this.

Electric cars may replace petrol vehicles if batteries are developed that give a greater range than those at present. Until then, car buyers are unlikely to be persuaded to buy electric cars.

Satellite tracking for purposes such as road pricing may be implemented without adequate trials because of pressure group influence.

The improved communication that digital electronics bring to society means that people can find out more easily what is happening and give their views. (For example, there are many petitions on the Downing Street website that you can sign online.) The range of information made available to decision-makers in industry, services, and government has increased now that information can be processed and presented using computers.

The expense of space travel is one area in which people have strong views. Some people regard it as a waste of money that could be used for building hospitals, for example. Others see the interest generated from space travel as beneficial and the technological spin-offs as worthwhile. Space travel contributes to global warming, but also to a greater understanding of climate on Earth and other planets.

Nuclear power, and the safe disposal of nuclear waste (see J6, J7, M5).

A1 Units and dimensions

Physical quantity

Say a plank is 2 metres long. This measurement is called a ***physical quantity***. In this case, it is a length. It is made up of two parts:

2 m — magnitude (number) and unit

Note:
- '2 m' really means '2 × metre', just as, in algebra, $2y$ means '$2 \times y$'.

SI base units

Scientific measurements are made using SI units (standing for Système International d'Unités). The system starts with a series of ***base units***, the main ones being shown in the table above right. Other units are derived from these.

SI base units have been carefully defined so that they can be accurately reproduced using equipment available to national laboratories throughout the world.

Physical quantity	Unit	
	Name	Symbol
length	metre	m
mass	kilogram	kg
time	second	s
current	ampere	A
temperature	kelvin	K
amount*	mole	mol

* In science, 'amount' is a measurement based on the number of particles (atoms, ions or molecules) present. One mole is 6.02×10^{23} particles, a number which gives a simple link with the total mass. For example, 1 mole (6.02×10^{23} atoms) of carbon-12 has a mass of 12 grams. 6.02×10^{23} is called the ***Avogadro constant***.

SI derived units

There is no SI base unit for speed. However, speed is defined by an equation (see B1). If an object travels 12 m in 3 s,

$$\text{speed} = \frac{\text{distance travelled}}{\text{time taken}} = \frac{12\ \text{m}}{3\ \text{s}} = 4\,\frac{\text{m}}{\text{s}}$$

The units m and s have been included in the working above and treated like any other numbers or algebraic quantities. To save space, the final answer can be written as 4 m/s, or 4 m s^{-1}. (Remember, in maths, $1/x = x^{-1}$ etc.)

The unit m s^{-1} is an example of a ***derived SI unit***. It comes from a defining equation. There are other examples below. Some derived units are based on other derived units. And some derived units have special names. For example, 1 joule per second (J s^{-1}) is called 1 watt (W).

Prefixes

Prefixes can be added to SI base and derived units to make larger or smaller units.

Prefix	Symbol	Value	Prefix	Symbol	Value
pico	p	10^{-12}	kilo	k	10^{3}
nano	n	10^{-9}	mega	M	10^{6}
micro	μ	10^{-6}	giga	G	10^{9}
milli	m	10^{-3}	tera	T	10^{12}

For example,

1 mm = 10^{-3} m 1 km = 10^{3} m

Note:
- 1 gram (10^{-3} kg) is written '1 g' and not '1 mkg'.

Physical quantity	Defining equation (simplified)	Derived unit	Special symbol (and name)
speed	distance/time	m s^{-1}	–
acceleration	speed/time	m s^{-2}	–
force	mass × acceleration	kg m s^{-2}	N (newton)
work	force × distance	N m	J (joule)
power	work/time	J s^{-1}	W (watt)
pressure	force/area	N m^{-2}	Pa (pascal)
density	mass/volume	kg m^{-3}	–
charge	current × time	A s	C (coulomb)
voltage	energy/charge	J C^{-1}	V (volt)
resistance	voltage/current	V A^{-1}	Ω (ohm)

Dimensions

Here are three measurements:

length = 10 m   area = 6 m^2 volume = 4 m^3

These three quantities have ***dimensions*** of length, length squared, and length cubed.

Starting with three basic dimensions – length [L], mass [M], and time [T] – it is possible to work out the dimensions of many other physical quantities from their defining equations. There are examples on the right and below.

Example 1

$$\text{speed} = \frac{\text{distance travelled}}{\text{time taken}} = \frac{[L]}{[T]} = [LT^{-1}]$$

So the dimensions of speed are $[LT^{-1}]$.

Example 2

$$\text{density} = \frac{\text{mass}}{\text{volume}} = \frac{[M]}{[L^3]} = [ML^{-3}]$$

So the dimensions of density are $[ML^{-3}]$.

Physical quantity	Defining equation (simplified)	Dimensions		In terms of base units
		from equation	reduced form	
length	–	–	[L]	m
mass	–	–	[M]	kg
time	–	–	[T]	s
speed	$\frac{\text{distance}}{\text{time}}$	$\frac{[L]}{[T]}$	$[LT^{-1}]$	$m\,s^{-1}$
acceleration	$\frac{\text{speed}}{\text{time}}$	$\frac{[LT^{-1}]}{[T]}$	$[LT^{-2}]$	$m\,s^{-2}$
force	mass × acceleration	$[M] \times [LT^{-2}]$	$[MLT^{-2}]$	$kg\,m\,s^{-2}$
work	force × distance	$[MLT^{-2}] \times [L]$	$[ML^2T^{-2}]$	$kg\,m^2\,s^{-2}$
power	$\frac{\text{work}}{\text{time}}$	$\frac{[ML^2T^{-2}]}{[T]}$	$[ML^2T^{-3}]$	$kg\,m^2\,s^{-3}$
pressure	$\frac{\text{force}}{\text{area}}$	$\frac{[MLT^{-2}]}{[L^2]}$	$[ML^{-1}T^{-2}]$	$kg\,m^{-1}\,s^{-2}$

Using base units to check equations

Each term in the two sides of an equation must always have the same units or dimensions. For example,

work = force × distance moved
J = Nm = N × m

An equation cannot be accurate if the base units on both sides do not match. It would be like claiming that '6 apples equals 6 oranges'.

Base units are a useful way of checking that an equation is reasonable.

Example *Check whether the equation PE = mgh is dimensionally correct.*

To do this, start by working out the base units of the right-hand side:

mgh units = kg × ms^{-2} × m = Kg $m^{-2}s^{-2}$

These are the base unit of work, and of energy. So the equation is dimensionally correct.

Note:

- A units check cannot tell you whether an equation is accurate. For example, both of the following are dimensionally correct, but only one is right:

  PE = mgh PE = $2mgh$

Dimensionless numbers

A pure number, such as 6, has no dimensions. Here are two consequences of this fact.

Dimensions and units of frequency The frequency of a vibrating source is defined as follows:

$$\text{frequency} = \frac{\text{number of vibrations}}{\text{time taken}}$$

As number is dimensionless, the dimensions of frequency are $[T^{-1}]$. The SI unit of frequency in the hertz (Hz):
1 Hz = 1 s^{-1}

Dimensions and units of angle

On the right, the angle θ in ***radians*** is defined like this:

$$\theta = \frac{s}{r}$$

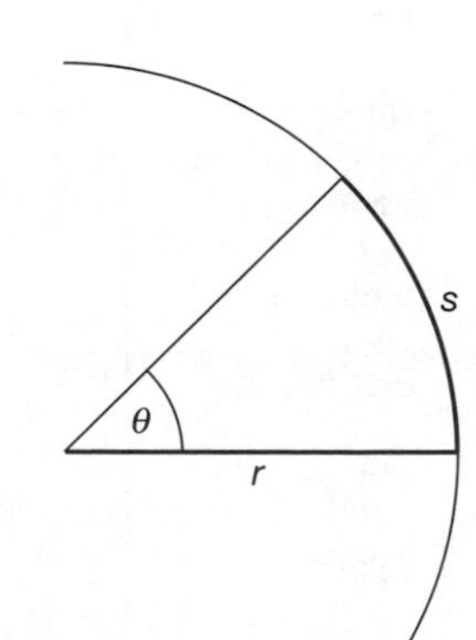

s/r has no dimensions because $[L] \times [L^{-1}] = 1$. However, when measuring an angle in radians, a unit is often included for clarity: 2 rad, for example.

A2 Measurements, uncertainties and graphs

Scientific notation

The average distance from the Earth to the Sun is 150 000 000 km.

There are two problems with quoting a measurement in the above form:

- the inconvenience of writing so many noughts,
- uncertainty about which figures are important (i.e. How approximate is the value? How many of the figures are significant?).

These problems are overcome if the distance is written in the form 1.50×10^8 km.

'1.50×10^8' tells you that there are three significant figures – 1, 5, and 0. The last of these is the least significant and, therefore, the most uncertain. The only function of the other zeros in 150 000 000 is to show how big the number is. If the distance were known less accurately, to two significant figures, then it would be written as 1.5×10^8 km.

Numbers written using powers of 10 are in ***scientific notation*** or ***standard form***. This is also used for small numbers. For example, 0.002 can be written as 2×10^{-3}.

Uncertainty

When making any measurement, there is always some ***uncertainty*** in the reading. As a result, the measured value may differ from the true value. In science, an uncertainty is sometimes called an ***error***. However, it is important to remember that it is *not* the same thing as a mistake.

In experiments, there are two types of uncertainty.

Systematic uncertainties These occur because of some inaccuracy in the measuring system or in how it is being used. For example, a timer might run slow, or the zero on an ammeter might not be set correctly.

There are techniques for eliminating some systematic uncertainties. However, this spread will concentrate on dealing with uncertainties of the random kind.

Random uncertainties These can occur because there is a limit to the sensitivity of the measuring instrument or to how accurately you can read it. For example, the following readings might be obtained if the same current was measured repeatedly using one ammeter:

2.4 2.5 2.4 2.6 2.5 2.6 2.6 2.5

Because of the uncertainty, there is variation in the last figure. To arrive at a single value for the current, you could find the mean of the above readings, and then include an estimation of the uncertainty:

current = 2.5 ± 0.1

(2.5 = mean; ± 0.1 = uncertainty)

Writing '2.5 ± 0.1' indicates that the value could lie anywhere between 2.4 and 2.6.

Note:

- On a calculator, the mean of the above readings works out at 2.5125. However, as each reading was made to only two significant figures, the mean should also be given to only two significant figures i.e. 2.5.
- Each of the above readings may also include a systematic uncertainty.

Uncertainty as a percentage

Sometimes, it is useful to give an uncertainty as a percentage. For example, in the current measurement above, the uncertainty (0.1) is 4% of the mean value (2.5), as the following calculation shows:

$$\text{percentage uncertainty} = \frac{0.1}{2.5} \times 100 = 4$$

So the current reading could be written as 2.5 ± 4%.

Combining uncertainties

Sums and differences Say you have to *add* two length readings, A and B, to find a total, C. If $A = 3.0 \pm 0.1$ and $B = 2.0 \pm 0.1$, then the minimum possible value of C is 4.8 and the maximum is 5.2. So $C = 5.0 \pm 0.2$.

Now say you have to subtract B from A. This time, the minimum possible value of C is 0.8 and the maximum is 1.2 . So $C = 1.0 \pm 0.2$, and the uncertainty is the same as before.

If $C = A + B$ or $C = A - B$, then

uncertainty in C = uncertainty in A + uncertainty in B

The same principle applies when several quantities are added or subtracted: $C = A + B - F - G$, for example.

Products and quotients If $C = A \times B$ or $C = A/B$, then

% uncertainty in C = % uncertainty in A + % uncertainty in B

For example, say you measure a current I, a voltage V, and calculate a resistance R using the equation $R = V/I$. If there is a 3% uncertainty in V and a 4% uncertainty in I, then there is a 7% uncertainty in your calculated value of R.

Note:

- The above equation is only an approximation – and a poor one for uncertainties greater than about 10%.
- To check that the equation works, try calculating the maximum and minimum values of C if, say, A is 100 ± 3 and B is 100 ± 4. You should find that $A \times B$ is 10 000 ± approximately 700 (i.e. 7%).
- The principle of adding % uncertainties can be applied to more complex equations: $C = A^2B/FG$, for example. As $A^2 = A \times A$, the % uncertainty in A^2 is twice that in A.

Calculated results

Say you have to calculate a resistance from the following readings:

voltage = 3.3 V (uncertainty ± 0.1 V, or ± 3%)
current = 2.5 A (uncertainty ± 0.1 A, or ± 4%)

Dividing the voltage by the current on a calculator gives a resistance of 1.32 Ω. However, as the combined uncertainty is ±7%, or ± 0.1 Ω, the calculated value of the resistance should be written as 1.3 Ω. As a general guideline, a calculated result should have no more significant figures than any of the measurements used in the calculation. (However, if the result is to be used in further calculations, it is best to leave any rounding up or down until the end.)

Choosing a graph

The general equation for a straight-line graph is

$$y = mx + c$$

In this equation, m and c are ***constants***, as shown below. y and x are ***variables*** because they can take different values. x is the ***independent variable***. y is the ***dependent variable***: its value depends on the value of x.

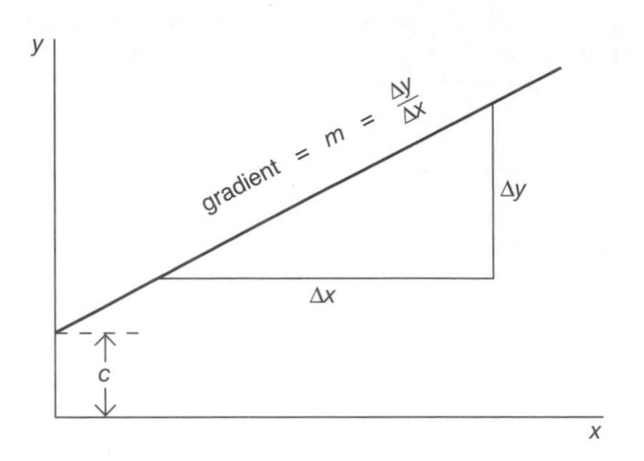

In experimental work, straight-line graphs are especially useful because the values of constants can be found from them. Here is an example.

Problem Theoretical analysis shows that the period T (time per swing) of a simple pendulum is linked to its length l, and the Earth's gravitational field strength g by the equation $T = 2\pi\sqrt{l/g}$. If, by experiment, you have corresponding values of l and T, what graph should you plot in order to work out a value for g from it?

Answer First, rearrange the equation so that it is in the form $y = mx + c$. Here is one way of doing this:

$$\underbrace{T^2}_{y} = \underbrace{\frac{4\pi^2}{g}}_{m}\,\underbrace{l}_{x} + \underbrace{0}_{c}$$

So, if you plot a graph of T^2 against l, the result should be a straight line through the origin (as $c = 0$). The gradient (m) is $4\pi^2/g$, from which a value of g can be calculated.

Showing uncertainties on graphs

In an experiment, a wire is kept at a constant temperature. You apply different voltages across the wire and measure the current through it each time. Then you use the readings to plot a graph of current against voltage.

The general direction of the points suggests that the graph is a straight line. However, before reaching this conclusion, you must be sure that the points' scatter is due to random uncertainty in the current readings. To check this, you could estimate the uncertainty and show this on the graph using short, vertical lines called uncertainty bars. The ends of each bar represent the likely maximum and minimum value for that reading. In the example below, the ***uncertainty bars*** show that, despite the points' scatter, it is reasonable to draw a straight line through the origin.

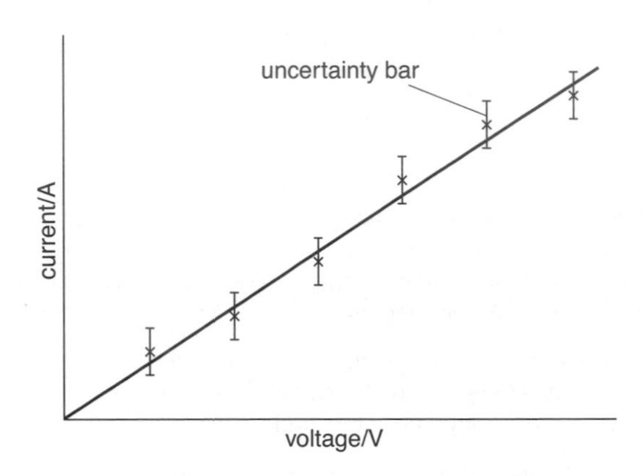

Labelling graph axes Strictly speaking, the scales on the graph's axes are pure, unitless numbers and not voltages or currents. Take a typical reading:

voltage = 10 V

This can be treated as an equation and rearranged to give:

voltage/V = 10

That is why the graph axes are labelled 'voltage/V' and 'current/A'. The values of these are pure numbers.

Reading a micrometer

The length of a small object can be measured using a micrometer screw gauge. You take the reading on the gauge like this:

Reading a vernier

Some measuring instruments have a vernier scale on them for measuring small distances (or angles). You take the reading like this:

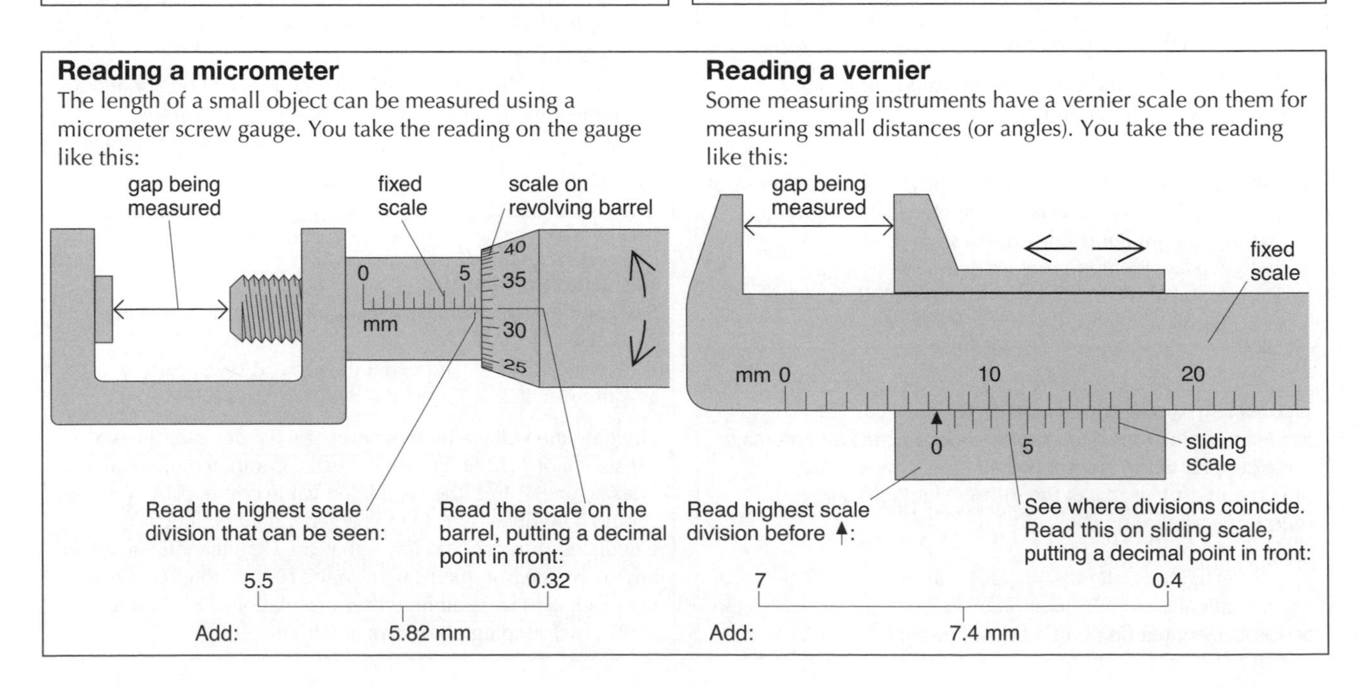

B1 Analysing motion

Units of measurement

Scientists make measurements using SI units such as the metre, kilogram, second, and newton. These and their abbreviations are covered in detail in A1. However, you may find it easier to appreciate the links between different units after you have studied the whole of section A.

For simplicity, units will be excluded from some stages of the calculations in this book, as in this example:

total length = 2 + 3 = 5 m

Strictly speaking, this should be written

total length = 2 m + 3 m = 5 m

Displacement

Displacement is distance moved in a particular direction. The SI unit of displacement is the ***metre*** (m).

Quantities, such as displacement, that have both magnitude (size) and direction are called ***vectors***.

A —— 12 m ——> B

The arrow above represents the displacement of a particle which moves 12 m from A to B. However, with horizontal or vertical motion, it is often more convenient to use a '+' or '–' to show the vector direction. For example,

Movement of 12 m *to the right*: displacement = +12 m
Movement of 12 m *to the left*: displacement = –12 m

Displacement is not necessarily the same as distance travelled. For example, when the ball below has returned to its starting point, its vertical displacement is zero. However, the distance travelled is 10 m.

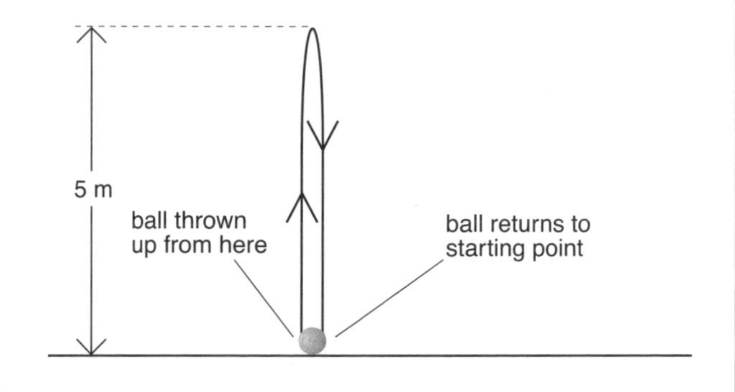

Speed and velocity

Average speed is calculated like this:

$$\text{average speed} = \frac{\text{distance travelled}}{\text{time taken}}$$

The SI unit of speed is the metre/second, abbreviated as m s^{-1}. For example, if an object travels 12 m in 2 s, its average speed is 6 m s^{-1}.

Average velocity is calculated like this:

$$\text{average velocity} = \frac{\text{displacement}}{\text{time taken}}$$

The SI unit of velocity is also the m s^{-1}. But unlike speed, velocity is a vector.

6 m s^{-1} ——>

The velocity vector above is for a particle moving to the right at 6 m s^{-1}. However, as with displacement, it is often more convenient to use a '+' or '–' for the vector direction.

Average velocity is not necessarily the same as average speed. For example, if a ball is thrown upwards and travels a total distance of 10 m before returning to its starting point 2 s later, its average speed is 5 m s^{-1}. But its average velocity is zero, because its displacement is zero.

Acceleration

Average acceleration is calculated like this:

$$\text{average acceleration} = \frac{\text{change in velocity}}{\text{time taken}}$$

The SI unit of acceleration is the m s^{-2} (sometimes written m/s^2). For example, if an object gains 6 m s^{-1} of velocity in 2 s, its average acceleration is 3 m s^{-2}.

3 m s^{-2} ——>>

Acceleration is a vector. The acceleration vector above is for a particle with an acceleration of 3 m s^{-2} to the right. However, as with velocity, it is often more convenient to use a '+' or '–' for the vector direction.

If velocity *increases* by 3 m s^{-1} every second, the acceleration is +3 m s^{-2}. If it *decreases* by 3 m s^{-1} every second, the acceleration is –3 m s^{-2}.

Mathematically, an acceleration of –3 m s^{-2} *to the right* is the same as an acceleration of +3 m s^{-2} *to the left*.

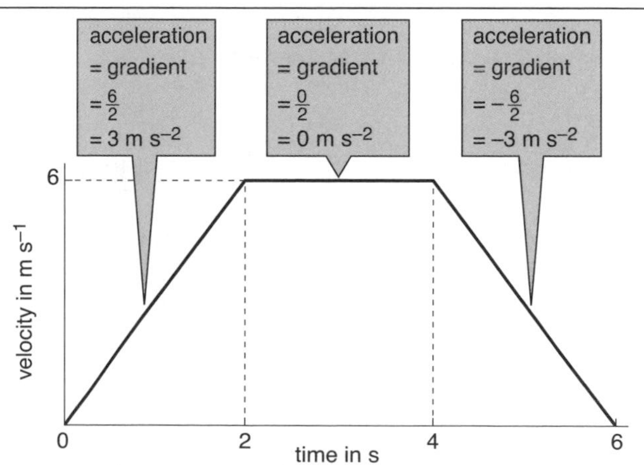

On the velocity–time graph above, you can work out the acceleration over each section by finding the *gradient* of the line. The gradient is calculated like this:

$$\text{gradient} = \frac{\text{gain along } y\text{-axis}}{\text{gain along } x\text{-axis}}$$

Vectors and scalars

Vectors are quantities that have magnitude and direction. Examples of these in this section are displacement, velocity, and acceleration. Force and moment are also vectors. When adding vectors you must allow for their direction. On page 24 there are diagrams of two 6 N forces being added. In one case the resultant is 12 N. In the other it is zero.

Scalars are quantities with magnitude only. Examples are distance and speed and mass. Scalar addition is simple: if 6 kg of mass is added to 6 kg of mass the total is always 12 kg. Another example of a scalar quantity is energy. If an object has 6 J of potential energy and 6 J of kinetic energy, the total energy is 12 J.

Velocity–time graphs

The graphs which follow are for three examples of ***linear*** motion (motion in a straight line).

Graph A below shows how the velocity of a stone would change with time, if the stone were dropped near the Earth's surface and there were no air resistance to slow it.
The stone has a *uniform* (unchanging) acceleration a which is equal to the gradient of the graph:

$$a = \frac{\Delta v}{\Delta t}$$

In this case, the acceleration is g (9.81 m s^{-2}).

If air resistance is significant, then the graph is no longer a straight line (see B8).

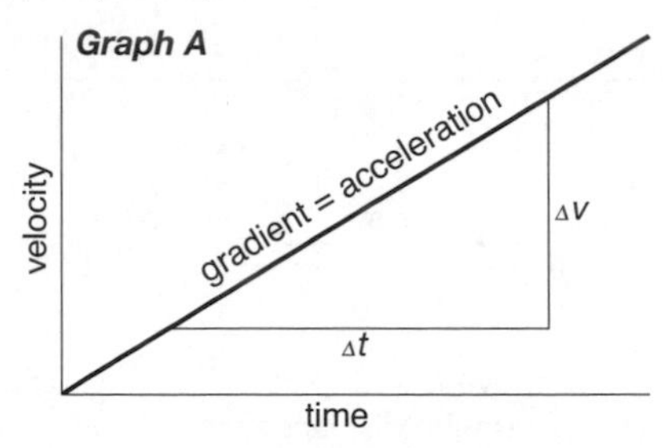

Graph B below is for a car travelling at a steady velocity of 30 m s^{-1}. In 2 s, the car travels a distance of 60 m. Numerically, this is equal to the area under the graph between the 0 and 2 s points. (Note: the area must be worked out using the scale numbers, not actual lengths.)

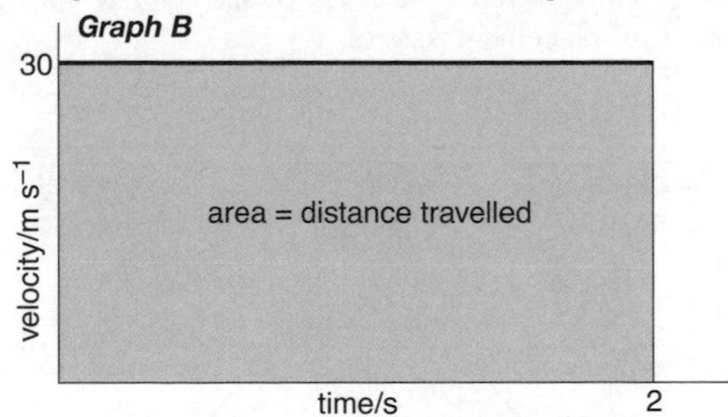

Graph C below is for a car with a changing velocity. However, the same principle applies as before: the area under the graph gives the distance travelled. (This is also true if the graph is not a straight line: see B8.)

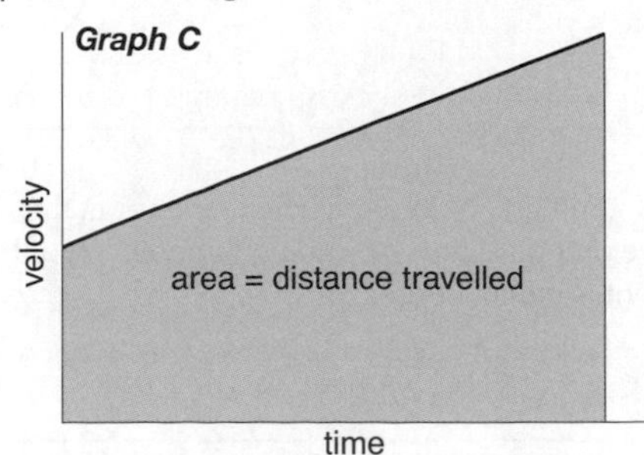

Displacement–time graphs

Uniform velocity The graph below describes the motion of a car moving with uniform velocity. The displacement and time have been taken as zero when the car passes a marker post. The gradient of the graph is equal to the velocity v:

$$v = \frac{\Delta s}{\Delta t}$$

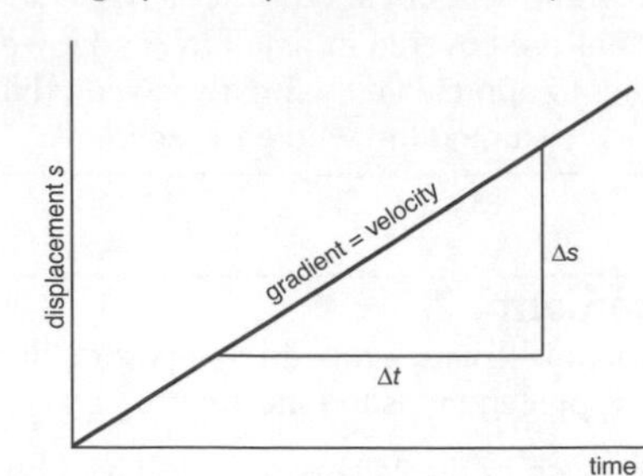

Changing velocity The gradient of this graph is increasing with time, so the velocity is increasing. The velocity v at any instant is equal to the gradient of the *tangent* at that instant.

In calculus notation $v = \frac{ds}{dt}$

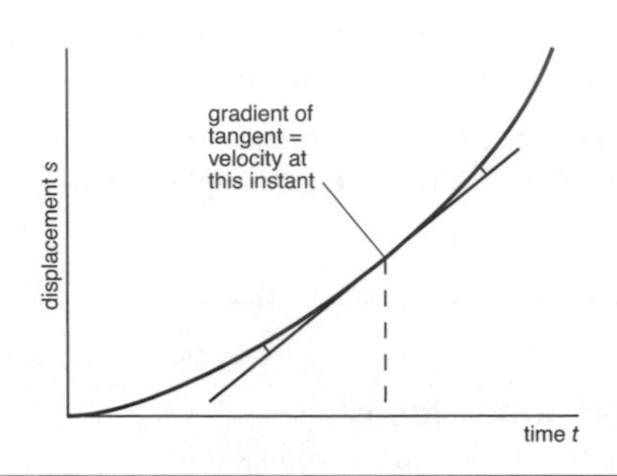

Uniform acceleration The graph below describes the motion of a car gaining velocity at a steady rate. The time has been taken as zero when the car is stationary. The gradient of the graph is equal to the acceleration.

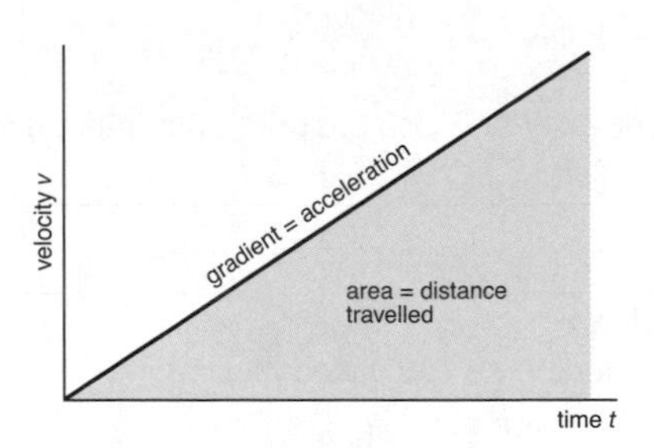

Changing acceleration The acceleration a at any instant is equal to the gradient of the *tangent* at that instant.

In calculus notation $a = \frac{dv}{dt}$

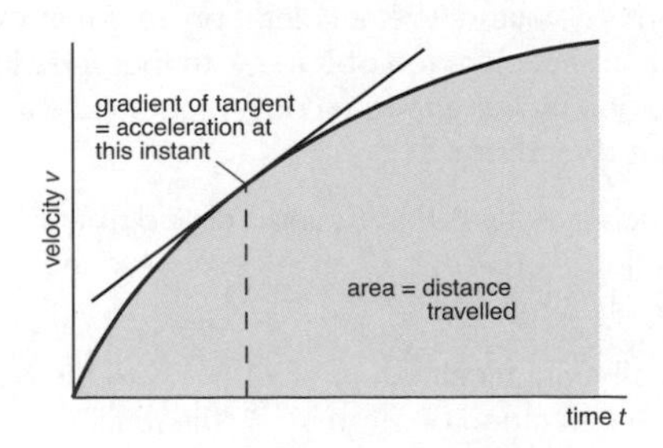

B2 Motion, mass and forces

Force

Force is a vector. The SI unit is the ***newton*** (N).

If two or more forces act on something, their combined effect is called the ***resultant force***. Two simple examples are shown below. In the right-hand example, the resultant force is zero because the forces are ***balanced***.

A resultant force acting on a mass causes an acceleration. The force, mass, and acceleration are linked like this:

resultant force = mass × acceleration $F = ma$

For example, a 1 N resultant force gives a 1 kg mass an acceleration of 1 m s^{-2}. (The newton is defined in this way.)

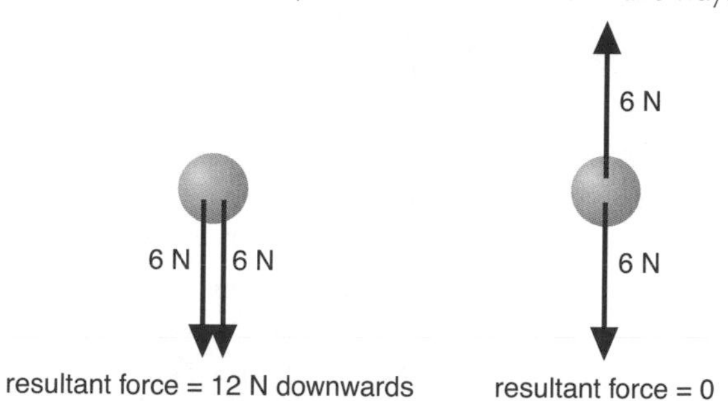

The more mass something has, the more force is needed to produce any given acceleration.

When balanced forces act on something, its acceleration is zero. This means that it is *either* stationary *or* moving at a steady velocity (steady speed in a straight line).

Weight and g

On Earth, everything feels the downward force of gravity. This gravitational force is called ***weight***. As for other forces, its SI unit is the newton (N).

Near the Earth's surface, the gravitational force on each kg is about 9.8 N; the ***gravitational field strength*** is 9.8 N kg^{-1}. This is represented by the symbol g.

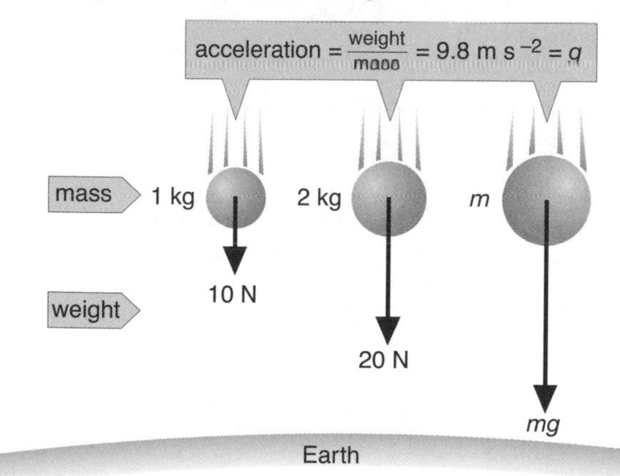

In the diagram above, all the masses are falling freely (gravity is the only force acting). From $F = ma$, it follows that all the masses have the same downward acceleration, g. This is the ***acceleration of free fall***.

You can think of g

either as a gravitational field strength of 9.8 N kg^{-1}

or as an acceleraton of free fall of 9.8 m s^{-2}.

Always use the value of g given on the question or formula sheet. It may be 10 or 9.81, depending on whether less, or more, accuracy is required.

Equations of motion

The car below has uniform acceleration. In the following analysis, only motion between X and Y will be considered.

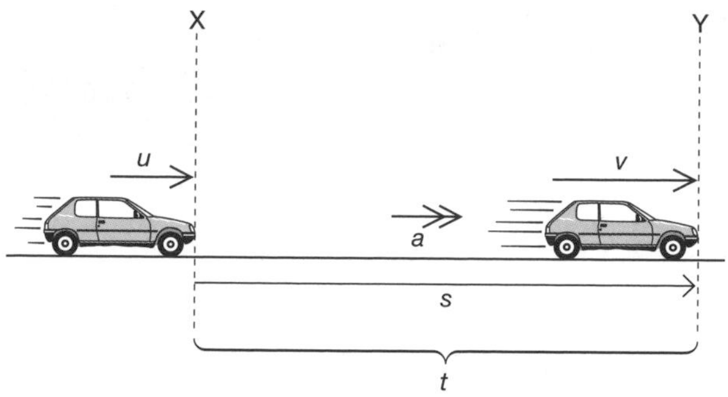

u = initial velocity (velocity on passing X)
v = final velocity (velocity on passing Y)
a = acceleration
s = displacement (in moving from X to Y)
t = time taken (to move from X to Y)

Here is a velocity–time graph for the car.

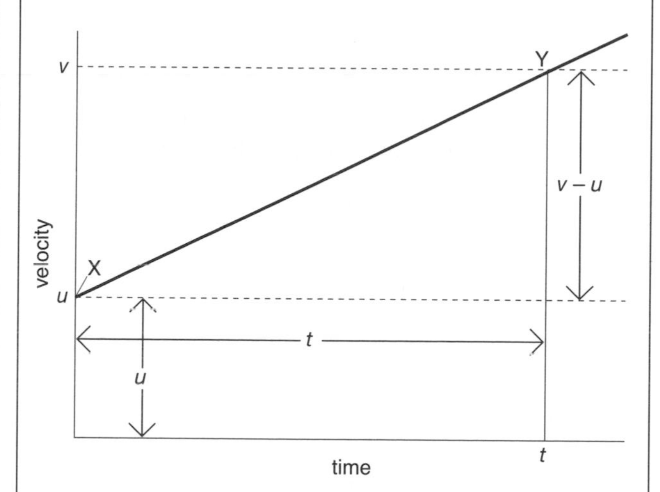

There are four equations (numbered 1–4 below) linking u, v, a, s, and t. They can be worked out as follows.

The acceleration is the gradient of the graph.
So $a = (v - u)/t$. This can be rearranged to give

$$v = u + at \quad (1)$$

The distance travelled, s in this case, is the area under the graph. This is the area of one rectangle (height × base) plus the area of one triangle ($\frac{1}{2}$ × height × base). So it is $u \times t$ plus $\frac{1}{2} \times (v - u) \times t$. But $v - u = at$ from equation (1), so

$$s = ut + \tfrac{1}{2}at^2 \quad (2)$$

As distance travelled = average velocity × time taken,

$$s = \tfrac{1}{2}(v + u)t \quad (3)$$

If equations (1) and (3) are combined so that t is eliminated,

$$v^2 = u^2 + 2as \quad (4)$$

Note:
- The equations are only valid for uniform acceleration.
- Each equation links a different combination of factors. You must decide which equation best suits the problem you are trying to solve.
- You must allow for vector directions. With horizontal motion, you might decide to call a vector to the right positive (+). With vertical motion, you might call a downward vector positive. So, for a stone thrown upwards at 30 m s^{-1}, $u = -30$ m s^{-1} and $g = +10$ m s^{-2}.

Motion problems

Here are examples of how the equations of motion can be used to solve problems. For simplicity, units will not be shown in some equations. It will be assumed that air resistance is negligible and that g is 10 m s^{-2}.

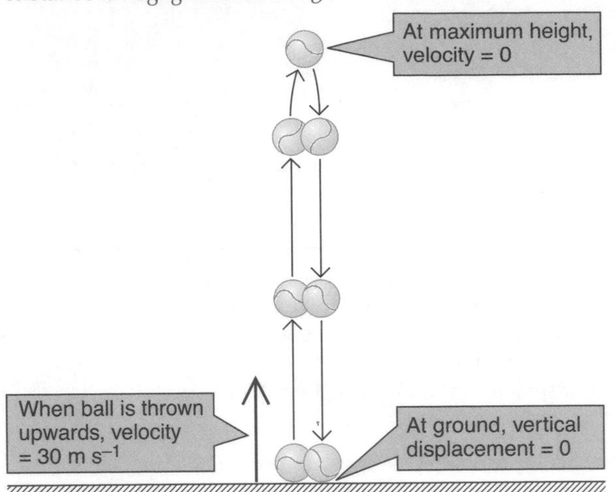

Example 1 *A ball is thrown upwards at 30 m s^{-1}. What time will it take to reach its highest point?*

The ball's motion only needs to be considered from when it is thrown to when it reaches its highest point. These are the 'initial' and 'final' states in any equation used.

When the ball is at it highest point, its velocity v will be zero. So, taking downward vectors as positive,

$u = -30$ m s^{-1} $\quad v = 0 \quad a = g = 10$ m s^{-2} $\quad t$ is to be found.

In this case, an equation linking u, v, a, and t is required. This is equation (1) on the opposite page:

$$v = u + at$$

So $0 = -30 + 10t$

Rearranged, this gives $t = 3.0$ s.

Example 2 *A ball is thrown upwards at 30 m s^{-1}. What is the maximum height reached?*

In this case,

$u = -30$ m s^{-1} $\quad v = 0 \quad a = g = 10$ m s^{-2} $\quad s$ is to be found.

This time, the equation required is (4) on the opposite page:

$$v^2 = u^2 + 2as$$

So $0 = (-30)^2 + (2 \times 10 \times s)$

This gives $s = -45$ m.

(Downwards is positive, so the negative value of s indicates an *upward* displacement.)

Example 3 *A ball is thrown upwards at 30 m s^{-1}. For what time is it in motion before it hits the ground?*

When the ball reaches the ground, it is back where it started, so its displacement s is zero. Therefore

$u = -30$ m s^{-1} $\quad s = 0 \quad a = g = 10$ m s^{-2} $\quad t$ is to be found.

This time, the equation required is (2) on the opposite page:

$$s = ut + \tfrac{1}{2}at^2$$

So $0 = (-30t) + (\tfrac{1}{2} \times 10 \times t^2)$

This gives $t = 6$ s.

(There is also a solution $t = 0$, indicating that the ball's displacement is also zero at the instant it is thrown.)

Measuring g

By measuring the time t it takes an object to fall through a measured height h, a value of g can be found (assuming that air resistance is negligible).

In the diagram on the right, $u = 0 \quad a = g \quad s = h$

Applying equation (2) on the opposite page gives

$$s = ut + \tfrac{1}{2}at^2$$

So $h = 0t + \tfrac{1}{2}gt^2$

This gives $g = \dfrac{2h}{t^2}$.

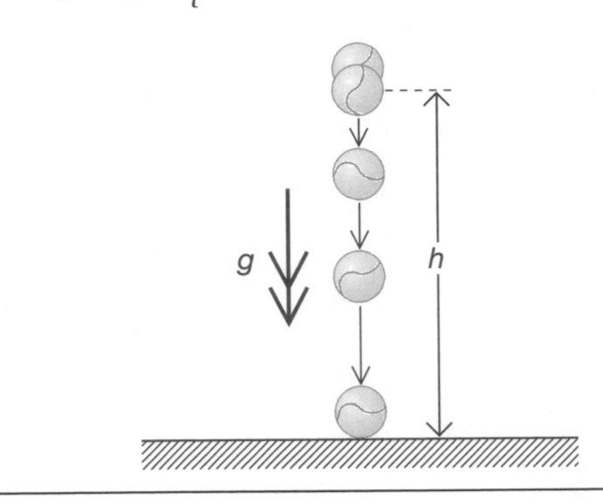

Upwards and downwards

A ball bounces upwards from the ground. The graph on the right shows how the velocity of the ball changes from when it leaves the ground until it hits the ground again.
Downward velocity has been taken as positive.
Air resistance is assumed to be negligible.

Initially, the ball is travelling upwards, so it has negative downward velocity. This passes through zero at the ball's highest point and then becomes positive.

The gradient of the graph is constant and equal to g.

Note:

- The ball has downward acceleration g, even when it is travelling upwards. (Algebraically, losing upward velocity is the same as gaining downward velocity.)
- The ball has downward acceleration g, even when its velocity is zero (at its highest point).

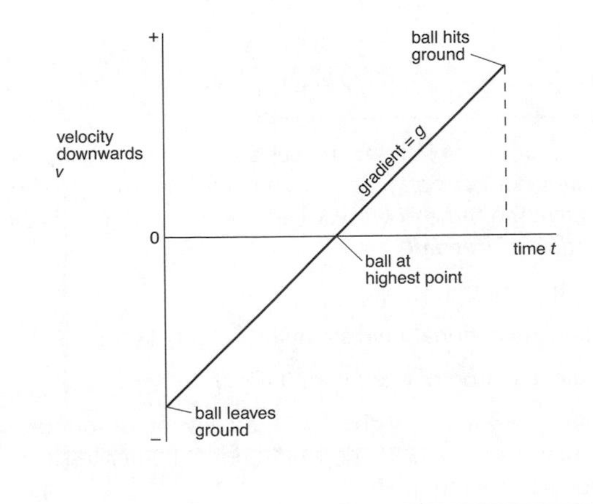

B3 Vectors

Vector arrows

Vectors are quantities which have both magnitude (size) and direction. Examples include displacement and force.

For problems in one dimension (e.g. vertical motion), vector direction can be indicated using + or –. But where two or three dimensions are involved, it is often more convenient to represent vectors by arrows, with the length and direction of the arrow representing the magnitude and direction of the vector. The arrowhead can either be drawn at the end of the line or somewhere else along it, as convenient. Here are two displacement vectors.

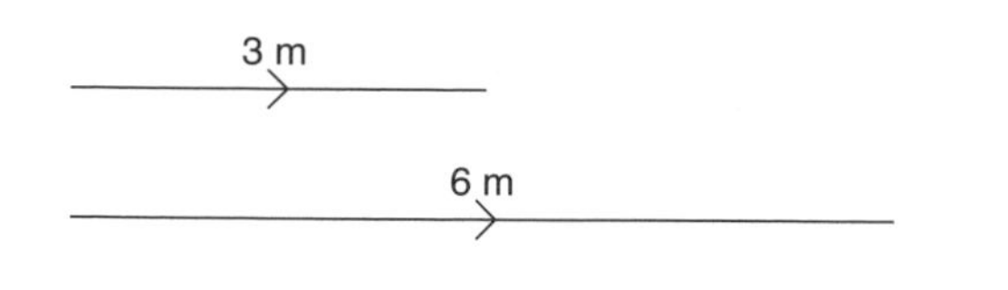

Adding vectors

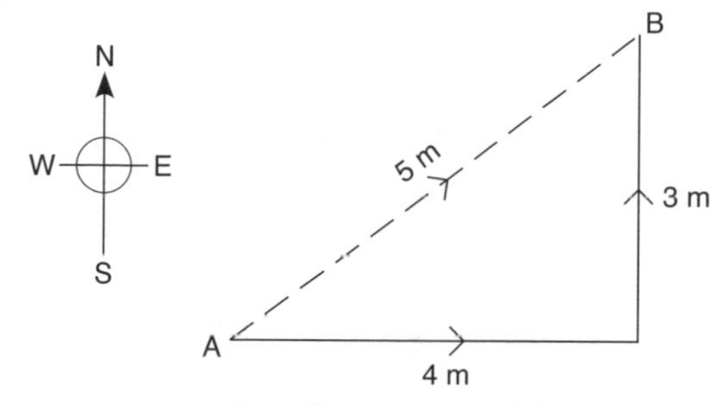

If someone starts at A, walks 4 m East and then 3 m North, they end up at B, as shown above. In this case, they are 5 m from where they started – a result which follows from Pythagoras' theorem. This is an example of vector addition. Two displacement vectors, of 3 m and 4 m, have been added to produce a ***resultant*** – a displacement vector of 5 m.

This principle works for any type of vector. Below, forces of 3 N and 4 N act at right-angles through the same point, O. The ***triangle of vectors*** gives their resultant. The vectors being added must be drawn head-to-tail. The resultant runs from the tail of the first arrow to the head of the second.

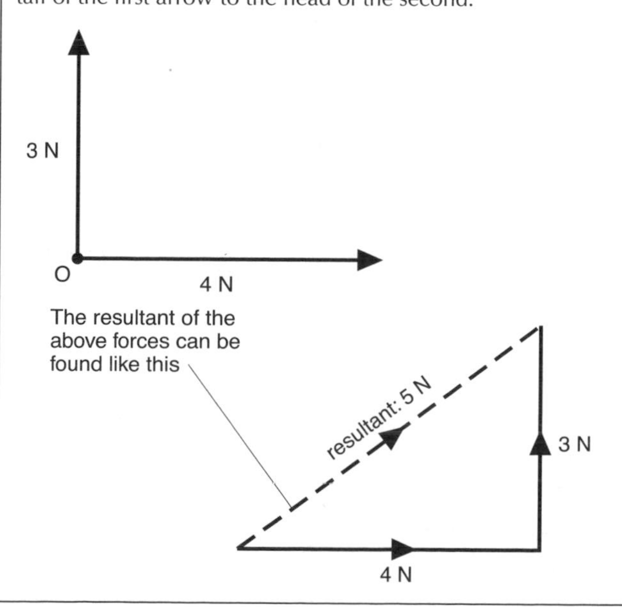

Parallelogram of vectors

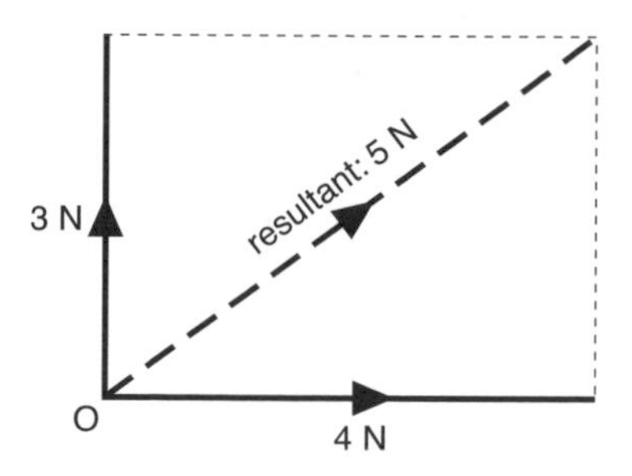

Above, you can see another way of finding the resultant of two forces, 3 N and 4 N, acting at right-angles through the same point. The vectors are drawn as two sides of a rectangle. The diagonal through O gives the magnitude and direction of the resultant. Note that the lines and angles in this diagram match those in the previous force triangle.

By drawing a parallelogram, the above method can also be used to add vectors which are not at right-angles. Here are two examples of a ***parallelogram of vectors***.

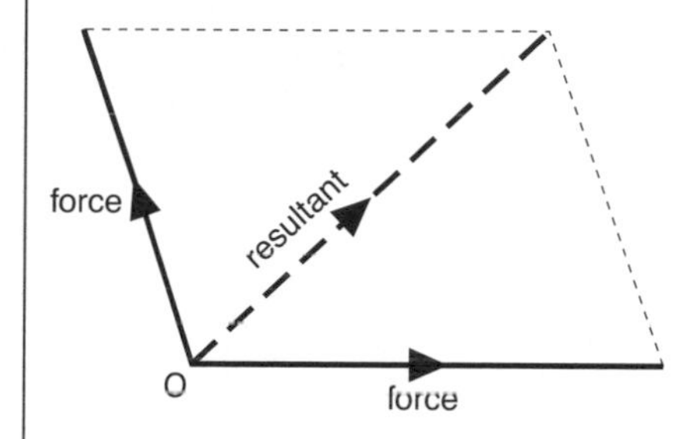

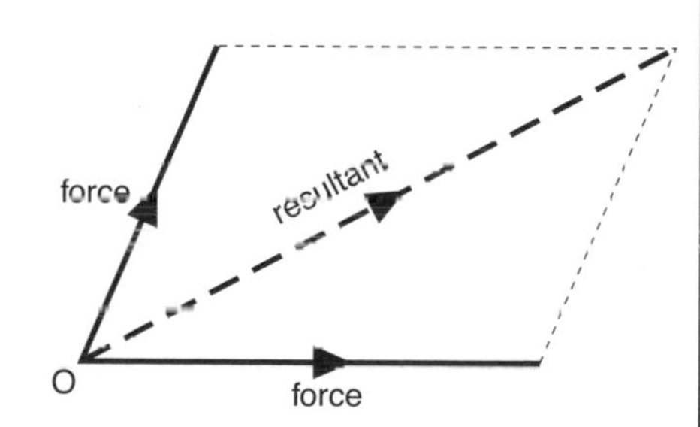

Note:

- The magnitude of the resultant depends on the relative directions of the vectors. For example, if forces of 3 N and 4 N are added, the resultant could be anything from 1 N (if the vectors are in opposite directions) to 7 N (if they are in the same direction).
- In the diagrams on this page, the resultant is always shown using a dashed arrow. This is to remind you that the resultant is a *replacement* for the other two vectors. There are *not* three vectors acting.

Multiplying vectors

When vectors are multiplied together, the product is not necessarily another vector. For example, work is the product of two vectors: force and displacement. But work is a scalar, not a vector. It has magnitude but no direction.

Methods of multiplying vectors have not been included in this book, other than for simple cases: for example, when force and displacement vectors are in the same direction.

Components

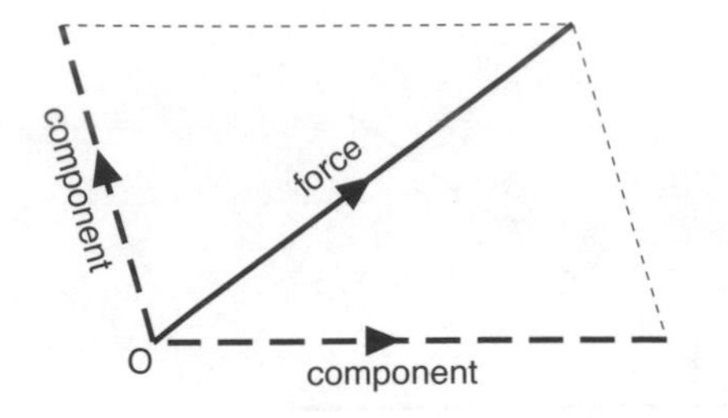

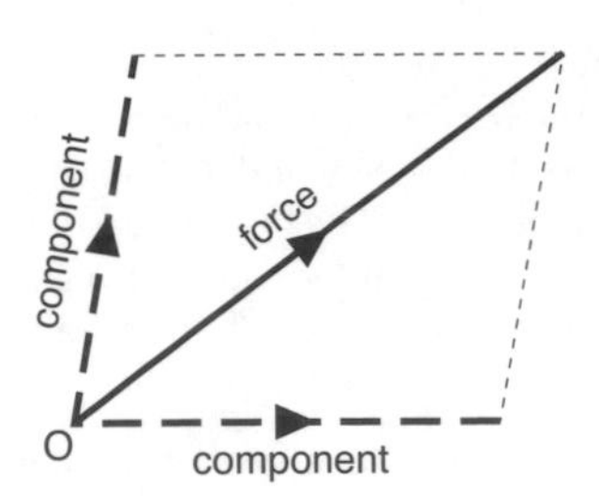

Two forces acting through a point can be replaced by a single force (the resultant) which has the same effect. Conversely, a single force can be replaced by two forces which have the same effect – a single force can be ***resolved*** into two ***components***. Two examples of the components of a force are shown above, though any number of other sets of components is possible.

Note:

- Any vector can be resolved into components.
- The components above are shown as dashed lines to remind you that they are a *replacement* for a single force. There are *not* three forces acting.

In working out the effects of a force (or other vector), the most useful components to consider are those at right-angles, as in the following example.

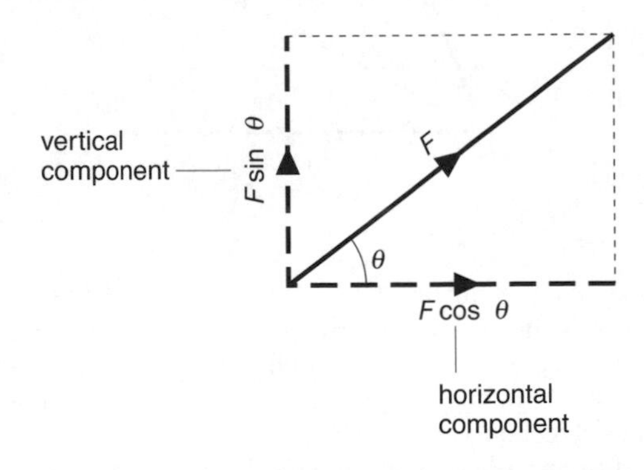

Below, you can see why the horizontal and vertical components have magnitudes of $F \cos \theta$ and $F \sin \theta$.

$$\cos \theta = \frac{F_x}{F}$$

So $F_x = F \cos \theta$

$$\sin \theta = \frac{F_y}{F}$$

So $F_y = F \sin \theta$

Equilibrium

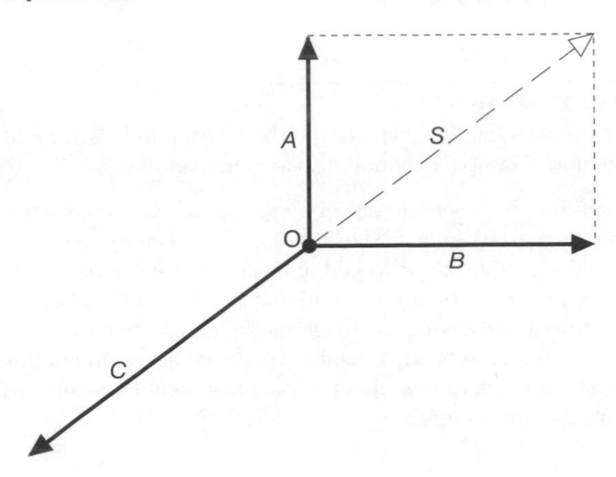

The particle O above has three forces acting on it – *A, B,* and *C*. Forces *A* and *B* can be replaced by a single force *S*. As force *C* is equal and opposite to *S*, the resultant of *A, B,* and *C*, is zero. This means that the three forces are in balance – the system is in ***equilibrium***.

If three forces are in equilibrium, they can be represented by the three sides of a triangle, as shown below. Note that the sides and angles match those in the previous force diagram. The forces can be drawn in any order, provided that the head of each arrow joins with the tail of another.

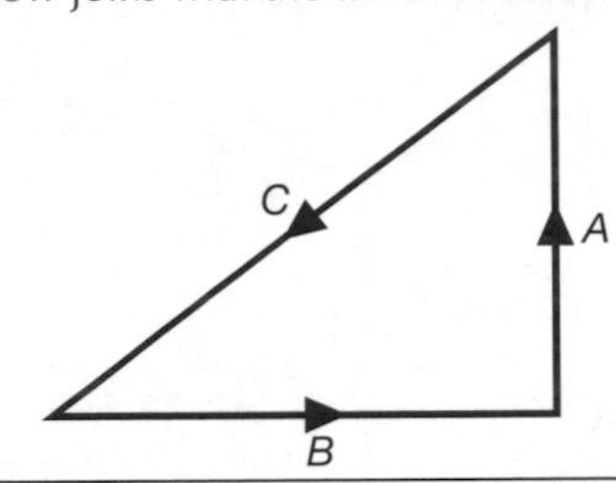

Resolving problem

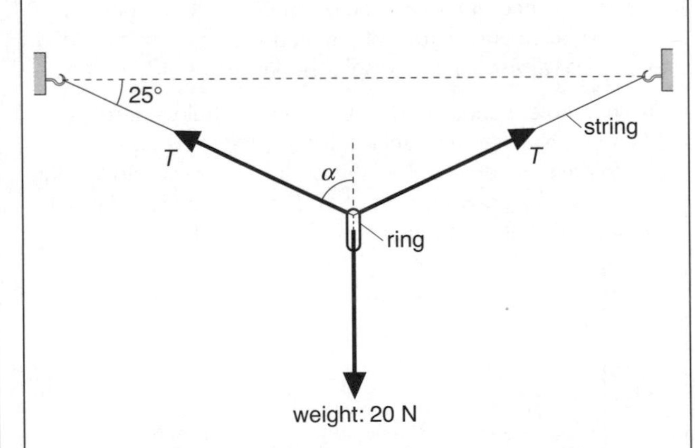

Example *Above, a ring is suspended from the middle of a piece of string. What is the tension in the string?*

Force T is the tension. It is present in both halves of the string. As angle α is 65°, this force has a component (upwards) of $T \cos 65°$. So

total of upward components on ring = $2T \cos 65°$

As the system is in equilibrium, the total of upward components must equal the downward force on the ring.

So $2T \cos 65° = 20$

This gives $T = 24$ N

B4 Moments and equilibrium

Moment of a force

The turning effect of a force is called its ***moment***.

moment of force about a point = force × perpendicular distance from point measured from the line of action of the force

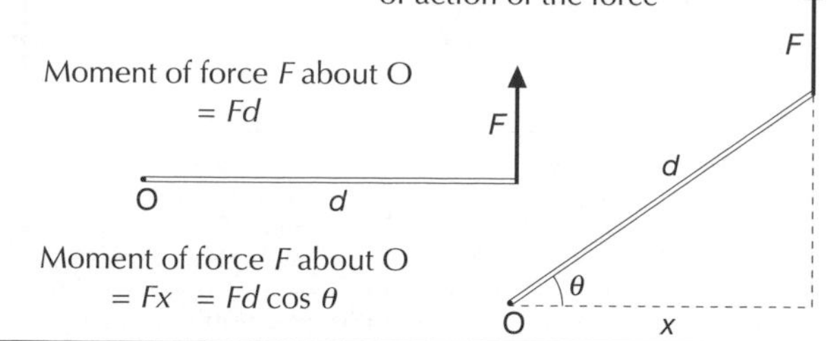

Note:

- In the diagram on the left, although O is shown as a point, it is really an *axis* going perpendicularly into the paper.
- Moments are measured in N m. However this is not the same unit as the N m, or J (joule), used for measuring energy.
- A moment can be *clockwise* or *anticlockwise*, depending on its *sense* (direction of turning). This can be indicated with a + or –. For example,

anticlockwise moment of 2 N m = +2 N m
clockwise moment of 2 N m = –2 N m

Principle of moments

The beam in the diagram on the right has weights on it. (The beam itself is of negligible weight.) The total weight is supported by an upward force R from the fulcrum.

The beam is in a state of balance. It is in equilibrium.

As the beam is not tipping to the left or right, the turning effects on it must balance. So, when moments are taken about O, as shown, the total clockwise moment must equal the total anticlockwise moment. (Note: R has zero moment about O because its distance from O is zero.)

As the beam is static, the upward force on it must equal the total downward force. So $R = 10 + 8 + 4 = 22$ N.

The beam is not turning about O. But it is not turning about any other axis either. So you would expect the moments about *any* axis to balance. This is exactly the case, as you can see in the next diagram. The beam and weights are the same as before, but this time, moments have been taken about point P instead of O. (Note: R does have a moment about P, so the value of R must be known before the calculation can be done.)

The examples shown on the right illustrate the ***principle of moments***, which can be stated as follows:

> If an object is in equilibrium, the sum of the clockwise moment about any axis is equal to the sum of the anticlockwise moments.

Here is another way of stating the principle. In it, moments are regarded as + or –, and the ***resultant moment*** is the algebraic sum of all the moments:

> If a rigid object is in equilibrium, the resultant moment about any axis is zero.

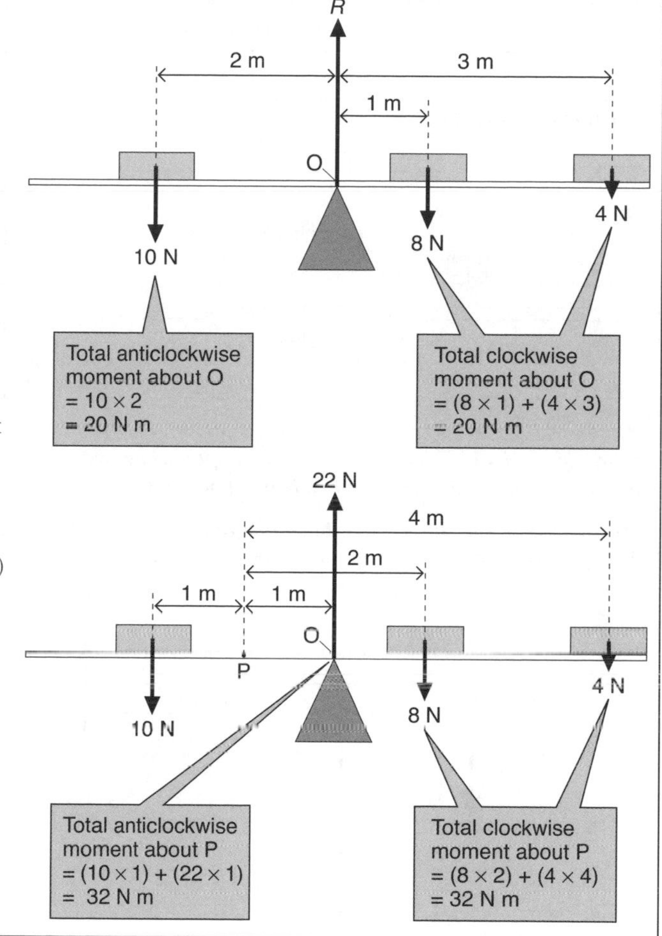

Centre of gravity

All the particles in an object have weight. The weight of the whole object is the resultant of all these tiny, downward gravitational forces. It appears to act through a single point called the ***centre of gravity***.

For a rectangular beam with an even weight distribution, the centre of gravity is in the middle. Unless negligible, the weight must be included when analysing the forces and moments acting on the beam.

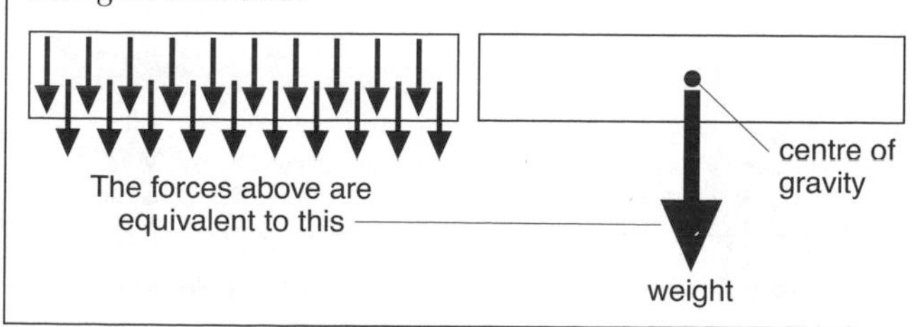

For a dumb-bell with uneven weight distribution, the centre of gravity is at point 0 because the two moments about 0 are equal and opposite.

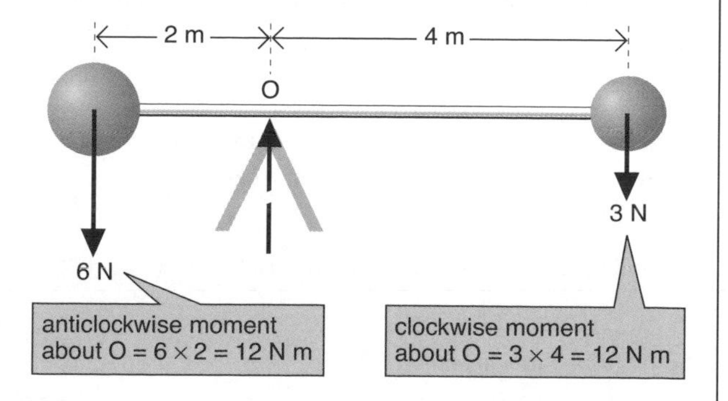

Conditions for equilibrium

There are two types of motion: ***translational*** (from one place to another) and ***rotational*** (turning). If a static, rigid object is in equilibrium, then

- the forces on it must balance, otherwise they would cause translational motion,
- the moments must balance, otherwise they would cause rotational motion.

The balanced beam on the opposite page is a simple system in which the forces are all in the same plane. A ***coplanar*** system like this is in equilibrium if

- the vertical components of all the forces balance,
- the horizontal components of all the forces balance,
- the moments about any axis balance.

To check for equilibrium, components can be taken in any two directions. However, vertical and horizontal components are often the simplest to consider. The balanced beam is especially simple because there are no horizontal forces.

Equilibrium problem

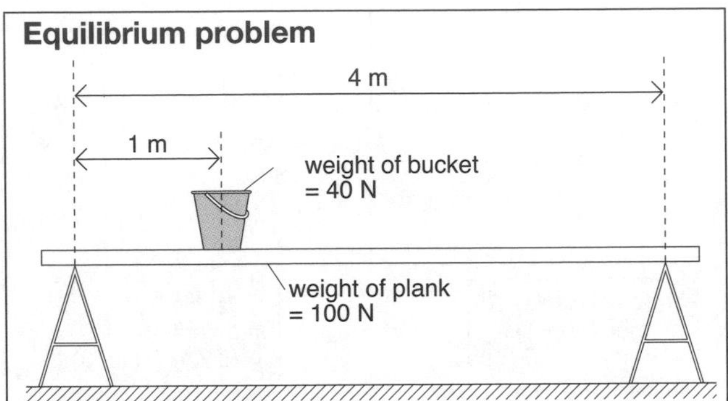

Example *A plank with a bucket on it is supported by two trestles. What force does each trestle exert on the plank?*

The first stage is to draw a ***free-body diagram*** showing just the rigid body (the plank) and the forces acting on it:

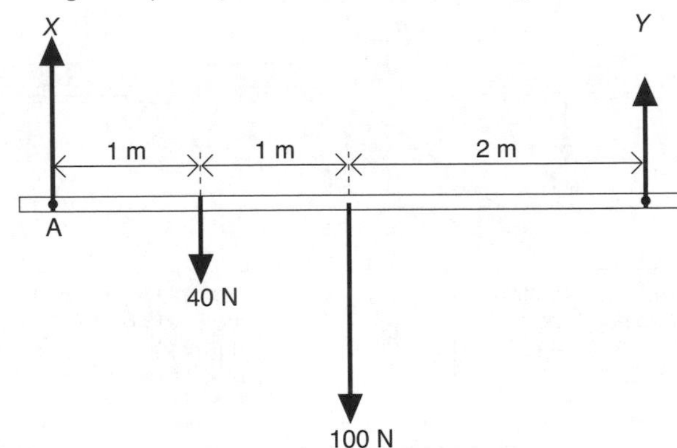

The body is in equilibrium, so the moments must balance, and the forces also. X and Y are the unknown forces.

Taking moments about A:

total clockwise moment = total anticlockwise moment

$$(40 \times 1) + (100 \times 2) = (Y \times 4)$$

This gives $Y = 60$ N.

Note the advantage of taking moments about A: X has a zero moment, so it does not feature in the equation.

Comparing the vertical forces:

total upward force = total downward force

$$Y + X = 40 + 100$$

As Y is 60 N, this gives $X = 80$ N.

Couples and torque

A pair of equal but opposite forces, as below, is called a ***couple***. It has a turning effect but no resultant force.

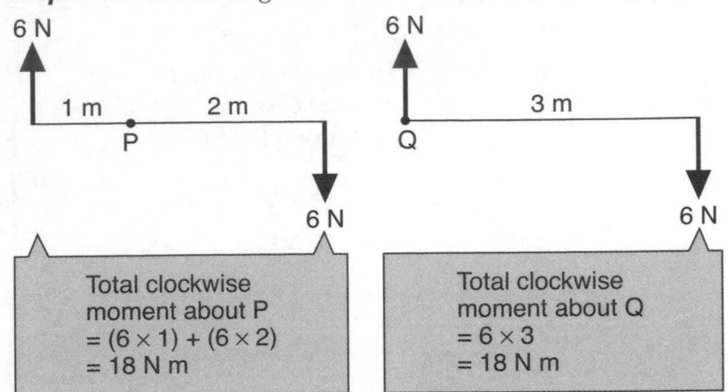

To find the total moment of a couple, you could choose any axis, work out the two moments and add them up. Whichever axis you choose, the answer is the same, so the simplest way of calculating the total moment is like this:

moment of couple = one force × perpendicular distance between forces

Note:

- The total moment of a couple is called a ***torque***.
- Strictly speaking, a couple is any system of forces which has a turning effect only i.e. one which produces rotational motion without translational (linear) motion.

Stability

For a static object, there are three types of equilibrium, as shown below. Whether the equilibrium is ***stable***, ***unstable***, or ***neutral*** depends on the couple formed by the weight and the reaction when the object is displaced.

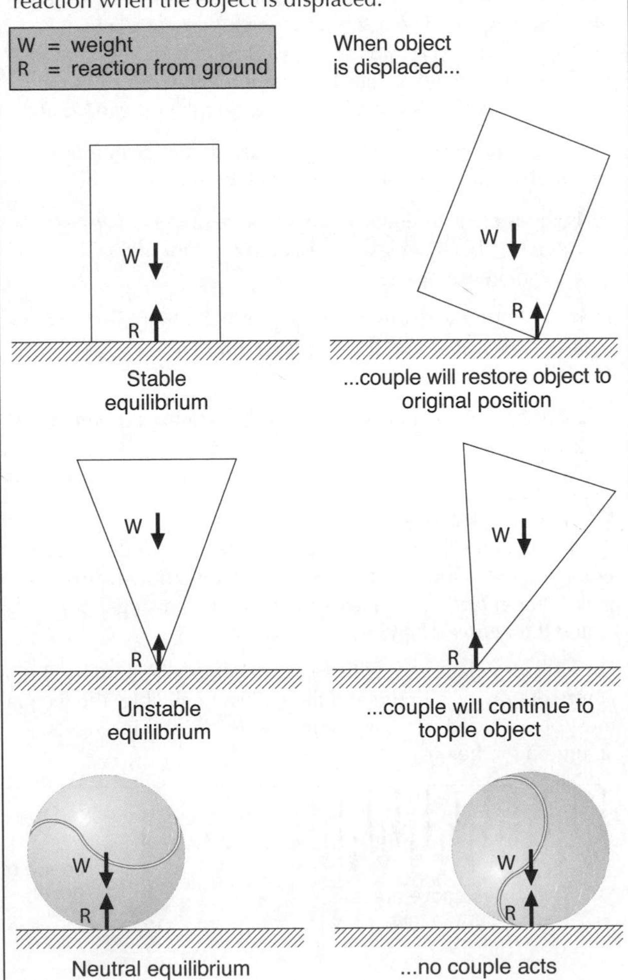

B5 Projectiles and theories

Projectiles

A projectile is an object that:

- has sideways (horizontal) movement as well as falling under gravity, but is not powered and so has no horizontal acceleration
- has negligible air resistance, which can be ignored.

Examples are balls and bullets.

Downwards and sideways

ball dropped

ball thrown sideways

Above, one ball is dropped, while another is thrown sideways at the same time. There is no air resistance. The positions of the balls are shown at regular time intervals.

- Both balls hit the ground together. They have the same downward acceleration *g*.
- As it falls, the second ball moves sideways over the ground at a steady speed.

Results like this show that the vertical and horizontal motions are independent of each other.

Example *Below, a ball is thrown horizontally at 40 m s^{-1}. What horizontal distance does it travel before hitting the water? (Assume air resistance is negligible and g = 10 m s^{-2}).*

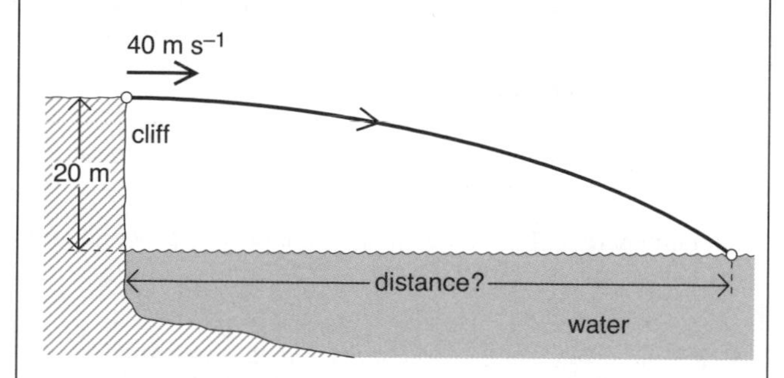

First, work out the time the ball would take to fall vertically to the sea. This can be done using the equation $s = ut + \frac{1}{2}at^2$, in which $u = 0$, s = 20 m, a = –10 m s^{-2}, and t is to be found. This gives t = 2.0 s.

Next, work out how far the ball will travel horizontally in this time (2 s) at a steady horizontal speed of 40 m s^{-1}.

As distance travelled = average speed × time, horizontal distance travelled = 40 × 2 = 80 m.

How Science Works

Theories of motion

Aristotle's ideas

Almost 2400 years ago, the Greek philosopher Aristotle developed theories of Physics that were very different to the ones we use today. For example, he said that arrows move forward after they have left the bowstring because as they move through the air they create an empty space behind them that propels them forward. He said that things could move in a substance only where they could create an empty space. The 'thinner' the substance the faster things would move, and a vacuum was impossible, because things would move at infinite speed.

To explain gravity he said that some things naturally belong to the centre of the Earth and would try to return there. However, other things, such as steam and gases, naturally belong to, and try to return to, the heavenly spheres.

These ideas lasted for almost 2000 years, although a number of Persian and European scientists showed that they did not work.

Galileo's ideas

About 400 years ago, the Italian scientist Galileo Galilei made a large number of scientific observations. He pioneered the use of standard units of length and time, so that his results could be repeated and checked. This was one of the reasons he was able to convince people that Aristotle's theories were wrong.

For example, Galileo said that two cannonballs, a light one and a heavy one, dropped from the tower of Pisa would reach the ground together, and this would be true of any objects as long as air resistance was negligible. This would have been too difficult for him to measure accurately. However, he built and used inclined planes to roll balls down, timing them to show that their acceleration did not depend on their mass, providing friction was negligible. Today in Florence you can still see the equipment he used for these experiments.

When an astronaut on the Moon dropped a feather and a hammer at the same time, because there was no air resistance they both reached the surface together. The same result can be shown on Earth if they are dropped in a vacuum tube.

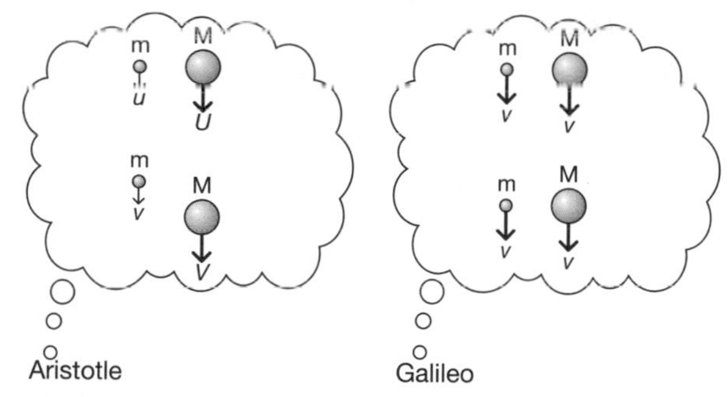

Newton's ideas

Isaac Newton continued to extend and improve our theories of motion and gravity (see B10, p. 36). Today we know that a force is required to change the motion of an object, and that resistive forces such as air resistance and friction slow down moving objects. Newton's laws have been tested and shown to work on the Moon and in space.

Einstein's ideas

Almost 100 years ago, Albert Einstein published his special theory of relativity, which says that nothing (except light) can travel at the speed of light, and so mass increases with speed. When an object approaches the speed of light, the equation $F = ma$ cannot be used, because the mass m increases. The Special and General Theories of Relativity have both been tested and shown to work.

B6 Work and energy

Work

Work is done whenever a force makes something move. It is calculated like this:

work done = force × distance moved in direction of force

The SI unit of work is the ***joule*** (J). For example, if a force of 2 N moves something a distance of 3 m, then the work done is 6 J.

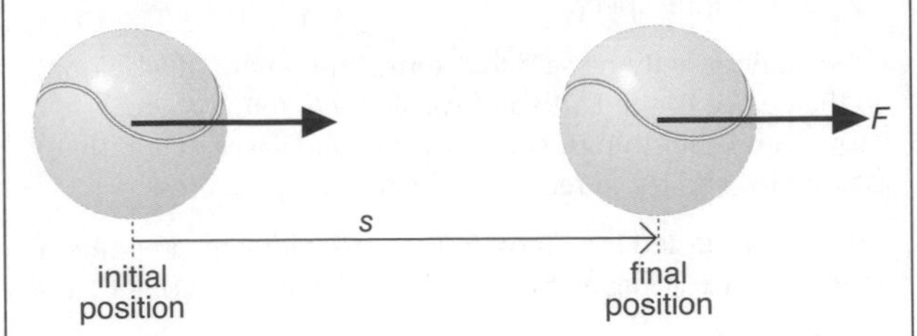

Above, F is the resultant force on an object. If W is the work done when the force has caused a displacement s, then

$W = Fs$

When the direction of the force is not in the same direction as the displacement, the displacement in the direction of the force is $s \cos\theta$.

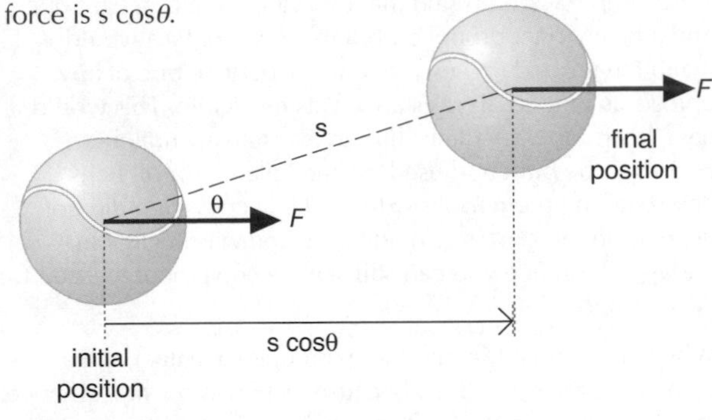

$W = F s \cos \theta$

Force-displacement graphs

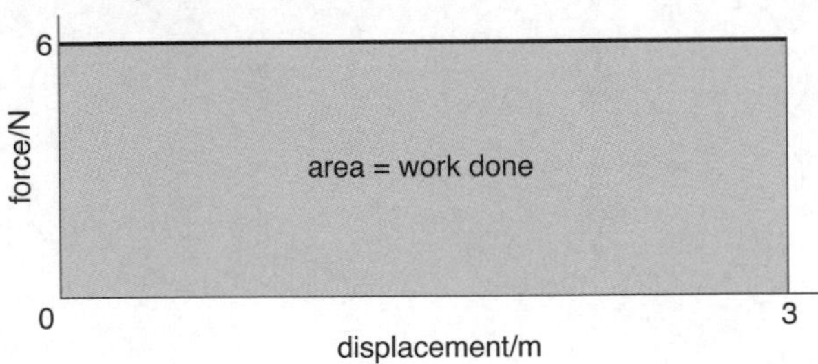

The graph above is for a uniform force of 6 N. When the displacement is 3 m, the work done is 18 J. Numerically, this is equal to the area under the graph between the 0 and 3 m points.

The area under a force-displacement graph is equal to the work done by the force. The following graph is for a non-uniform force, for example the force used to stretch a spring.

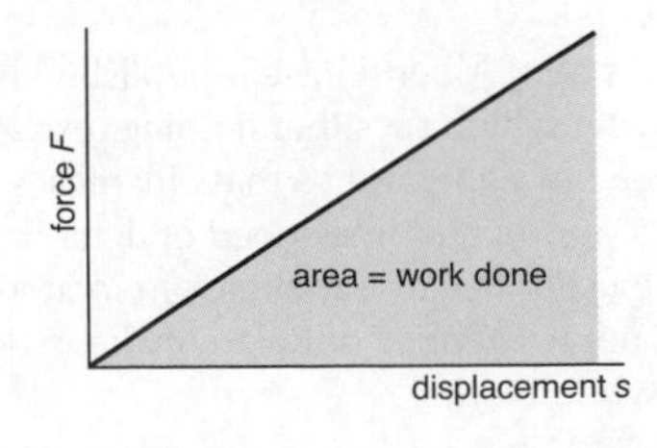

Energy

Things have energy if they can do work. The SI unit of energy is also the joule (J). You can think of energy as a 'bank balance' of work which can be done in the future.

When something does work, its energy decreases. When work is done on something, its energy increases.

A useful model of energy is to say that energy exists in different forms and can be transformed from one to another. In this model the forms are:

Kinetic energy This is energy which something has because it is moving.

Potential energy This is energy which something has because of its position, shape, or state. A stone about to fall from a cliff has ***gravitational*** potential energy. A stretched spring has ***elastic*** potential energy. Foods and fuels have ***chemical*** potential energy. Charge from a battery has ***electrical*** potential energy. Particles from the nucleus (centre) of an atom have ***nuclear*** potential energy.

Internal energy Matter is made up of tiny particles (e.g. atoms or molecules) which are in random motion. They have kinetic energy because of their motion, and potential energy because of the forces of attraction trying to pull them together. An object's internal energy is the total kinetic and potential energy of its particles.

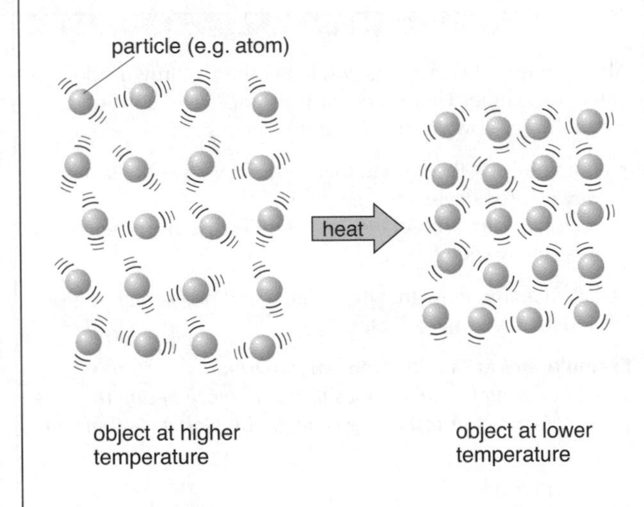

Heat (thermal energy) This is the energy transferred from one object to another because of a temperature difference. Usually, when heat is transferred, one object loses internal energy, and the other gains it.

Radiant energy This is often in the form of waves. Sound and light are examples.

Note:

- Kinetic energy, and gravitational and elastic potential energy are sometimes known as ***mechanical energy***. They are the forms of energy most associated with machines and motion.
- Gravitational potential energy is sometimes just called potential energy (or PE), even though there are other forms of potential energy as described above.

B7 Energy and power

Energy changes

According to the ***law of conservation of energy***,

energy cannot be made or destroyed, but it can be changed from one form to another.

The diagram below shows the sequence of energy changes which occur when a ball is kicked along the ground. At every stage, energy is lost as heat. Even the sound waves heat the air as they die away. As in other energy chains, all the energy eventually becomes internal energy.

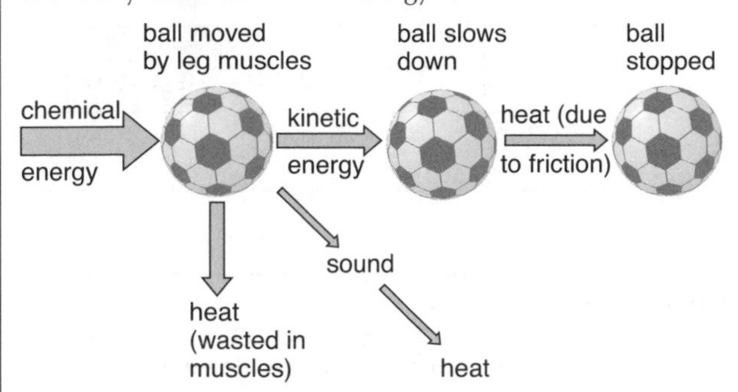

Whenever there is an energy change, work is done – although this may not always be obvious. For example, when a car's brakes are applied, the car slows down and the brakes heat up, so kinetic energy is being changed into internal energy. Work is done because tiny forces are making the particles of the brake materials move faster.

An energy change is sometimes called an energy transformation. Whenever it takes place,

work done = energy transformed

So, for each 1 J of energy transformed, 1 J of work is done.

Calculating kinetic energy (KE)

The stone below has kinetic energy. This is equal to the work done in increasing the velocity from zero to v.
The result is

kinetic energy $= \frac{1}{2}mv^2$

For example, if a 2 kg stone has a speed of 10 m s^{-1},
its KE $= \frac{1}{2} \times 2 \times 10^2 = 100$ J

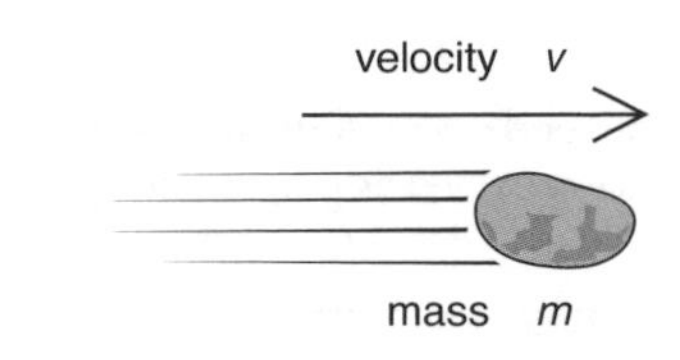

Calculating potential energy (PE)

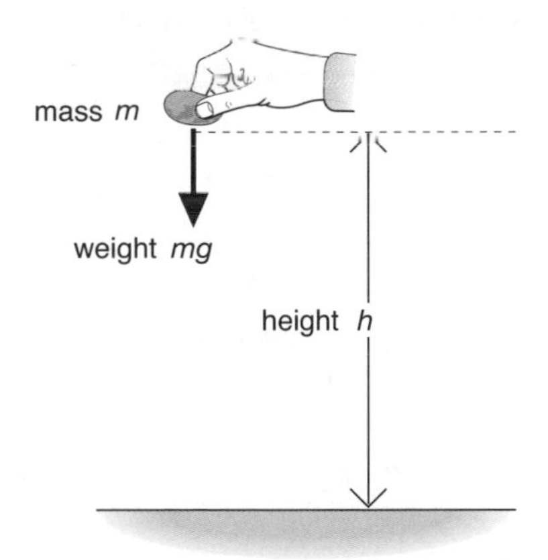

The stone above has potential energy. This is equal to the work done in lifting it to a height h above the ground.

The stone, mass m, has a weight of mg.
So the force needed to overcome gravity and lift it is mg.

As the stone is lifted through a height h,

work done = force × distance moved = $mg \times h$

So potential energy = mgh

For example, if a 2 kg stone is 5 m above the ground, and g is 10 N kg^{-1}, then the stone's PE $= 2 \times 10 \times 5 = 100$ J.

Finding an equation for kinetic energy (KE)

Below, an object of mass m is accelerated from velocity u to v by a resultant force F. While gaining this velocity, its displacement is s and its acceleration is a.

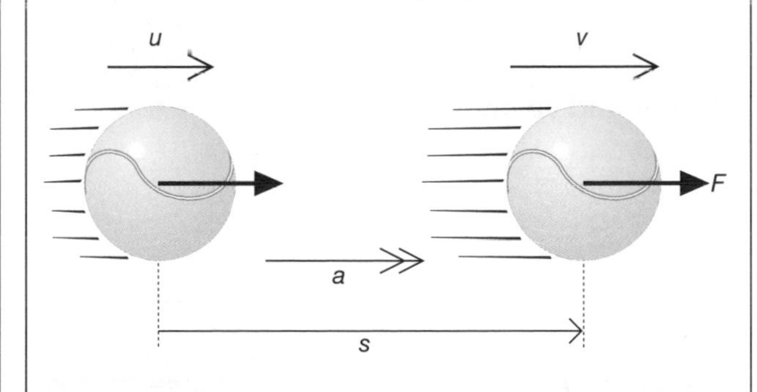

From the law of conservation of energy, the KE gained by the object is equal to the work done on it, Fs.

From equation (4) in B3, $v^2 - u^2 = 2as$

∴ $as = \frac{1}{2}v^2 - \frac{1}{2}u^2$

So $mas = \frac{1}{2}mv^2 - \frac{1}{2}mu^2$

But $mas = Fs$ (because $F = ma$)

So $Fs = \frac{1}{2}mv^2 - \frac{1}{2}mu^2$.

As Fs is the work done, the right-hand side of the equation represents the KE gained. So, when the object's velocity is v, its KE $= \frac{1}{2}mv^2$.

PE to KE

The diagram on the right shows how PE is changed into KE when something falls. The stone in this example starts with 100 J of PE. Air resistance is assumed to be zero, so no energy is lost to the air as the stone falls.

By the time the stone is about to hit the ground (with velocity v), all of its potential energy has been changed into kinetic energy. So

$$\tfrac{1}{2}mv^2 = mgh$$

Dividing both sides by m and rearranging gives

$$v = \sqrt{2gh}$$

In this example, $v = \sqrt{2 \times 10 \times 5} = 10\ \text{m s}^{-1}$.

Note that v does not depend on m. A heavy stone hits the ground at exactly the same speed as a light one.

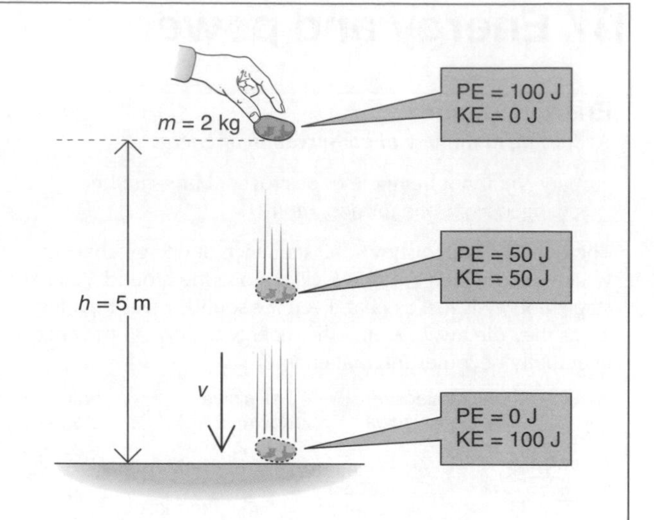

Power

Power is calculated like this:

$$\text{power} = \frac{\text{energy transferred}}{\text{time taken}} \quad \text{or} \quad \text{power} = \frac{\text{work done}}{\text{time taken}}$$

The SI unit of power is the **watt** (W). A power of 1 W means that energy is being transformed at the rate of 1 joule/second (J s^{-1}), so work is being done at the rate of 1 J s^{-1}.

Below, you can see how to calculate the power output of an electric motor which raises a mass of 2 kg through a height of 12 m in 3 s:

$$\text{PE gained} = mgh = 2 \times 10 \times 12 = 240\ \text{J}$$

$$\text{power} = \frac{\text{energy transferred}}{\text{time taken}} = \frac{240}{3} = 80\ \text{W}$$

motor

Efficiency

Energy changers such as motors waste some of the energy supplied to them. Their ***efficiency*** is calculated like this:

$$\text{efficiency} = \frac{\text{useful energy output}}{\text{energy input}} = \frac{\text{useful power output}}{\text{power input}}$$

power input
100 W
electric motor: efficiency 80%
80 W
useful power output
20 W power wasted as heat

For example, if an electric motor's power input is 100 W, and its useful power output (mechanical) is 80 W, then its efficiency is 0.8. This can be expressed as 80%.

Power and velocity

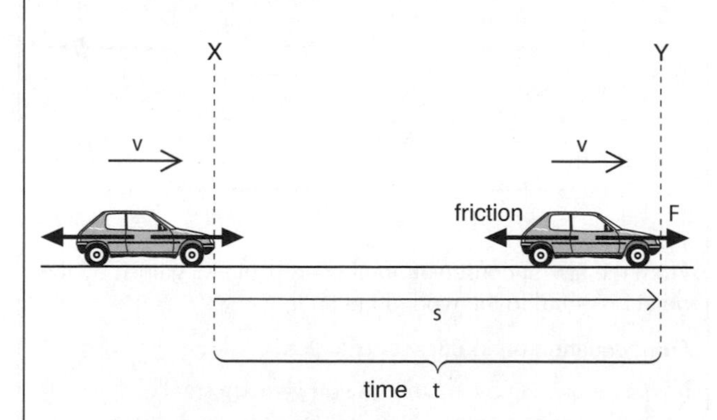

Above, the car's engine provides (via the driven wheels) a forward force F which balances the total frictional force (mainly air resistance) on the car. As a result, the car maintains a steady velocity v. The displacement of the car is s in time t. P is the power being delivered to the wheels.

In moving from X to Y, work done (by F) $= Fs$.

$$\text{power} = P = \frac{\text{work done}}{\text{time taken}} = \frac{Fs}{t}$$

But $v = \frac{s}{t}$ so $P = Fv$

i.e. power delivered = force × velocity

For example, if a force of 200 N is needed to maintain a steady velocity of 5 m s^{-1} against frictional forces,

power delivered = 200 × 5 = 1000 W

All of this power is wasted as heat in overcoming friction. Without friction, no forward force would be needed to maintain a steady velocity, so no work would be done.

B8 Motion in fluids

Drag

When objects move through a fluid – a gas or a liquid – there is a resistive force that acts to slow the object down. This is called drag (or sometimes air resistance, if the motion is in air). The force depends on what fluid the object is moving through, the shape of the object and the velocity of the object through the fluid (or the velocity squared in many cases).

Terminal velocity

Air resistance on a falling object can be significant. As the velocity increases, the air resistance increases, until it eventually balances the weight. The resultant force is then zero, so there is no further gain in velocity. The object has reached its ***terminal velocity***.

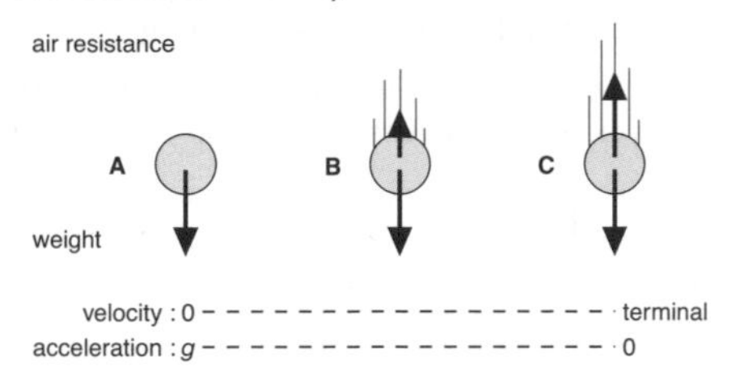

An object falling at terminal velocity loses potential energy but does not gain kinetic energy. Instead, the energy heats up the object and the surrounding fluid. Spacecraft and asteroids re-entering the Earth's atmosphere at very high speeds can burn up, so the space shuttle is covered with special heat-resistant tiles.

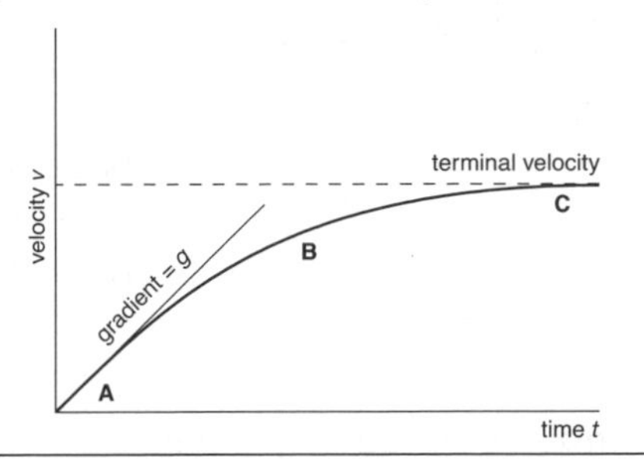

How Science Works

Making use of drag

Increasing drag

Parachutes increase the drag force to slow objects down. They are used on some jet aircraft when landing, and on some racing cars.

Forces on a skydiver:

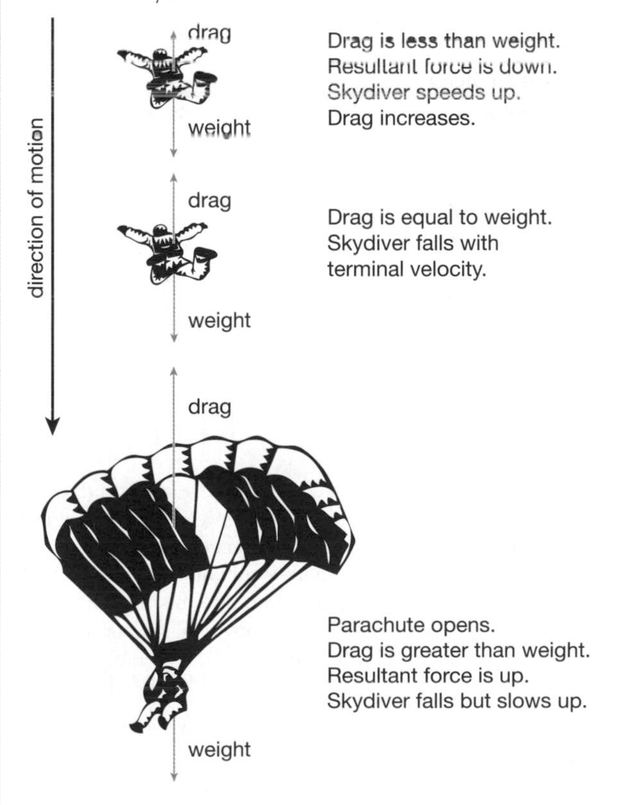

Reducing drag

In the 2008 Beijing Olympic Games, a number of swimming world records were broken. The swimmers will have been helped by the drag-reducing swimsuits that covered more skin and were made of a new low-drag, water-repellent polyurethane membrane. Car designers try to achieve as low a drag as possible by reducing the front area of a vehicle and by shaping it so that it is streamlined, which means that the air flows smoothly over the surface.

Maximum speed of a car

Provided that the maximum frictional force between the tyres and the road is not exceeded the maximum traction force F between the road and a car is fixed by the size of the engine. When the traction force is equal to the sum of all the resistive forces acting the maximum speed is reached.

The frictional force is approximately constant when motion has started. Initially the drag force will be small and the car is able to accelerate. As speed builds up eventually a speed will be reached when

$$F_{traction} = F_{friction} + F_{drag}$$

$$F_{drag} \propto v^2$$

Higher maximum speeds can be achieved by

- increasing the traction force that the engine can provide
- reducing the drag coefficient by streamlining the car by reducing the frontal area of the vehicle and drag coefficient.

Maximum speed and power of an engine

At the maximum speed the output power of the engine

$$P_{max} = \text{traction force} \times \text{velocity}$$

The power used to overcome friction is much smaller than that used to overcome the drag , so the drag is the major limiting factor. Hence

$$P_{max} = F_{drag} \times v$$

$$P_{max} \propto v^3 \text{ (approximately)}$$

This means that to increase the speed by about 25% the engine power output has to be doubled ($1.25^3 \approx 2$).
Note: The above analysis applies to any object (from people to aircraft) moving against air resistance.

B9 Fluid flow

Streamlined/laminar flow

When a fluid flows in layers that slide over each other so that the movement of the layers can be represented by smooth lines that do not cross, the flow is said to be streamlined, or laminar ('laminar' means 'layered').

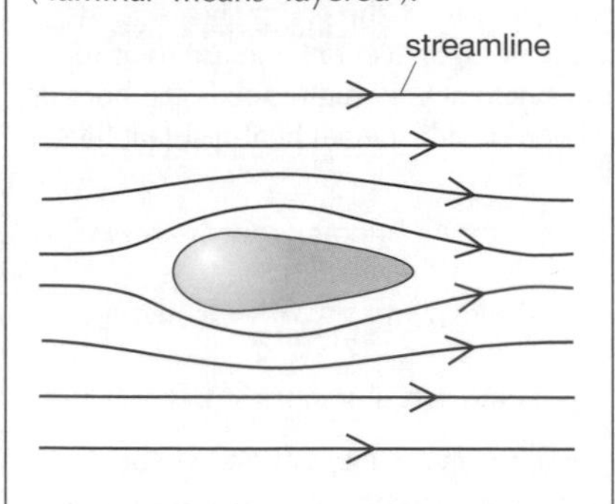

Viscosity

When a fluid flows through a pipe, one of the factors affecting the speed is how viscous it is. Treacle, for example, is very viscous. Small spherical objects moving slowly through fluids are slowed by a force called ***viscous drag***.

Viscous drag is due to the ***viscosity*** of the fluid. For large, high-speed, objects there is a much larger contribution to the drag, which depends on the density of the fluid.

To measure the ***coefficient of viscosity***, η, George Stokes timed the fall of a ball bearing, travelling at terminal velocity, through columns of different fluids (see Stokes' Law). The same idea is used in the falling-ball viscometer, an instrument used to measure the viscosity of different fluids.

Viscous drag slows a ball moving through the fluid at velocity v. If, instead, the fluid is moving at velocity v past the stationary ball, then the viscous drag is the same.

At any given temperature most fluids have a constant η. These fluids are called ***Newtonian fluids***. Values for η are given at a particular temperature, because η is very dependent on temperature.

Thixotropic liquids are very viscous until they are stirred. When v increases, η decreases. Examples are some paints and glues, and tomato ketchup.

Upthrust and Archimedes' principle

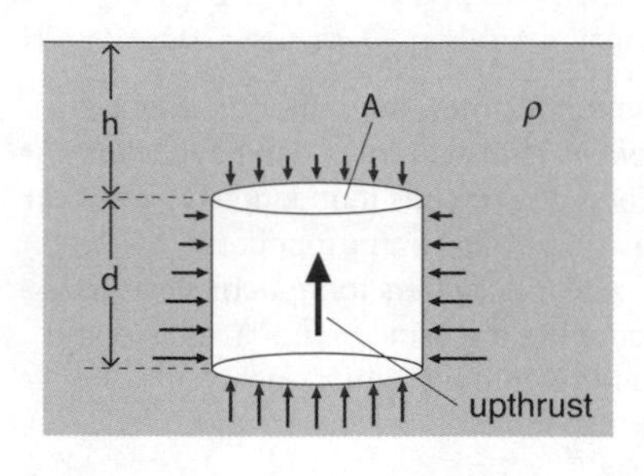

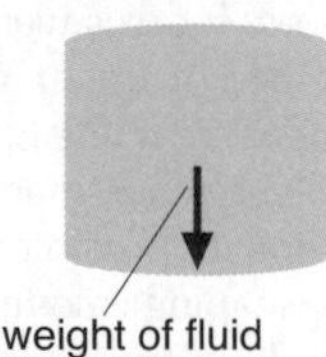

The cylinder above is immersed in a fluid (in this case, a liquid). As a result, fluid is ***displaced***. The mass of fluid displaced is ρdA.

The pressure on the bottom of the cylinder is greater than on the top, so the upward force on the cylinder is greater than the downward force. The *resultant* upward force is the difference between the two. It is called the **upthrust**.

	upward force	$= \rho g(h + d)A$
	downward force	$= \rho g h A$
So	upthrust	$= \rho g(h + d)A - \rho g h A = \rho g d A$
But	$\rho g d A$	= weight of fluid displaced
So	upthrust	= weight of fluid displaced

This is known as ***Archimedes' principle***.

Note:

- The principle applies to all fluids (liquids *and* gases).
- The principle applies to an immersed object of any shape, including one which is only partly immersed.

Stokes' law

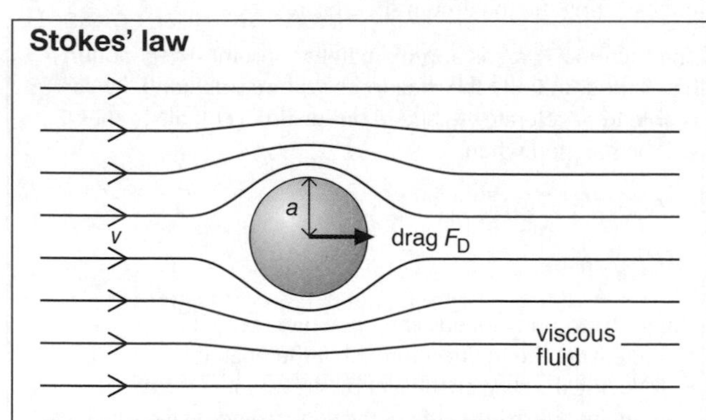

Above, a sphere with radius a is moving through a fluid at a speed v (for simplicity, in this and later diagrams, the air is shown moving, rather than the object). If the flow is streamline, as shown, then the ***viscous drag*** F_D (resisting force from the fluid) is given by this equation, called ***Stokes' law***:

$$F_D = 6\pi\eta a v$$

Note:

- In this case, drag ∝ speed

A falling sphere will reach its ***terminal velocity*** when the forces on it balance (see right, and also B8), i.e.

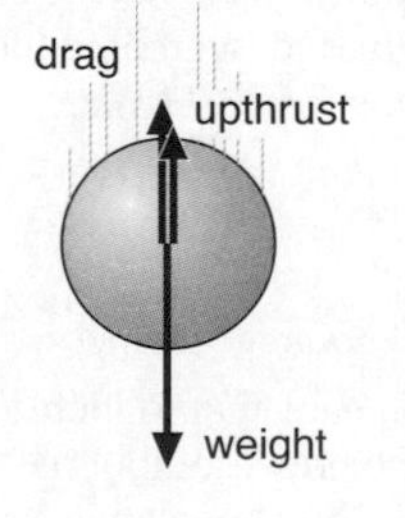

weight = drag + upthrust

Turbulent flow

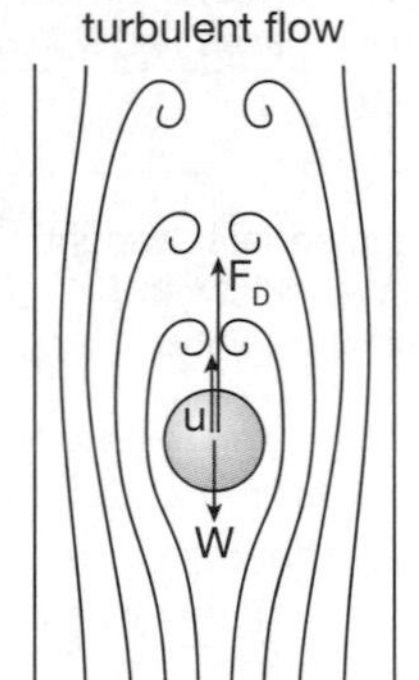

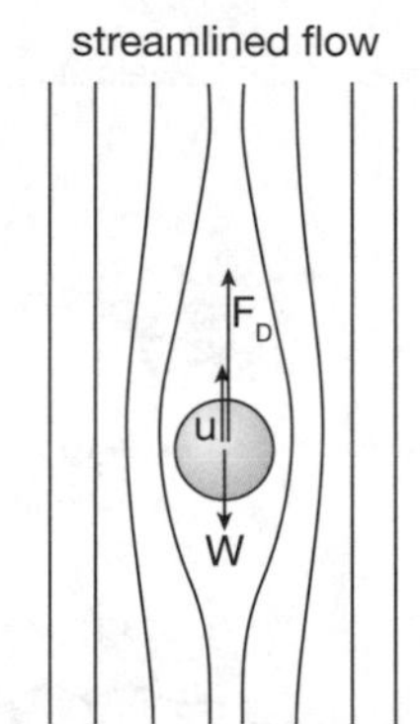

When a sphere (or other object) moves through a fluid, or a fluid flows through a pipe, the flow is only streamline beneath a certain ***critical speed***. Beyond this speed, it becomes *turbulent*, as shown above. There is mixing of the layers and eddies are formed. Energy is transferred to the fluid, which heats up. Turbulence arises in most practical situations involving fluid flow and results in reduced efficiency.

Note:

- Stokes' law only applies at low speeds, where the flow is non-turbulent.

B10 Newton's laws

Newton's first law

The equation $F = ma$ implies that, if the resultant force on something is zero, then its acceleration is also zero. This idea is summed up by ***Newton's first law of motion***:

> If there is no resultant force acting,
> - a stationary object will stay at rest,
> - a moving object will maintain a constant velocity (a steady speed in a straight line).

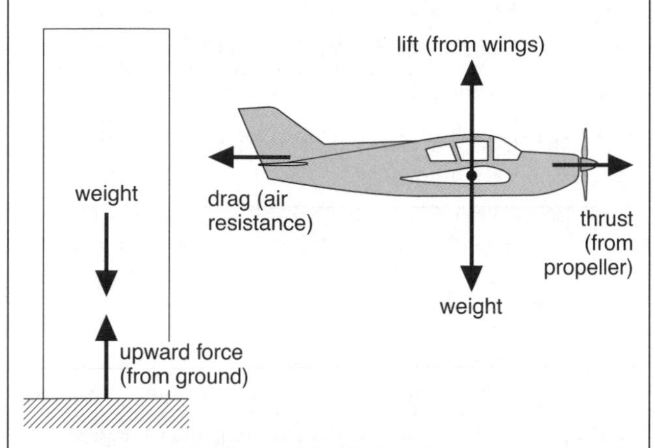

From Newton's first law, it follows that if an object is at rest or moving at constant velocity, then the forces on it must be balanced, as in the examples above.

The more mass an object has, the more it resists any change in motion (because more force is needed for any given acceleration). Newton called this resistance to change in motion ***inertia***.

Aircraft propulsion

To move forward, an aircraft pushes a mass of gas backwards so that, by Newton's third law, there is an equal forward force on the aircraft. Here are two ways of producing a backward flow of gas:

Jet engine Air is drawn in at the front by a large fan, and pushed out at the back. Exhaust gases are also ejected, at a higher speed.

Propeller This is driven by the shaft of a jet engine or piston engine. Its blades are angled so that air is pushed backwards as it rotates.

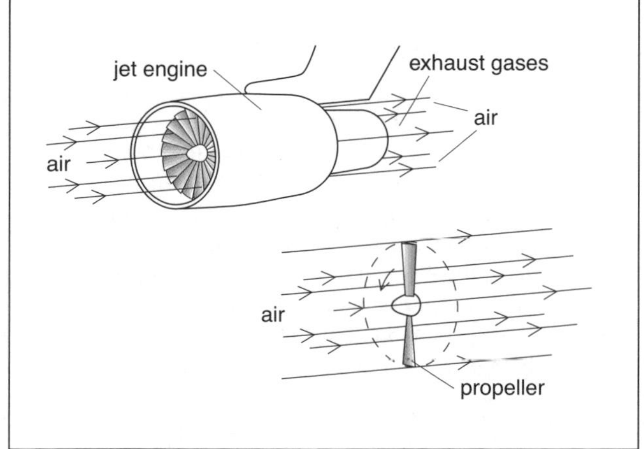

Newton's third law

A single force cannot exist by itself. Forces are always pushes or pulls between *two* objects, so they always occur in pairs. One force acts on one object; its equal but opposite partner acts on the other. This idea is summed up by ***Newton's third law of motion***:

> If A is exerting a force on B, then B is exerting an equal but opposite force on A.

The law is sometimes expressed as follows:

> To every action, there is an equal but opposite reaction.

Examples of action–reaction pairs are given below.

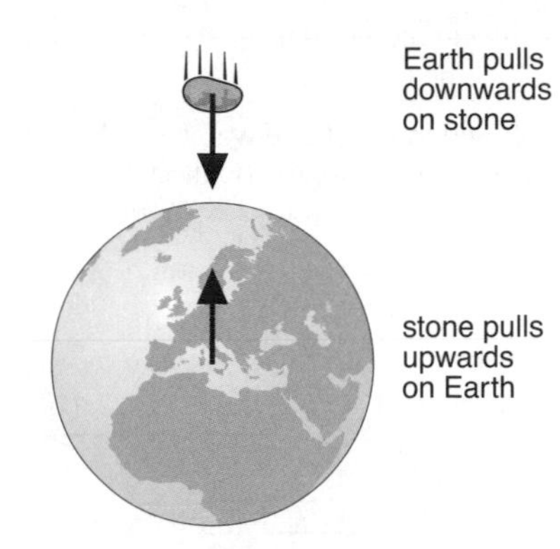

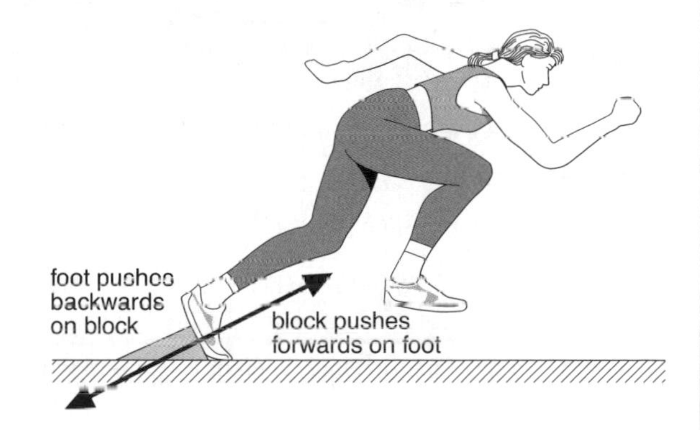

Note:
- It does not matter which force you call the action and which the reaction. One cannot exist without the other.
- The action and reaction do not cancel each other out because they are acting on *different* objects.

Newton's second law – for constant mass

When a resultant force acts on an object, its velocity changes. Newton's second law applies to situations in which the mass and velocity both change – for example, a rocket using fuel so that the mass reduces as the rocket accelerates.

In the special case of the mass staying constant, Newton's Law can be written as

Resultant force = mass × acceleration

$F = m\,a$

This equation comes from Newton's second law, but it is not correct to say it is Newton's second law – because the equation can only be used when the mass is constant.

B11 Momentum and impulse

Momentum and Newton's second law

The product of an object's mass m and velocity v is called its ***momentum, p***:

momentum = mass × velocity

$p = mv$

Momentum is measured in kg m s^{-1}. It is a vector.

According to ***Newton's second law of motion***:

The rate of change of momentum of an object is proportional to the resultant force acting.

This can be written in the following form:

$$\text{resultant force} \propto \frac{\text{change in momentum}}{\text{time taken}}$$

With the unit of force defined in a suitable way (as in SI), the above proportion can be changed into an equation:

$$F = \frac{mv - mu}{t} \qquad (1)$$

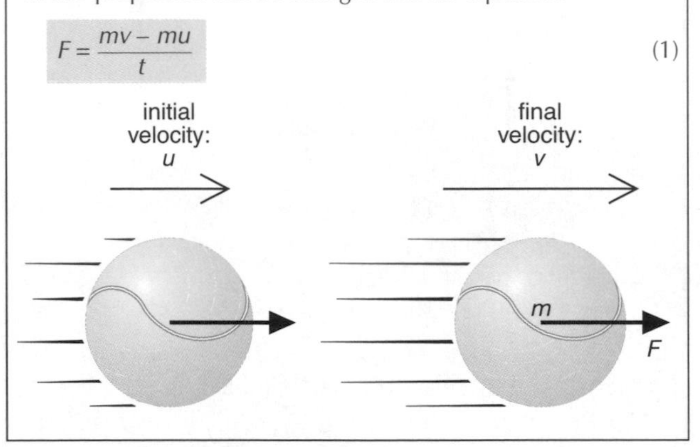

Linked equations

Equation (1) can be rewritten $F = \dfrac{m(v - u)}{t}$

But acceleration $a = \dfrac{(v - u)}{t}$. So, when mass is constant, $F = ma$ (2)

Equations (1) and (2) are therefore different versions of the same principle.

Note:

- According to Einstein, mass increases with velocity (though insignificantly for velocities much below that of light). This means that $F = ma$ is really only an approximation, though an acceptable one for most practical purposes.
- When using equations (1) and (2), remember that F is the resultant force acting. For example, on the right, the resultant force is 26 – 20 = 6 N upwards. The upward acceleration a can be worked out as follows:

$$a = \frac{F}{m} = \frac{6}{2} = 3 \text{ m s}^{-2}$$

force F

area = impulse

time t

mass: 2 kg

weight: 20 N

engine thrust: 26 N

Impulse

Equation (1) can also be rewritten $Ft = mv - mu$

In words force × time = change in momentum

The quantity 'force × time' is called an ***impulse***.

A given impulse always produces the same change in momentum, irrespective of the mass. For example, if a resultant force of 6 N acts for 2 s, the impulse delivered is 6 × 2 = 12 N s.

This will produce a momentum change of 12 kg m s^{-1}

So a 4 kg mass will gain 3 m s^{-1} of velocity

or a 2 kg mass will gain 6 m s^{-1} of velocity, and so on.

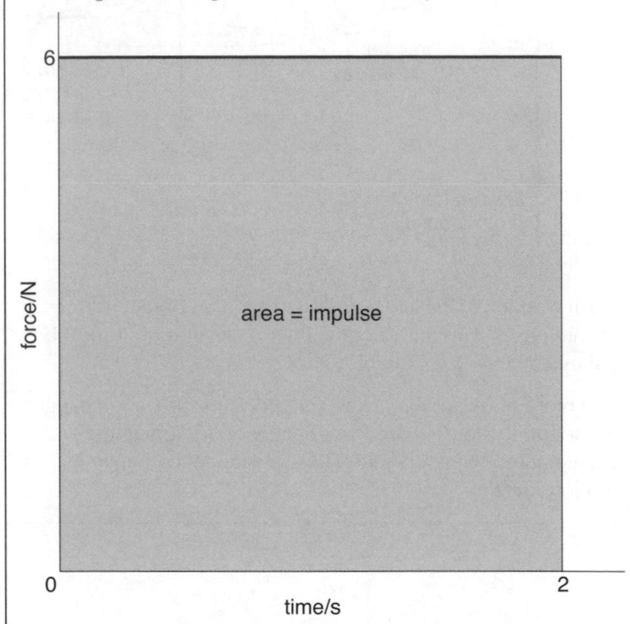

The graph above is for a uniform force of 6 N. In 2 s, the impulse delivered is 12 N s. Numerically, this is equal to the area of the graph between the 0 and 2 s points.

Momentum problem

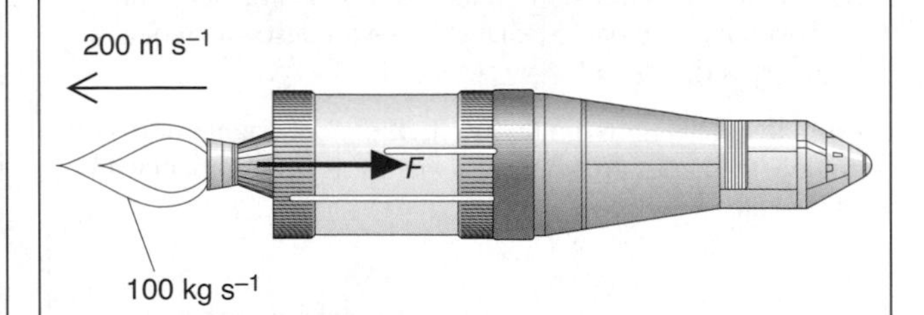

Example *A rocket engine ejects 100 kg of exhaust gas per second at a velocity (relative to the rocket) of 200 m s^{-1}. What is the forward thrust (force) on the rocket?*

By Newton's third law, the forward force on the rocket is equal to the backward force pushing out the exhaust gas. By Newton's second law, this force F is equal to the momentum gained per second by the gas, so it can be calculated using equation (1) with the following values:

$m = 100$ kg $\quad t = 1$ s $\quad u = 0 \quad v = 200$ m s^{-1}

$$\text{So } F = \frac{mv - mu}{t} = \frac{(100 \times 200) - (100 \times 0)}{1} = 20\,000 \text{ N}$$

B12 Collisions – momentum and energy

Conservation of momentum

Trolleys A and B below are initially at rest. When a spring between them is released, they are pushed apart.

By Newton's third law, the force exerted by A on B is equal (but opposite) to the force exerted by B on A. These equal forces also act for the same time, so they deliver equal (but opposite) impulses. As a result, A gains the same momentum to the left as B gains to the right.

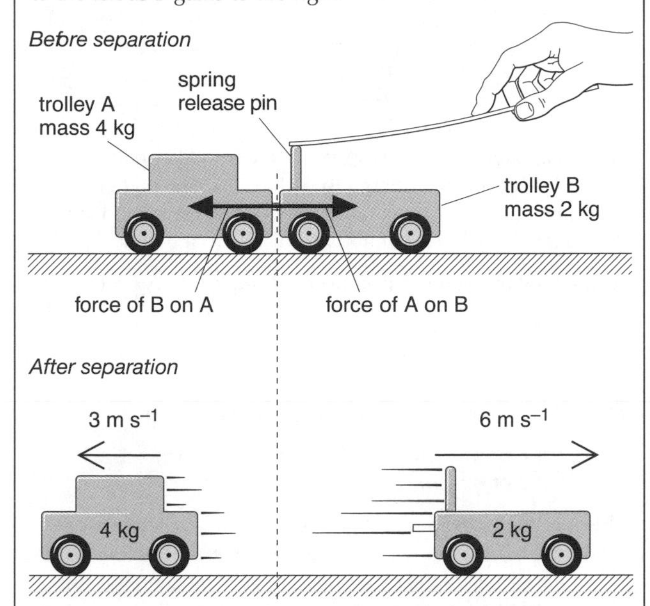

Momentum is a vector, so its direction can be indicated using + or –. If vectors to the right are taken as +,

before the trolleys separate
total momentum = 0

after the trolleys separate
momentum of A = 4 × (–3) = –12 kg m s^{-1}
momentum of B = 2 × (+6) = +12 kg m s^{-1}
so total momentum = 0 kg m s^{-1}

Together, trolleys A and B make up a ***system***. The total momentum of this system is the same (zero) before the trolleys push on each other as it is afterwards. This illustrates the ***law of conservation of momentum***:

> When the objects in a system interact, their total momentum remains constant, provided that there is no external force on the system.

Below, the separating trolleys are shown with velocities of v_1 and v_2 instead of actual values. In cases like this, it is always best to choose the same direction as positive for all vectors. It does not matter that A is really moving to the left. If A's velocity is 3 m s^{-1} to the left, then $v_1 = -3$ m s^{-1}.

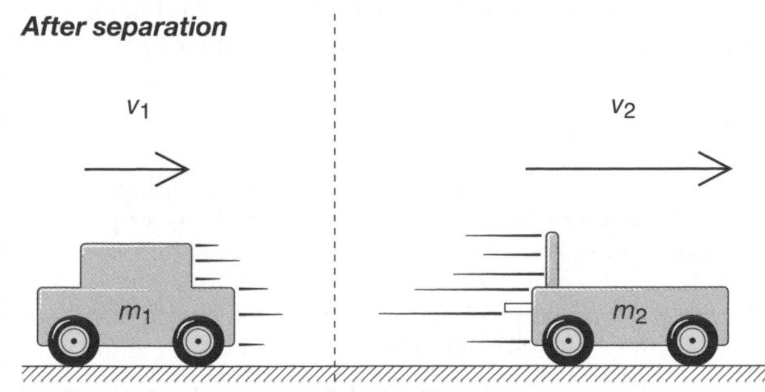

As the total momentum of the trolleys is zero,

$$m_1v_1 + m_2v_2 = 0$$

So, if v_2 is positive, v_1 must be negative.

Momentum in collisions

Whenever objects collide, their total momentum is conserved, provided that there is no external force acting.

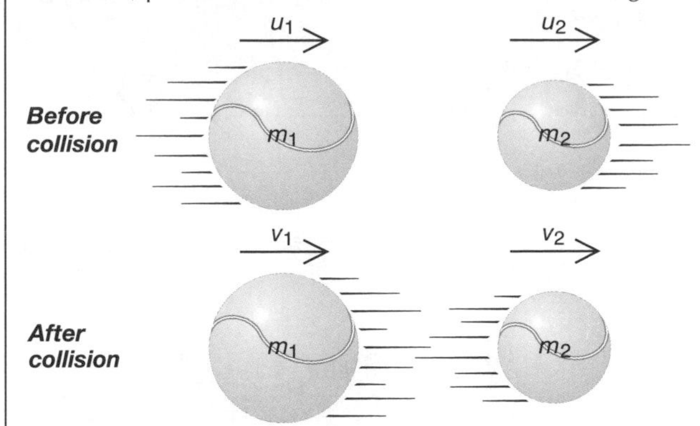

Above, two balls collide and then separate. All vectors have been defined as positive to the right. As the total momentum is the same before and after,

$$m_1u_1 + m_2u_2 = m_1v_1 + m_2v_2$$

Kinetic energy in collisions

Elastic collision An elastic collision is one in which the total kinetic energy of the colliding objects remains constant. In other words, no energy is converted into heat (or other forms). If the above collision is elastic,

$$\tfrac{1}{2}m_1u_1^2 + \tfrac{1}{2}m_2u_2^2 = \tfrac{1}{2}m_1v_1^2 + \tfrac{1}{2}m_2v_2^2$$

One consequence of the above is that the speed of separation of A and B is the same after the collision as before:

$$u_1 - u_2 = -(v_1 - v_2)$$

Inelastic collision In an inelastic collision, kinetic energy is converted into heat. The total amount of *energy* is conserved, but the total amount of *kinetic energy* is not.

Energy or momentum

Energy and momentum are completely different quantities. They have different dimensions. Energy has dimensions of force × distance. It is measured in joules. One joule of kinetic energy is gained when a force of 1 newton acts on an object over a distance of 1 m.

Momentum has dimensions of force × time. It is measured in newton seconds (or kg m s^{-1}, which is the same). One newton second of momentum is gained when a force of 1 newton acts on an object for 1 second. Momentum is a vector and has direction as well as magnitude. Energy is a scalar and has magnitude only.

The equation KE = $\tfrac{1}{2}mv^2$ and $p = mv$, so

$$\text{KE} = \frac{p^2}{2m}$$

When considering collisions remember the following.

The ***principle of conservation of momentum*** says that the total momentum is conserved if there is no external force acting. This also results from Newton's third law (the force on each of the two objects is equal and opposite) and the fact that the forces must act for the same time on each of the objects. In an ***elastic collision*** kinetic energy is conserved. In an inelastic collision kinetic energy is not conserved. You must specify that it is what happens to the kinetic energy that determines whether a collision is elastic or inelastic, because the ***principle of conservation of energy*** says that energy is always conserved. In collisions energy may be transferred, for example to heat and sound, but the total energy is unchanged.

Collision problems

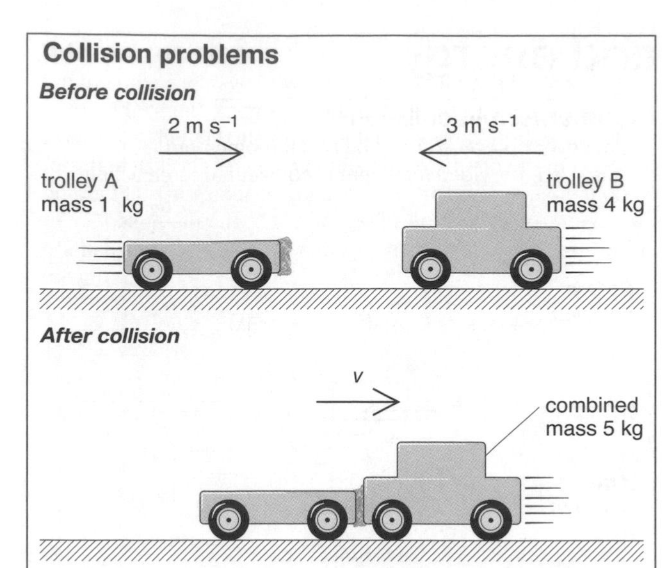

Example 1 *The trolleys above collide and stick together. What is their velocity after the collision? (Assume no friction.)*

All vectors to the right will be taken as positive.
The unknown velocity is v (to the right).

momentum = mass × velocity

before the collision

momentum of A = $1 \times 2 = 2\ \text{kg m s}^{-1}$

momentum of B = $4 \times (-3) = -12\ \text{kg m s}^{-1}$

∴ total momentum = $-10\ \text{kg m s}^{-1}$ (2)

After the collision

A and B have a combined mass of 5 kg, and a combined velocity of v. So total momentum = $5 \times v$.

As the total momentum is the same before and after,

$5v = -10$ which gives $v = -2\ \text{m s}^{-1}$

So the trolleys have a velocity of $2\ \text{m s}^{-1}$ to the *left*.

Example 2 *When the trolleys collide, how much of their total kinetic energy is lost (converted into other forms)?*

$KE = \frac{1}{2}mv^2$

before the collision

KE of A = $\frac{1}{2} \times 1 \times 2^2 = 2$ J

KE of B = $\frac{1}{2} \times 4 \times (-3)^2 = 18$ J

∴ total KE = 20 J (3)

after the collision

total KE = $\frac{1}{2} \times 5 \times (-2)^2 = 10$ J

Comparing the total KEs before and after, 10 J of KE is lost.

Example 3 *If the collision had been elastic, what would the velocities of the trolleys have been after separation?*

Let v_1 be the final velocity of A and v_2 be the final velocity of B (both defined as positive to the right).

As both total momentum and total KE are conserved,

total momentum after collision $= -10\ \text{kg m s}^{-1}$ (from 2)
total KE after collision = 20 J (from 3)

So $(1 \times v_1) + (4 \times v_2) = -10$
And $(\frac{1}{2} \times 1 \times v_1^2) + (\frac{1}{2} \times 4 \times v_2^2) = 20$

Solving these equations for v_1 and v_2 gives

$v_1 = -6\ \text{m s}^{-1}$ and $v_2 = -1\ \text{m s}^{-1}$

Note:

- There is an alternative solution which gives the velocities before the collision: $2\ \text{m s}^{-1}$ and $-3\ \text{m s}^{-1}$, which is not possible, as it means the trolleys have passed through each other.

Recoiling particles

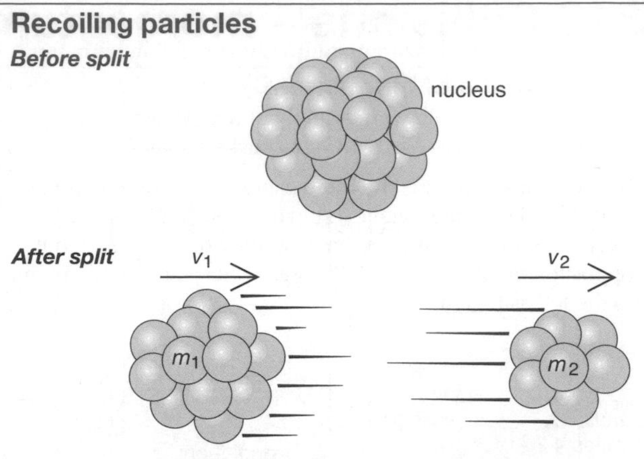

Above, an atomic nucleus splits into two smaller particles with a loss of nuclear energy (see also G3). The particles share the energy released (as kinetic energy) and shoot apart. All vectors have been defined as positive to the right.

As the total momentum is conserved, $m_1v_1 + m_2v_2 = 0$ (4)

Also KE of A = $\frac{1}{2}m_1v_1^2$ (5)

and KE of B = $\frac{1}{2}m_2v_2^2$ (6)

From (4), (5), and (6), the following can be obtained:

$$\frac{\text{KE of A}}{\text{KE of B}} = \frac{m_2}{m_1}$$

This means, for example, that if A has 9 times the mass of B, then B will shoot out with 9 times the KE of A. In other words, it will have 90% of the available energy. The energy is only shared equally if A and B have the same mass.

How Science Works

Reducing collision damage

When a vehicle comes to a sudden stop, the momentum and kinetic energy of the vehicle and the occupants are reduced to zero very rapidly. There are two ways of looking at this.

To reduce the kinetic energy, work is done on the person over a very small distance, which means the force is very large. ($\Delta KE = W = Fs$)

To reduce the momentum there is an impulse on the person for a very short time, which means the force is very large. ($\Delta p = Ft$)

The force on the occupants can be reduced, while still transferring the same amount of kinetic energy, by increasing the distance over which they are brought to a stop.

The force on the occupants can be reduced, while still reducing the momentum to zero, by increasing the time over which the change of momentum occurs.

Safety features, such as air bags, seat belts and crumple zones, make use of these ways of reducing the force (see B13, p. 41).

Other examples are:

- bending your knees when you land from jumping, to increase the distance and time over which your upper body is brought to a stop
- moving your hands with a ball when you catch it to prevent it stinging your hands
- train buffers compressing and bringing the train to a stop
- safety helmets containing padding that compresses and stops the head hitting a hard surface.

B13 Cars in motion

Stopping distances

When a driver notices a hazard that requires an emergency stop, he or she has to react to the emergency and apply the brakes. The brakes then have to bring the car to rest.

stopping distance = thinking distance + braking distance

thinking distance = speed of vehicle × driver's reaction time = vt

braking distance = distance travelled between brakes being applied and car stopping (for a constant deceleration of the vehicle, a using $v^2 = u^2 + 2as$ gives $s = \frac{v^2}{2a}$)

The stopping distance d_m is given by

$$d_m = vt + \frac{v^2}{2a}$$

A typical reaction time is 0.7 s and a typical deceleration is 7.5 m s^{-2}.

The graph shows the stopping distances for this data.

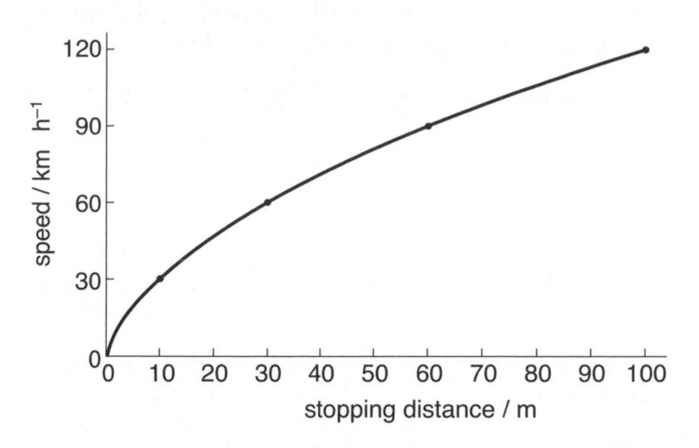

Stopping distance will increase if thinking distance is increased because the reaction time is increased, owing to driver fatigue or drugs including alcohol, for example. Braking distance will increase if:

- a car is heavily loaded
- the condition of the car's brakes is poor, so the braking force is less
- road conditions are poor, because of rain or ice for example, so friction is reduced.

How Science Works

Speed and speed cameras

When the speed of a car is doubled, its kinetic energy goes up by a factor of four. If the maximum braking force is applied, the braking distance will be four times as far at double the speed. The thinking distance will be doubled at double the speed.

Speed limits are set at the maximum safe speed for a road in good conditions, and vehicles should travel at lower than the speed limit in poor weather conditions, which reduce visibility or increase braking distance.

Speed cameras are sited where crashes have occurred when vehicles have exceeded the speed limit. They are one way of trying to ensure that vehicles travel within the speed limits and so reduce the number and severity of crashes. They are unpopular with many drivers. Speed can be shown to be a factor in some crashes, but you can never prove that reducing the speed prevented a particular crash that did not happen.

How Science Works

Satellite navigation systems

Satellite navigation systems use the Global Positioning System (GPS) to pinpoint the position of a vehicle within 10 m.

At any point on the Earth's surface there are four satellites in range. Each satellite has an accurate clock and broadcasts a radio signal with the current time. The receiver compares the signal broadcast time with the arrival time and uses the difference, and the speed of the signal, to work out the distance, d, to the satellite.

This places the receiver on a sphere with the satellite at the centre and a radius d.

The position is the found using trilateration. Using two satellites narrows the position to somewhere on the circle where the two spheres intersect. The third satellite gives a third sphere that intersects the circle at two points. The fourth satellite gives a fourth sphere that will go through only one of the points – and this is the position of the receiver.

Trilateration in two dimensions:

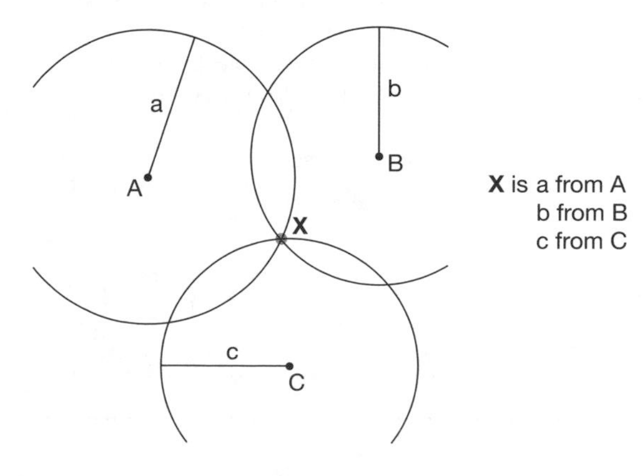

The radio signal travels at the speed of light, and the satellites orbit at 20 000 km, so the satellites have atomic clocks accurate to within a few nanoseconds.

Car safety features

In the event of a crash occurring, cars and equipment in them are designed to protect the occupants.

Large forces can kill or injure the occupants of a vehicle. There are several safety features that work on the principle that the force on a passenger is reduced by:

- reducing the velocity over a longer period of time, because $F = m\Delta(v)/t$
- losing kinetic energy over a longer distance, because $F = \Delta(E_K)/s$.

The design is such that the kinetic energy is reduced to zero in as long a time as possible but before the passenger hits a rigid part of the car.

The energy of the motion is converted into thermal energy and elastic energy.

Seat belts

When front seat occupants hit the windscreen, or rear seat passengers hit the front seats, they stop very quickly in a small distance and the force is large. When they wear seat belts, they move forward a small distance as the seat belt stretches slightly. This increases the distance (and time) during which they stop, so the force is reduced. After a crash the seat belt may have stretched beyond its elastic limit (see C1) and should be replaced.

It is also better to be stopped by the large surface area of the seat-belt strap, than a hard part of the car with a small surface area.

Air bags

An air bag is a flexible nylon bag used for cushioning the impact with hard rigid parts of a vehicle, especially the steering wheel or dashboard. It slows the body and prevents it hitting a hard surface at high speed.

The steps in the deployment of the air bag are:

- a sensor (or sensors) detect the collision and produce an electric signal
- an initiator burns when it receives the electrical signal and ignites the inflator, which provides the gas
- the gas inflates the bag
- the bag gradually deflates.

Sensors

An accelerometer is a sensor that detects sudden deceleration of the vehicle. A simple design is a mass on a spring. When the vehicle decelerates, the mass continues to move. The spring compression depends on the deceleration. When compressed enough, it switches on an electric circuit; for example, a magnetized mass can close a reed switch.

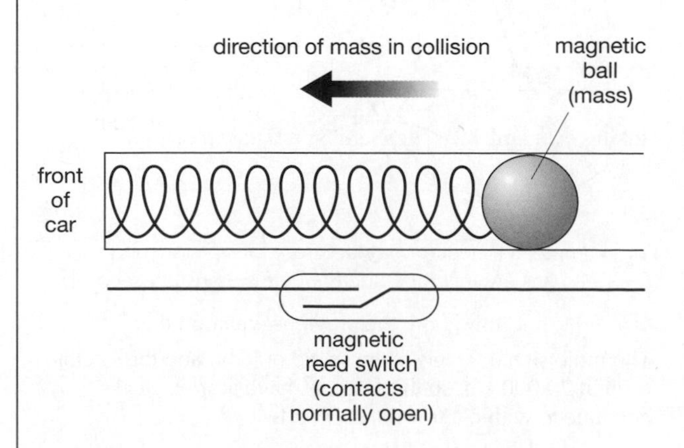

Much smaller, improved accelerometers are now used. For example, some MEMS (micro electromechanical systems) accelerometers use a change in electrical properties in part of an integrated circuit. Some designs use a small moving or tilting mass on part of the chip to change the separation of metal plates, which changes the capacitance of the circuit (see H2).

Other sensors include those that detect the wheel speed and those that detect whether the passenger seat is occupied, to avoid the air bag being deployed unnecessarily.

Inflators

Older inflators contain sodium azide and potassium nitrate, which produce nitrogen gas. Sodium azide is toxic. Today other gas-generating chemicals are used, and there are also designs using compressed gas, and designs that use a mixture of both methods.

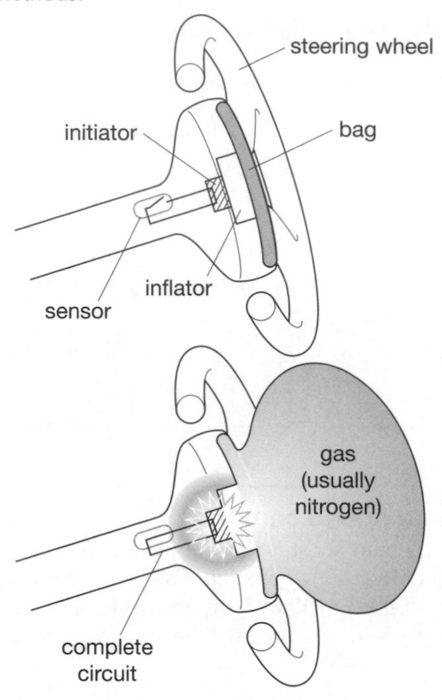

The bag is deployed and inflates in about 0.05 s. It must be fully inflated before the body hits it, otherwise it could do damage as it is moving at high speed towards the body. For this reason, it is important to wear a seat belt to slow the body down and give the air bag time to inflate. The air bag then deflates over the next 0.25 s, through vents in the side, bringing the body slowly to a stop.

Crumple zones

Cars are designed so that there is a rigid central safety cage for the occupants, surrounded by crumple zones at the front and rear of the car. These zones are designed to crumple on impact, so that kinetic energy is transformed by deforming the metal and heating it. While this happens, the occupants can continue to move forward, increasing the time and distance over which they stop and reducing the stopping force.

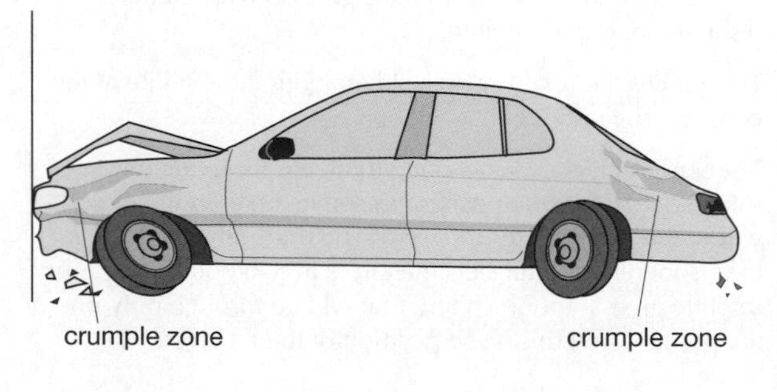

B14 Circular motion

Angular displacement

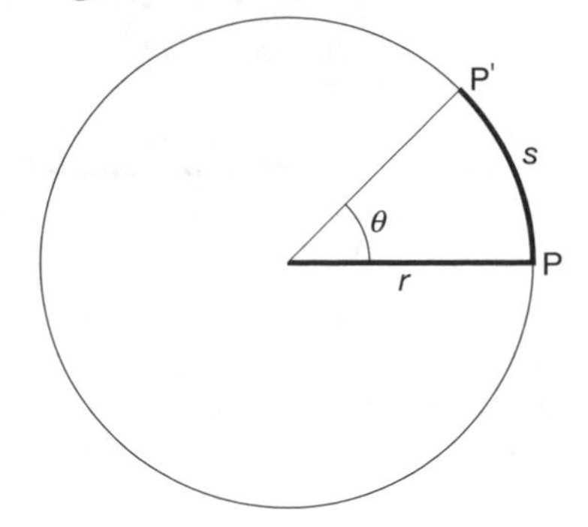

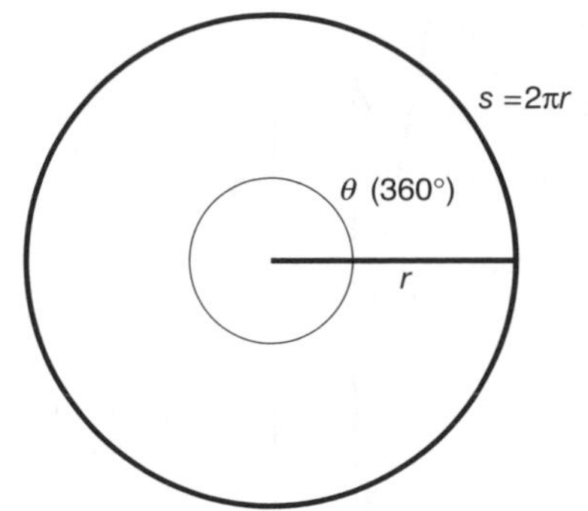

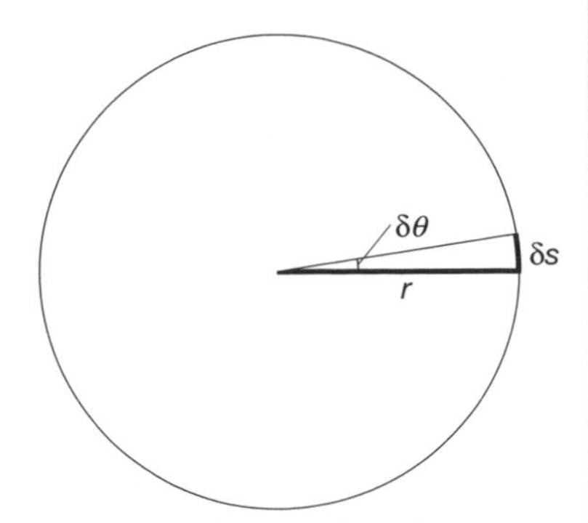

If point P moves to P′, then the angle θ is called the ***angular displacement***. It is measured in ***radians*** (see also A1):

$$\theta = \frac{s}{r}$$

If s is the full circumference of the circle,

$$\theta = \frac{2\pi r}{r} = 2\pi$$

So $\quad 2\pi$ radians = 360°

$$\therefore \; 1 \text{ radian} = \frac{360}{2\pi} = 57.3°$$

Above, $\delta\theta$ is a very small angle. ($\delta\theta$ counts as one symbol.) δs is so small that it can either be the arc of a circle or the side of a triangle. So

$$\sin \delta\theta = \frac{s}{r} = \delta\theta$$

i.e. for *small* angles $\sin \delta\theta = \delta\theta$.

Rate of rotation

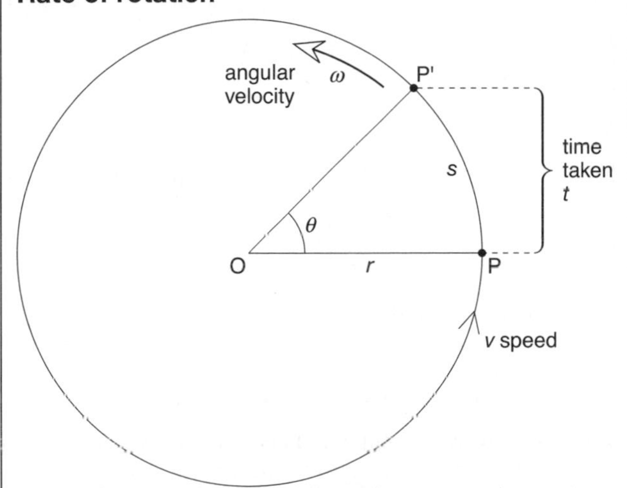

P is a point on a wheel which is turning at a steady rate. In time t, it moves to P′. The rate of rotation can be measured either as an ***angular velocity*** or as a ***frequency***.

Angular velocity ω

$$\text{angular velocity} = \frac{\text{angular displacement}}{\text{time taken}}$$

In symbols $\qquad \omega = \dfrac{\theta}{t}$

For example, if a wheel turns through 10 radians in 2 seconds, then $\omega = 5$ rad s^{-1}. Angular frequency is also measured in rad s^{-1}. It is the magnitude of the vector angular velocity.

Frequency f

$$\text{frequency} = \frac{\text{number of rotations}}{\text{time taken}}$$

Frequency is measured in hertz (Hz). For example, if a wheel completes 12 rotations in 4 seconds, then $f = 3$ Hz.

Period T This is time taken for one rotation. If a wheel makes 3 complete rotations per second ($f = 3$ Hz), then the time taken for one rotation is $\frac{1}{3}$ second. So

$$T = \frac{1}{f}$$

Linking v, ω, and r

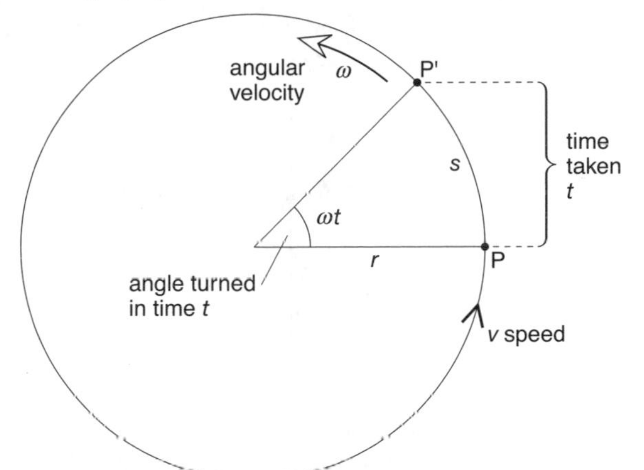

Above, a particle is moving in circle with a steady speed v. (It is not a steady velocity because the direction of the velocity vector is changing.) The particle moves a distance s in time t, so

$$v = \frac{s}{t}$$

As the angular velocity is ω, the angle turned in time t is ωt (from the equation on the left). But $\omega t = s/r$. So $s = \omega t r$. Substituting this in the above equation gives

$$v = \omega r$$

Linking ω, f, and T As there are 2π radians in one full rotation (360°),

$$\omega = 2\pi f$$

For example, a wheel turning at 3 rotations per second ($f = 3$ Hz) has an angular velocity of 6π radians per second.

As $T = 1/f$, it follows from the previous equation that

$$T = \frac{2\pi}{\omega}$$

Centripetal acceleration

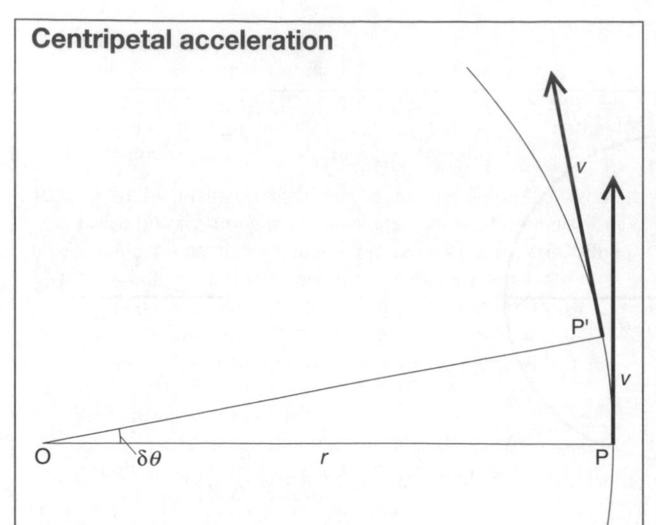

Above, a particle is moving in a circle with a steady speed v. The diagram shows how the velocity vector changes direction as the particle moves from P to P′ in time δt.

Below, the velocity vectors from the previous diagram have been used in a triangle of vectors (see B4).

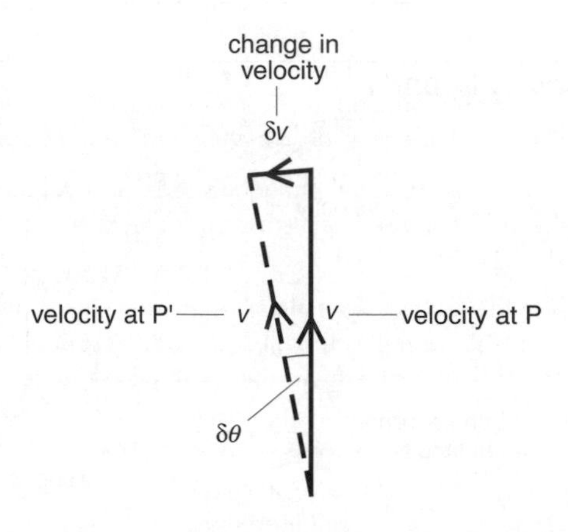

The δv vector represents the *change* in velocity because it is the velocity vector which must be *added* to the velocity at P to produce the new velocity (the resultant) at P′. Note that the change in velocity is towards O. In other words, the particle has an *acceleration* towards the centre of the circle. This is called ***centripetal acceleration***.

If a is the centripetal acceleration, $a = \dfrac{\delta v}{\delta t}$

But, from the triangle above, $\delta\theta = \dfrac{\delta v}{v}$. So $\delta v = v\delta\theta$

Substituting this in the previous equation, $a = \dfrac{v\delta\theta}{\delta t}$

But $\delta\theta = \omega\delta t$. So $a = v\omega$

Using $v = \omega r$, two more versions of the above equation can be obtained. So

$a = v\omega$ $\qquad a = \dfrac{v^2}{r}$ $\qquad a = \omega^2 r$

For example, if a particle is moving at a steady speed of 3 m s^{-1} in a circle of radius 2 m, its centripetal acceleration a is found using the middle equation: $a = 3^2/2 = 4.5$ m s^{-2}.

Note:

- When something accelerates, its velocity changes. As velocity is a vector, this can mean a change in *speed* or *direction* (or both). Centripetal acceleration is produced by a change in direction, not speed.

Centripetal force

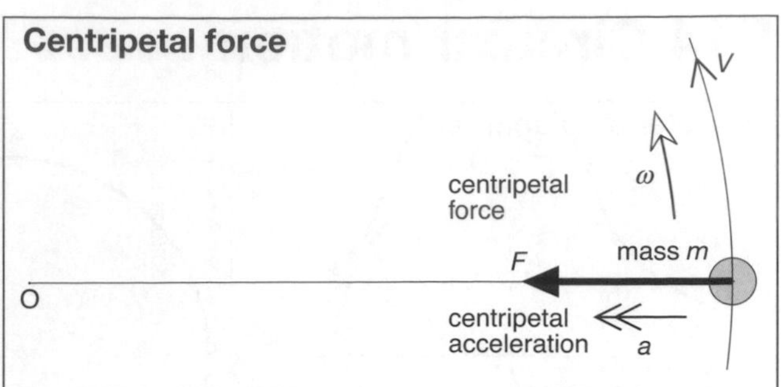

To produce centripetal acceleration, a ***centripetal force*** is needed. It must act towards the centre of the circle. The centripetal force F, mass m, and centripetal acceleration a are linked by the equation $F = ma$. So, using the equations for a in the previous column,

$F = mv\omega$ $\qquad F = \dfrac{mv^2}{r}$ $\qquad F = m\omega^2 r$

Note:

- Centripetal force is *not* produced by circular motion. It is the force *needed* for circular motion. Without it, the object would travel in a straight line.

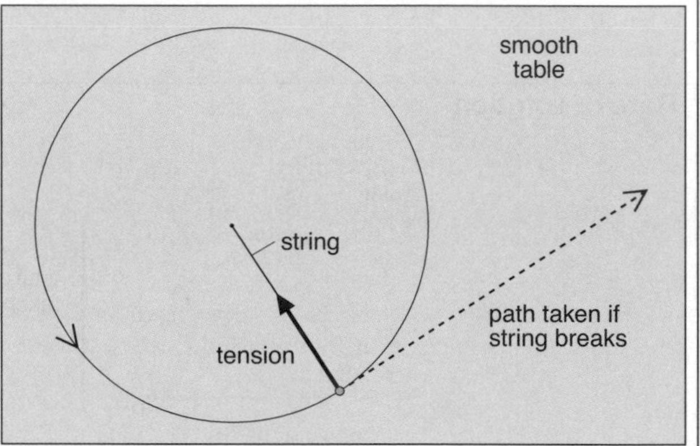

Above, a mass moves in a circle on a smooth table. The tension in the string provides the centripetal force needed. There is no outward 'centrifugal force' on the mass. If the string breaks, the mass travels along a tangent.

Angle of bank An aircraft must bank to turn. This is so that the lift L (from the wings) and the weight mg can produce a resultant to provide the centripetal force F, where

$$F = \frac{mv^2}{r}$$

In the triangle of vectors (below right):

$L \cos\theta = mg$ and $L \sin\theta = \dfrac{mv^2}{r}$

Dividing the second equation by the first gives $\tan\theta = v^2/r$, where θ is the angle of bank required for the turn.

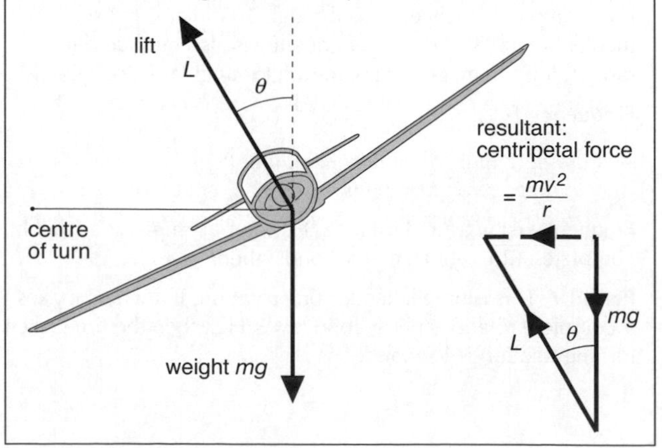

B15 Cycles, oscillations and SHM

Periodic motion

This is motion in continually repeating ***cycles***. Here are two examples:

Circular motion Particle P moving at a steady speed in a circle (see the diagram below).

T is the ***period*** (the time for one cycle).
f is the ***frequency*** (the number of cycles per second).
ω is the ***angular velocity*** (measured in rad s^{-1}).

T, f, and ω are linked by the equations below:

$$T = \frac{1}{f} \qquad \omega = 2\pi f \qquad T = \frac{2\pi}{\omega}$$

(see B11)

Oscillatory motion (e.g. a swinging pendulum)
T, f, and ω are also used when describing oscillatory motion, although ω has no direct physical meaning. They are linked by the same equations as for circular motion.

Linking circular motion and SHM

Below, particle P is moving in a circle with a steady angular velocity ω. Particle B is oscillating about O along the horizontal axis so that it is always vertically above or beneath P. The amplitude A of the oscillation is equal to the radius of the circle, r.

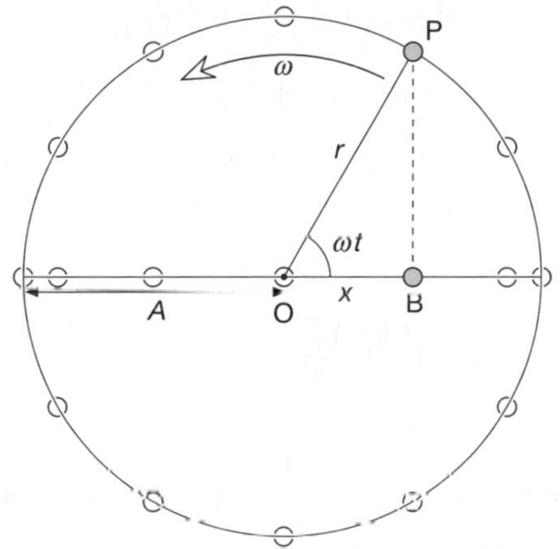

For particle B, $\quad x = r\cos\omega t \qquad (2)$

Using calculus, B's velocity v and acceleration a can be found from the above equation. These are the results:

$$v = -r\omega\sin\omega t \qquad (3)$$

$$a = -r\omega^2\cos\omega t \qquad (4)$$

From equations (4) and (2), it follows that

$$a = -\omega^2 x \qquad (5)$$

From the diagram, $\sin\omega t = \dfrac{\sqrt{r^2 - x^2}}{r}$

Using equation (3) and remembering that $r = A$, the **amplitude** of the oscillation, the velocity at a distance x from the centre of oscillation can be calculated from

$$v = \omega\sqrt{A^2 - x^2} = 2\pi f\sqrt{(A^2 - x^2)} \qquad (6)$$

$v_{max} = \omega A$ (when x = 0)

$a_{max} = -\omega^2 A$ (when x = A)

Note:

- Equation (5) has the same form as equation (1). So particle B is moving with SHM.
- The constant in equation (1) is equal to ω^2.
- Using calculus notation, the equation for SHM can be written in the following form:

$$\frac{d^2x}{dt^2} = -\omega^2 x$$

Defining simple harmonic motion

One commonly occurring type of oscillatory motion is called ***simple harmonic motion*** (**SHM**).

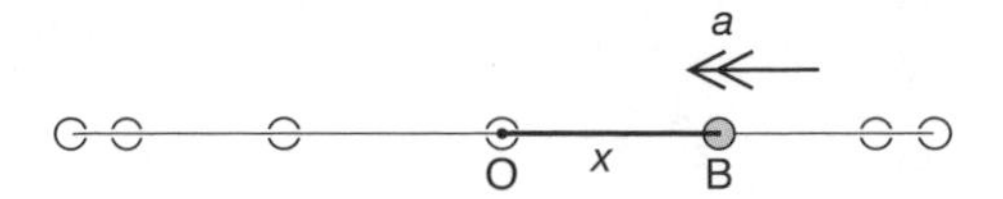

If particle B (above) oscillates about O with SHM, its acceleration is proportional to its displacement from O, and directed towards O.

If x is the displacement, and a is the acceleration (in the x direction), then this can be expressed mathematically:

$$a = -\text{(positive constant) } x \qquad (1)$$

The minus sign indicates that a is always in the opposite direction to x.

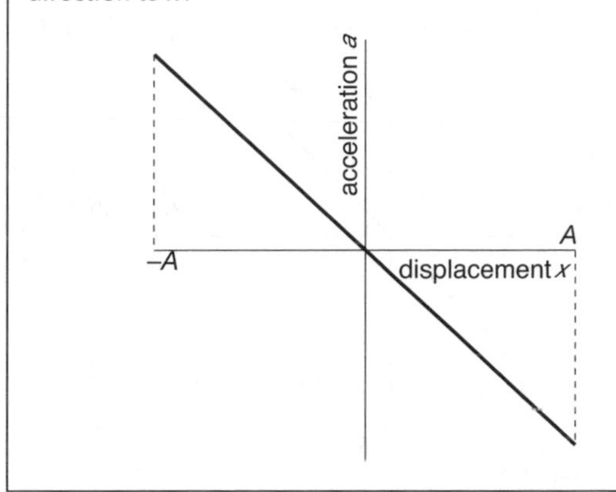

Displacement–time graph for SHM

The following graph shows how the displacement varies with time for one complete oscillation starting from the centre of oscillation.

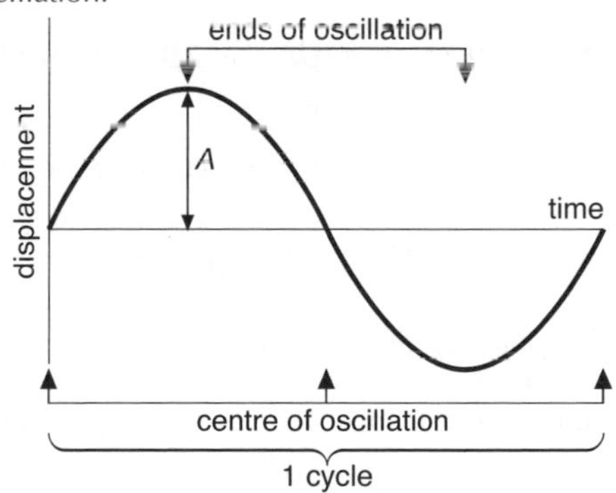

A is the **amplitude** of the oscillation.

Phase of the oscillation

An oscillation that has the same period but reaches its peak at a different time to that shown above is said to ***oscillate out of phase***. The phase difference between oscillators is quoted as an angle not a time. Two important examples are shown below:

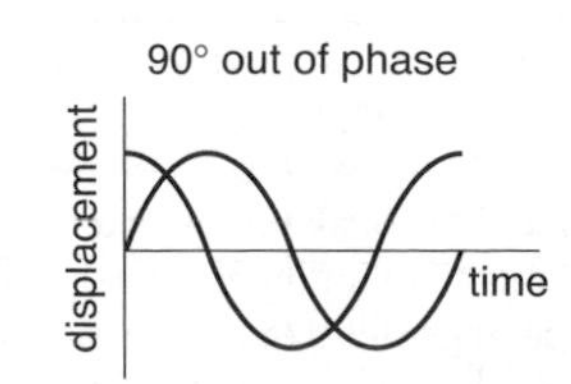

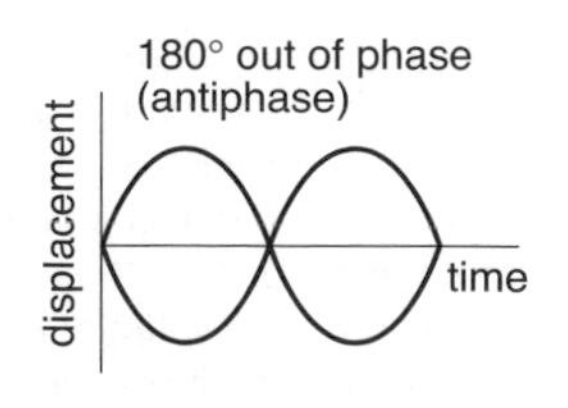

SHM and a mass on a spring

resultant upward force

mass m

At rest ***After stretch and release***

Above, a mass hangs from a spring. When pulled down and released, the mass makes small, vertical oscillations.

Springs obey Hooke's law. This means that extension x of a spring is directly proportional to the applied force (or load) F.

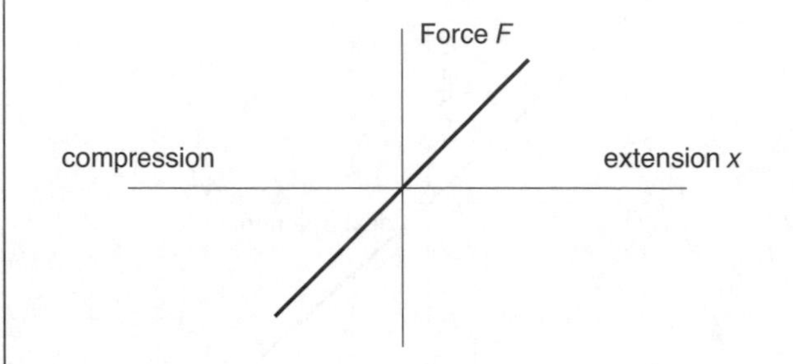

The graph of applied force against extension is a straight line through the origin.

The gradient of this graph is equal to the ***spring constant*** or ***stiffness of the spring***, k:

$$k = \frac{F}{x}$$

So, if the mass m is pulled down by x and then released,

resultant upward force on mass $= kx$

But force = mass × acceleration

So acceleration (upwards) $= \frac{kx}{m}$

So acceleration (in x direction) $a = -\frac{kx}{m}$

Comparing this with equation (5) shows that the motion is SHM and that

$$\frac{k}{m} = \omega^2$$

As $T = \frac{2\pi}{\omega}$ $\quad T = 2\pi\sqrt{\frac{m}{k}}$

In any oscillating system to which Hooke's law applies, the motion is SHM.

SHM and the simple pendulum

Provided its swings are small, and air resistance is neglible, a simple pendulum moves with SHM. The following analysis shows why.

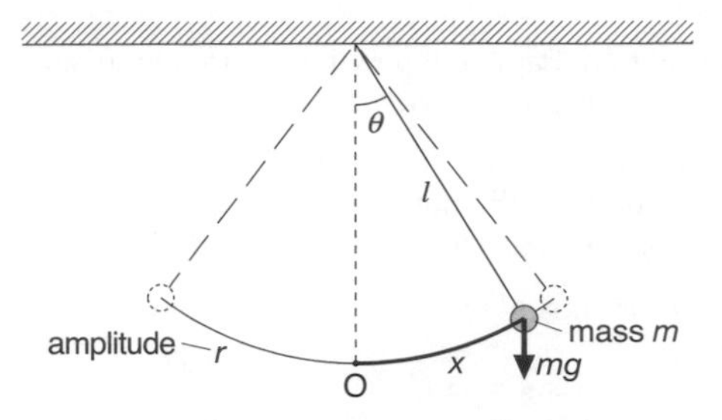

The mass m (above) has been displaced by x. It is being pulled towards O by a component of its weight:

force (towards O) $= mg \sin\theta$

But for very small angles $\sin\theta = \frac{x}{l}$ (see B11)

So force (towards O) $= \frac{mgx}{l}$

But force = mass × acceleration

So acceleration (towards O) $= \frac{gx}{l}$

So acceleration (in x direction) $a = \frac{-gx}{l}$

Comparing this with equation (5) shows that the motion is SHM and that

$$\frac{g}{l} = \omega^2$$

As $T = \frac{2\pi}{\omega}$ $\quad T = 2\pi\sqrt{\frac{l}{g}}$

Note:

- T does not depend on the amplitude (for smaller swings, the period stays the same). This is true for *all* SHM.

The graphs on the right are for an object moving with SHM: for example, a pendulum making small swings.
At the ends of each oscillation, the velocity is zero. The displacement and acceleration have their peak values, but when one is positive, the other is negative, and vice versa.
At the centre, the velocity has its peak positive or negative value, but the displacement and acceleration are both zero.

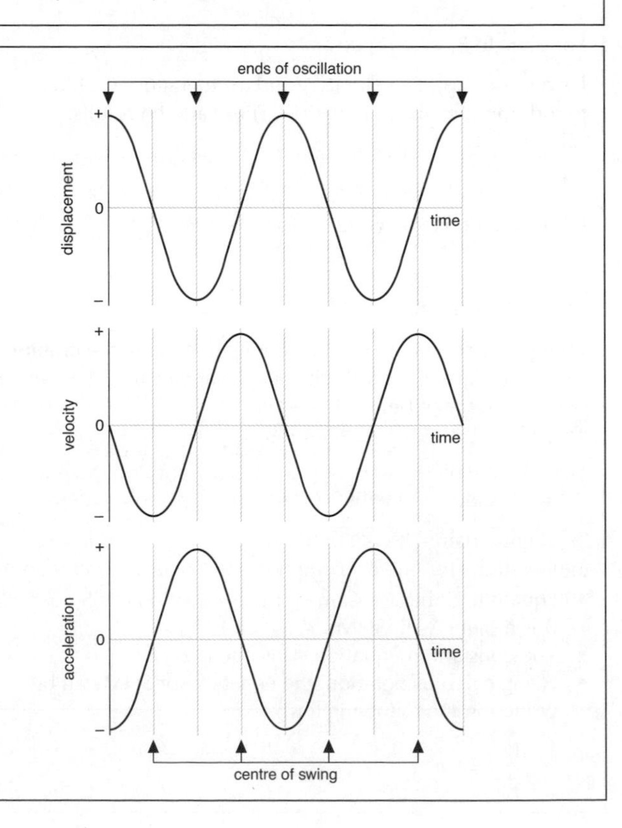

B16 Energy changes in oscillators

Mass–spring system

The elastic (or strain energy) stored in a mass–spring system is the work done in stretching the spring. This is the area under the force–displacement graph.

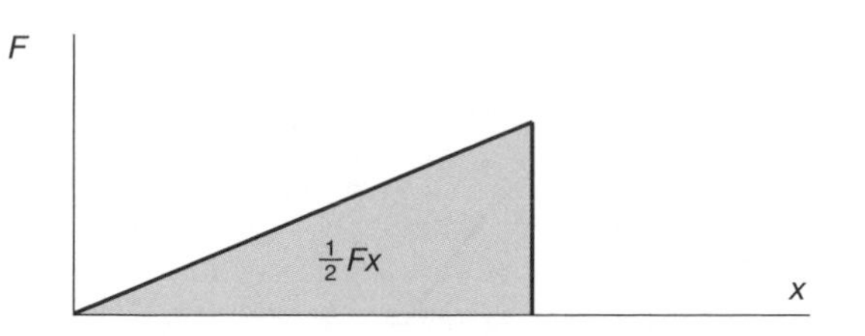

$$\text{elastic energy stored} = \tfrac{1}{2}Fx$$

Since $F = kx$, another useful equation for stored energy is

$$\text{elastic energy stored} = \tfrac{1}{2}kx^2$$

As the spring moves toward the equilibrium position it loses elastic stored energy and gains kinetic energy. But in the absence of any damping the total energy remains constant.

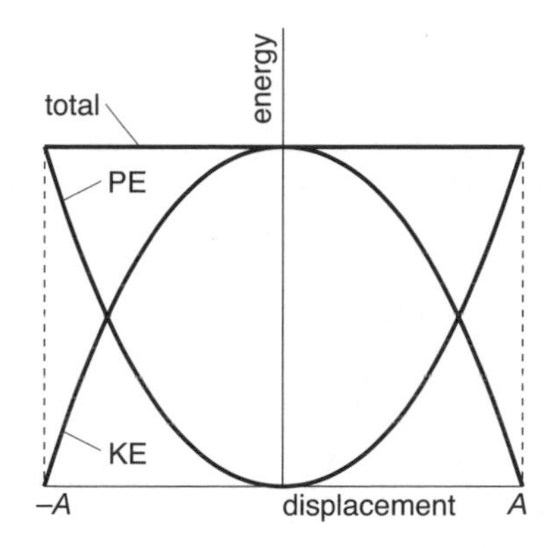

$$\text{maximum elastic stored energy} = \tfrac{1}{2}kA^2$$
$$\text{maximum kinetic energy} = \tfrac{1}{2}m(A\omega)^2$$

Pendulum

For a pendulum the mass loses potential energy (PE) as it swings downwards and gains kinetic energy (KE).

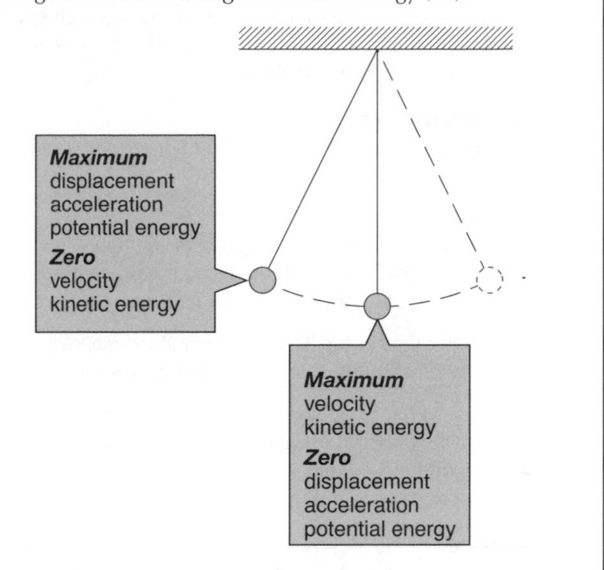

If there is no air resistance the total PE + KE is constant.

$$\text{total energy} = \text{maximum KE} = \tfrac{1}{2}m(A\omega)^2$$

Notice that in all the equations for total energy the total energy is proportional to the *square* of the amplitude (A^2).

Damping

A mass–spring system or a pendulum will not go on swinging for ever. Energy is gradually lost to the surroundings due to air resistance or some other resistive force and the oscillations die away. This effect is called **damping**.

In road vehicles, dampers (wrongly called 'shock absorbers') are fitted to the suspension springs so that unwanted oscillations die away quickly. Some systems (for example moving-coil ammeters and voltmeters) have so much damping that no real oscillations occur. The minimum damping needed for this is called **critical damping**.

The rate at which the amplitude falls depends on the fraction of the existing energy that is lost during each oscillation.

In a **lightly damped** system only a small fraction is lost so that the amplitude of one oscillation is only slightly lower than the one before.

The graphs here are for oscillations with different degrees of damping.

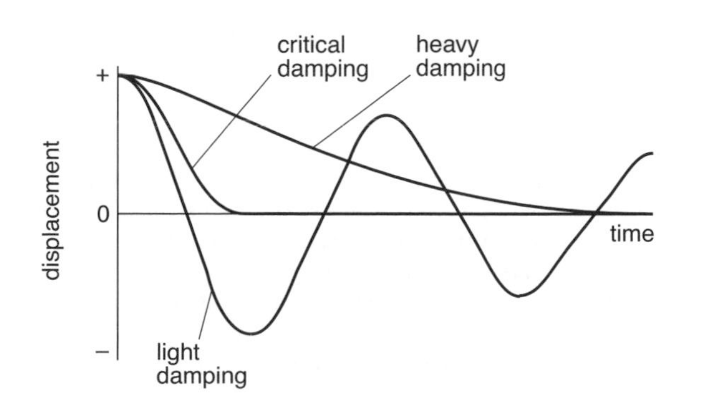

Other uses for dampers include fitting them to buildings in earthquake zones to reduce vibrations so the building can withstand earthquakes. The Millennium Bridge in London has dampers fitted to stop it oscillating from side to side.

B17 Forced oscillations and resonance

Natural frequency

This is the frequency of the oscillation that occurs when the mass of an oscillator (such as a pendulum bob or mass of a mass–spring system) is displaced and then released. The only forces acting are the internal forces of the oscillating system. These oscillations are **free oscillations**.

Forced oscillations

Forced oscillations occur when an external periodic force acts on an object that is free to oscillate.

Examples include:

- engine vibrations making bus windows oscillate
- the spinning drum causing vibrations in a washing machine
- the body of a guitar vibrating when a string is plucked.

The body that is forced to oscillate vibrates at the same frequency as that of the external source that is providing the energy.

The amplitude of the oscillations produced depends on

- how close the external frequency is to the natural frequency of the oscillator
- the degree of damping of the oscillating system.

Resonance

Resonance occurs when the frequency of the external source that is driving the oscillation is equal to the natural frequency of the oscillator that is being driven into oscillations.

When resonance occurs the amplitude of the resulting oscillations is a maximum.

The following graph shows how the amplitude of an oscillator with natural frequency f_0 varies with the frequency of the frequency of the source that is driving the oscillations.

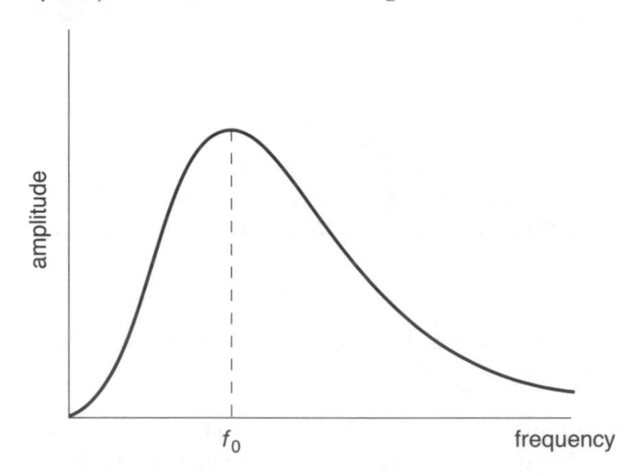

Graphs such as these are called ***frequency response graphs***.

Effect of damping on resonance

When a system is lightly damped, it loses very little energy during an oscillation. If it is being forced to oscillate it will retain most of the energy put into it so that the energy stored builds up and the amplitude becomes very large.

When a system is heavily damped, energy is lost quickly so that the amplitude is lower.

The graph shows the frequency response for lightly and heavily damped oscillators that have the same natural frequency.

Note:
The amplitude of an oscillation stops increasing when the energy put in each cycle is equal to that lost during the cycle.

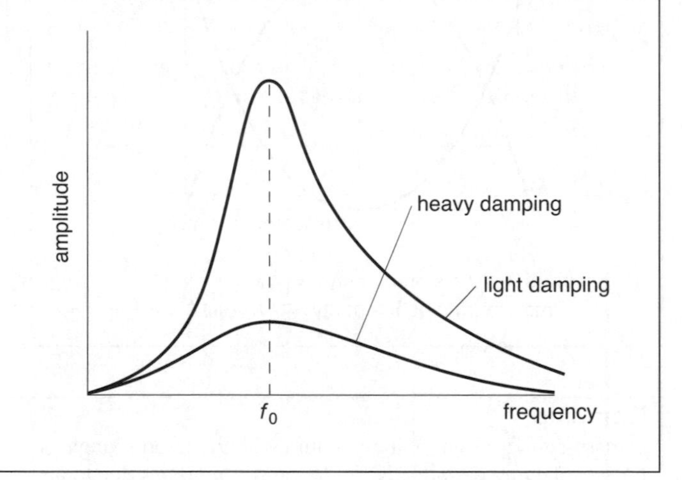

How Science Works

Electrical resonance

Electrical oscillators made from combinations of capacitors and inductors (coils) can also be forced into oscillations or be made to resonate. This effect is used in the tuner of a radio receiver. The current in the circuit reaches a maximum at the resonant frequency. By tuning the circuit so that the natural frequency matches the frequency of the radio channel, only the frequencies in a small range around this frequency are selected.

The natural frequency of an electrical oscillator depends on the capacitance and inductance of the coil used. By varying the capacitance you can tune in to different channels.

The range of frequencies selected depends on the damping which in turn depends on the resistance in the circuit.

How Science Works

Examples of resonance

Useful resonance

In microwave ovens, the microwaves are produced by a cavity magnetron. There is a resonant frequency that depends on the size of the cavity.

The microwaves are emitted at the resonant frequency and directed into the oven. The frequency is chosen to be one that vibrates the water, fat, and sugar molecules, which heats the food.

Resonance is also useful in musical instruments, for example the air in a clarinet when the reed vibrates.

Resonance to be avoided

Suspension bridges must be designed so that resonance is not caused by the wind vibrating the bridge. The Tacoma Narrows Bridge in Washington State, USA collapsed because of this effect.

C1 Density, pressure and elasticity

Density

The density of an object is calculated like this:

$$\text{density} = \frac{\text{mass}}{\text{volume}}$$

The SI unit of density is the kilogram/cubic metre (kg m^{-3}).

For example, 2000 kg of water occupies a volume of 2 m^3. So the density of water is 1000 kg m^{-3}.

Density values, in kg m^{-3}

alcohol	800	iron	7 900
aluminium	2 700	lead	11 300

Pressure

Pressure is calculated like this:

$$\text{pressure} = \frac{\text{force}}{\text{area}}$$

The SI unit of pressure is the newton/square metre, also called the ***pascal*** (**Pa**). For example, if a force of 12 N acts over an area of 3 m^2, the pressure is 4 Pa.

Hookes's law

If a material obeys **Hooke's Law** then, for an elastic deformation the extension, Δl or x, (or the compression) is proportional to the Force applied F.

$F \propto \Delta l$ or $F \propto x$ (1)

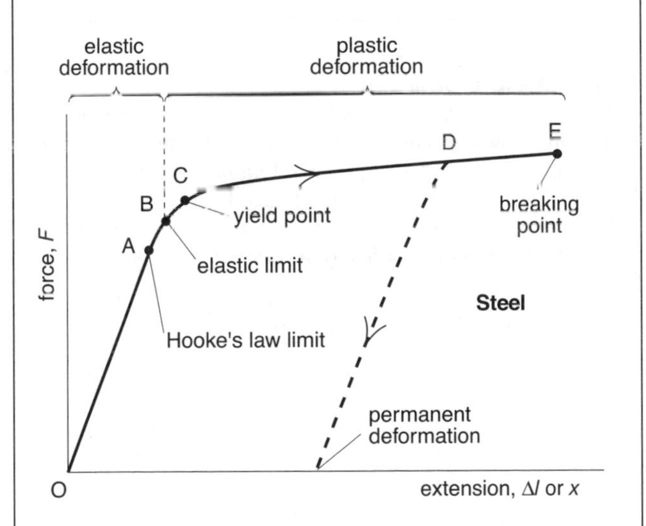

OA is the region where Hooke's Law is obeyed.

Equation 1 is used to define k, the ***stiffness*** of the material

$F = \frac{k}{\Delta l}$ or $F = \frac{k}{x}$ (2)

Deformation

The particles of a solid may be atoms, or molecules (groups of atoms), or ions (see J1). They are held closely together by electric forces of attraction.

When external forces are applied to a solid, its shape changes: ***deformation*** occurs. This alters the relative positions of its particles.

Forces on solids can be:

tension - the solid is stretched and you can measure the extension

compression – the solid is squashed and you can measure the compression

shear – the force is greater on one side than the other.

Bending is a combination of tension and compressive strain.

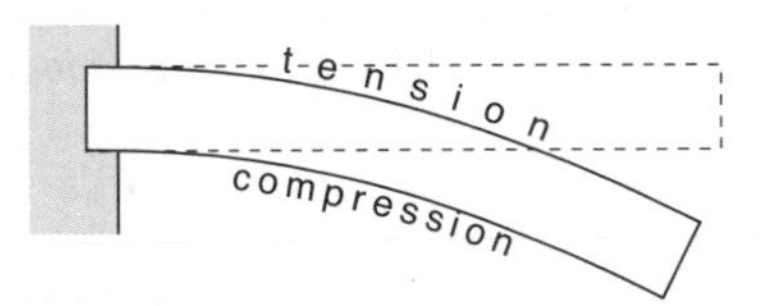

There are two types of deformation, as described below.

Elastic deformation If the deformation is ***elastic***, then the material returns to its original shape when the forces on it are removed.

Plastic deformation If the deformation is ***plastic***, then the material does not return to its original shape when the forces on it are removed. For example, Plasticine takes on a new shape when stretched.

Stress and strain

On the right, a wire of cross sectional area A is under tension from a force F (at each end). The ***tensile stress*** σ on the wire is defined like this:

$$\text{tensile stress} = \frac{\text{force}}{\text{area}} \qquad \sigma = \frac{F}{A}$$

The unit of tensile stress is the N m^{-2}.

The wire stretches so that its length l_0 increases by Δl, called its ***extension***. The ***tensile strain*** ε is defined like this:

$$\text{tensile strain} = \frac{\text{extension}}{\text{original length}} \qquad \varepsilon = \frac{\Delta l}{l_0}$$

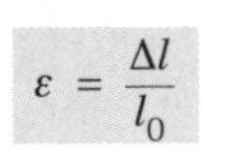

Tensile strain has no units.

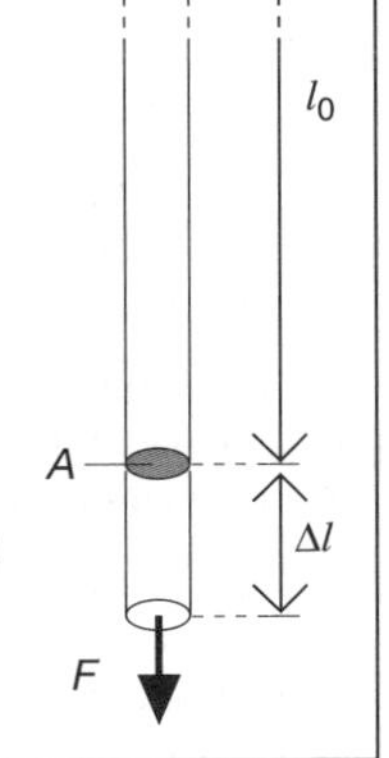

Note:

- There are stresses and strains linked with compression and twisting. On these pages however, the word stress or strain by itself will imply the tensile type.

Stress-strain graphs

The graph below shows how stress varies with strain when a metal wire (steel) is stretched until it breaks. The force-extension graph on the left will change according to the size and shape of the material. The advantage of the stress-strain graph is that it depends only on the material and not, for example, on its thickness or length.

Note:

- By convention, strain is plotted along the horizontal axis.
- The sequence O to E is described in detail on the next page.

> If a material obeys ***Hooke's law*** then, for an *elastic* deformation, the strain is proportional to the stress.

The wire obeys Hooke's law up to point A.

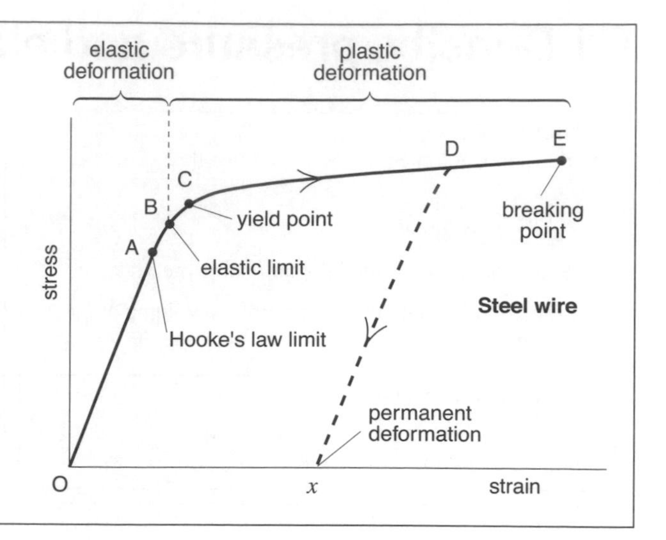

Young modulus

For a material which obeys Hooke's law, stress/strain is a constant. This constant is called ***the Young's modulus***, E:

$$\text{Young modulus} = \frac{\text{tensile stress}}{\text{tensile strain}} \qquad E = \frac{\sigma}{\varepsilon}$$

Using the equations for σ and ε,

$$E = \frac{Fl_0}{A\Delta l}$$

Note:

- l_0 is constant. If E and A are also constant, $\Delta l \propto F$. So the extension is proportional to the stretching force.
- The above equations can also be used when the material is being compressed.

Typical values for the Young modulus, in Nm^{-2}
steel 21×10^{10} aluminium 7×10^{10}

So, steel is proportionately three times as difficult to stretch as aluminium.

Stretching glass

elastic deformation

A

stress

glass

O strain

The graph above shows what happens if increasing tensile stress is applied to a glass thread. Elastic deformation occurs until, at point A, a crack suddenly grows, and the glass breaks. A material which behaves like this is said to be ***brittle***. The break is called a ***brittle fracture***.

Stretching a metal

Unlike glass, most metals do not experience brittle fracture when stretched because dislocations tend to stop cracks growing and spreading. The following descriptions refer to the graph for a steel wire on this page.

O to B The deformation of the wire is elastic.

B This is the ***elastic limit***. Beyond it, the deformation becomes plastic as layers of particles slide over each other. If the stress were removed at, say, point D, the wire would be left with a permanent deformation (strain x on the axis).

C This is the ***yield point***. Beyond it, little extra force is needed to produce a large extra extension. If a material can be stretched like this, it is said to be ***ductile***.

E The wire develops a thin 'neck', then a ***ductile fracture*** occurs. The highest stress just before the wire breaks is called the ***ultimate tensile stress***.

Fatigue If a metal is taken through many cycles of *changing* stress, a fatigue fracture may occur before the ultimate tensile stress is reached. Fatigue fractures are caused by the slow spread of small cracks.

Creep This is the deformation which goes on happening in some materials if stress is maintained. For example, unsupported lead slowly sags under its own weight.

Stretching a polymer

Polymers have long-chain molecules, each of which may contain many thousands of atoms. The molecules are formed from the linking of short units called ***monomers***. In a polymer, the chains may be coiled up and tangled like spaghetti. Depending on the amount of tangling, a polymer may be described as ***semi-crystalline*** or ***amorphous***.

Rubber and wool are natural polymers. Plastics, such as nylon and artificial rubber, are synthetic polymers.

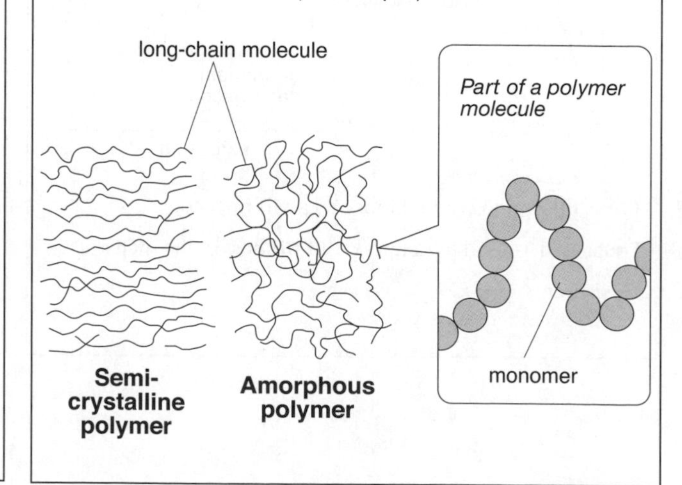

C2 Properties of materials

Strain energy

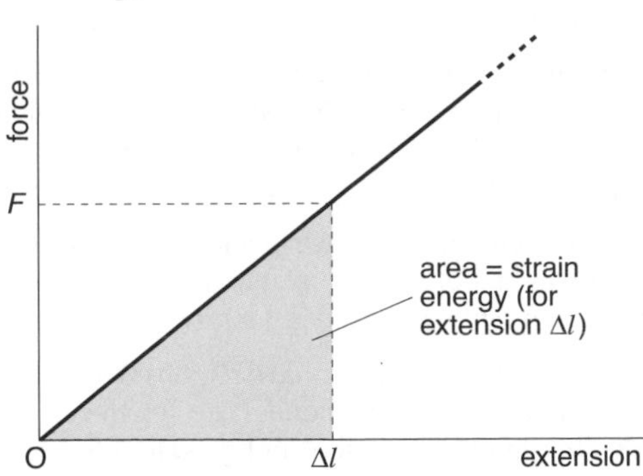

The graph above shows how the extension varies with the stretching force for a material which obeys Hooke's law. The work done for an extension Δl is given by the shaded area (see C1). The area of a triangle = $\frac{1}{2}$ × base × height. So the work done = $\frac{1}{2} F\Delta l$.

As work is done *on* the material, energy is stored *by* the material. This is its ***strain energy***. So

strain energy = $\frac{1}{2} F\Delta l$

using equation (2)

strain energy = $k\,(\Delta l)^2$

Note:

- If Hooke's law is *not* obeyed, the work done is still equal to the area under a force–extension graph. However, the above equation does not apply.

Stretching rubber

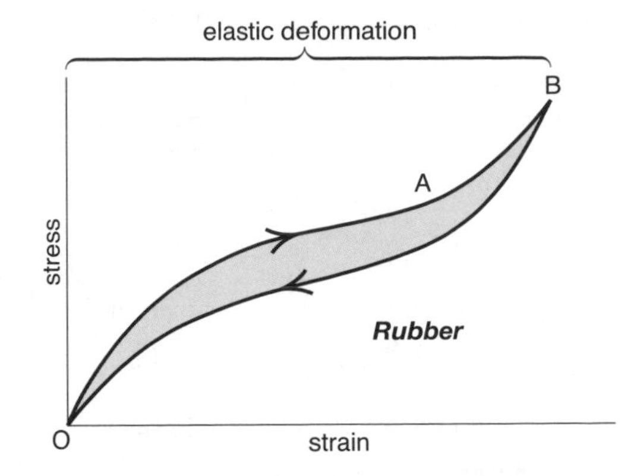

The graph shows what happens if increasing stress is applied to a rubber cord, and then released before the breaking point. Rubber does not obey Hooke's law. Also, much higher strains are possible than in steel or glass. For example, if the extension is twice the original length, the strain is 2.

O to A The molecular chains in the rubber are being uncoiled and straightened.

A to B The chains are almost straight, so the rubber is becoming proportionately more difficult to stretch.

B to A The rubber contracts when the stress is removed.

During this cycle of extension and contraction, energy is lost as heat. The effect is called ***elastic hysteresis***. The shaded area represents the energy lost per unit volume.

Mechanical properties

These describe how a material behaves when forces are applied. They depend on its structure, the strength of the bonds, and the type and number of defects present.

Strength A *strong* material has a high ultimate tensile stress i.e. a high stress is needed to break it.

Ductility A *ductile* material can be drawn into wires.

Malleability A *malleable* material can be hammered into different shapes.

Stiffness A *stiff* material has a high Young's modulus, i.e. a high stress produces little strain.

Toughness A *tough* material will deform plastically before it breaks. It is not brittle, i.e. cracks do not easily spread. It can withstand dynamic loads such as shock or impact.

Hardness A *hard* material cannot be easily scratched or indented. There are no absolute values for hardness. One material is harder than another if it will scratch it. Tests such as the Brinell Hardness Test use a machine to measure the indentation caused in controlled conditions and give the material a BHN (Brinell hardness number).

Durability A *durable* material can be repeatedly loaded and unloaded without its properties deteriorating.

Brittleness A *brittle* material cracks and breaks without plastic deformation.

Smoothness A *smooth* material has a low friction surface. This means you can treat friction as negligible.

C3 Material structure

Solid structures

Solids can be classified into three main types, according to how their particles are arranged.

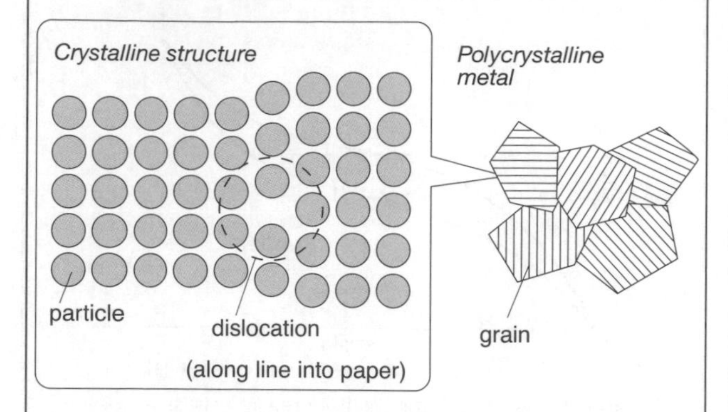

Polymers are covered on the right, the other two are:

Crystalline solids The particles are in a regular, repeating pattern. They may form a single crystal, as with a diamond. However, there may be millions of tiny crystals joined together. Most metals have this ***polycrystalline*** structure. Their crystals, called ***grains***, can be as small as 10^{-2} mm.

Crystal structures normally have imperfections in them called ***dislocations***. These allow particles to change their relative positions, so the solid is more easily deformed.

Amorphous (glassy) solids The particles have no regular pattern (except over very short distances). Glass and wax have structures like this.

Ceramics

Ceramics are non-metallic chemical compounds, which are crystalline, partly crystalline, or glass. They are un-reactive and have high melting points. They are often brittle, which makes them difficult to form. They are either formed from a molten mass that cools to a solid, or they are mixed with water, shaped and then heated (fired) to form a solid. Clay is used to make bricks and pottery. Other examples are porcelain, and glass.

Metals

Metals may be a single element, so all the atoms are the same. Examples are copper, gold, and aluminium. Alloys are a mixture of metals, (or of metals and non-metals) For example, steel is an alloy of iron (about 99% and carbon, and sometimes other elements as well. It is stronger, stiffer, and tougher than iron.

Polymers

Rubber and wool are natural polymers. Plastics are synthetic polymers. They are made up of long-chain molecules whose atoms are linked by strong covalent bonds. There are two main classes of plastics.

Thermoplastics These soften when heated and harden on cooling. Resoftening is possible because thermal activity can overcome the weak bonds between the polymer chains. Thermoplastics creep under stress: they are ***viscoelastic***.

Amorphous thermoplastics have tangled chains. They are *glassy* when cold, but *rubbery* (soft and flexible) above their ***glass transition temperature*** (e.g. 100 °C for Perspex).

Semicrystalline thermoplastics have regions where the chains are parallel and close, so the bonds between the chains are stronger. This produces stiffness and good tensile strength. Amorphous regions add flexibility.

Note:

- Stretching a thermoplastic makes it more crystalline, i.e. its chains uncoil and become less tangled.

Thermosets These do not soften when warmed, so they cannot be remoulded. During manufacture, they develop strong and permanent cross-links (bonds) between their chains.

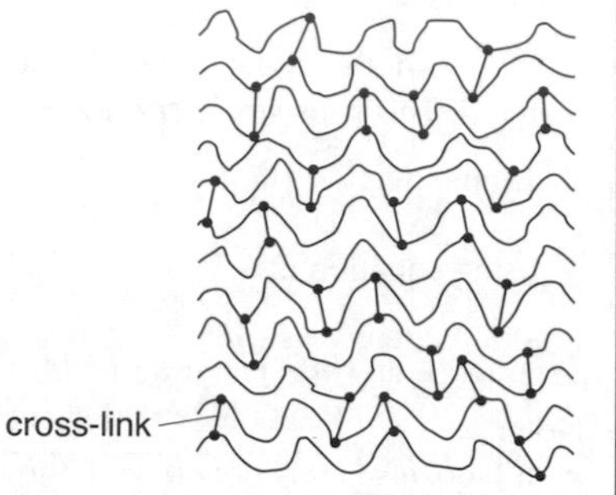

	Plastics			
	Thermoplastics		Thermosets	
Property	resoften on warming		permanent set	
Type	amorphous	semi-crystalline	elastomer	rigid
Structure	tangled chains	many chains parallel	some cross links	many cross links
Examples	Perspex	polythene nylon	artificial rubber	epoxy resins Melamine

C4 More about materials

Composites

These are combinations of materials, produced to make use of the best properties of each.

Wood and bone are natural composites.

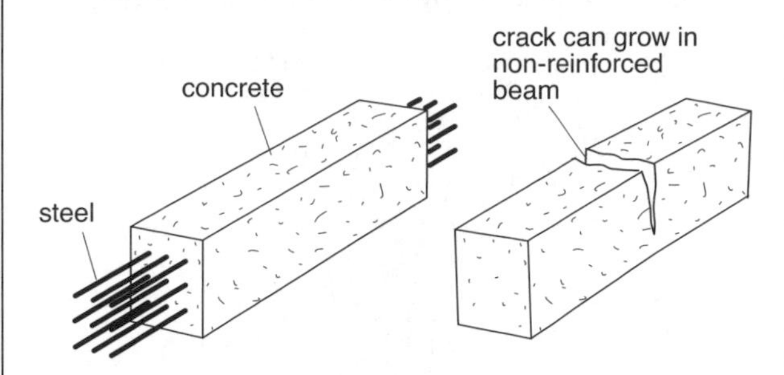

Reinforced concrete Concrete has high compressive strength but is brittle and weak in tension, so will crack and break if bent (see H4). To prevent this, it can be reinforced with steel rods. In ***pre-stressed concrete***, the rods are stretched elastically before the concrete sets. This gives even greater strength and stiffness. (Concrete is itself a composite of sand, chippings, and cement.)

Glass-reinforced plastic (GRP) ('fibre glass') Glass fibres are embedded in plastic resin. The fibres provide tensile strength. The resin gives stiffness, by bonding the fibres together, and toughness, by stopping crack growth.

Carbon-fibre-reinforced plastic (CFRP) ('carbonfibre') This is similar to GRP, but with stronger, stiffer carbon fibres instead of glass.

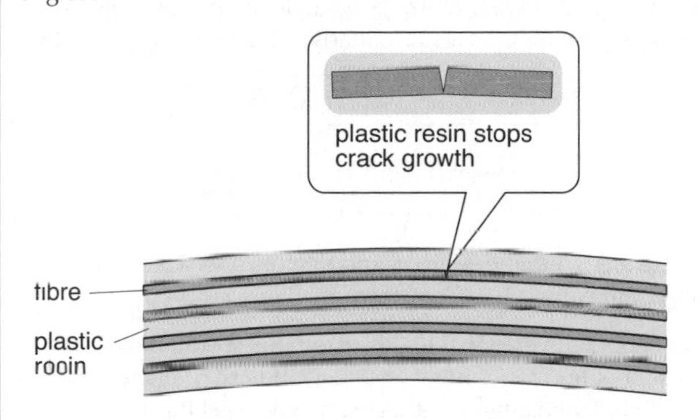

Electrical conduction

Read F1 and F2 first.

When atoms are close (e.g. in a solid), their energy levels broaden into ***bands***, i.e. lots of levels close together. To take part in conduction, charge carriers, usually electrons, must be able to move between atoms. This requires an energy transfer, so it can only happen if there are unoccupied levels for electrons to go to.

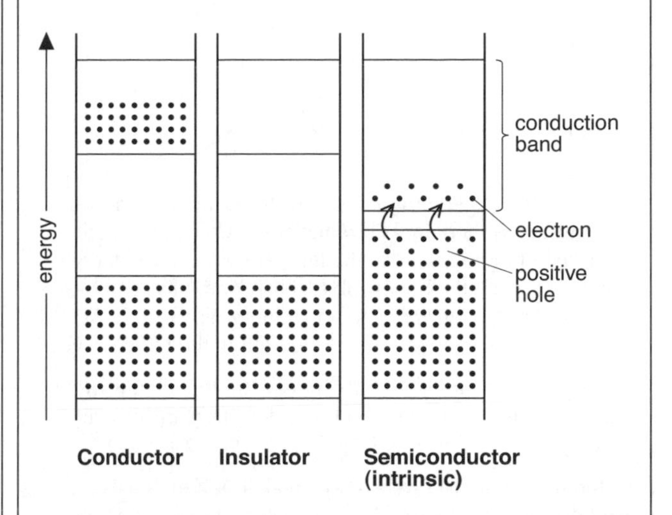

Conductors (e.g. copper) The outer band (the ***conduction band***) is only partly filled. It has unoccupied energy levels, so its electrons are free to move between atoms. In many metals, the conduction band overlaps with the band below.

Insulators (e.g. nylon) All the electrons are in full bands, so they are unable to change energy and move between atoms.

Semiconductors (e.g. silicon) When cold, these are insulators because the conduction band is empty. However, it is so close to the band below that a temperature rise can give electrons enough energy to jump the gap. Conduction can then take place both by electron movement in the conduction band and by movement of the 'gaps' (called ***positive holes***) created in the band below.

The conductivity depends on the number of mobile charge carriers – electrons or positive holes.

C5 The material for the job

How Science Works

Hip replacements

The hip joint is a ball-and-socket joint (see diagram). This type of joint has a wide range of movement.

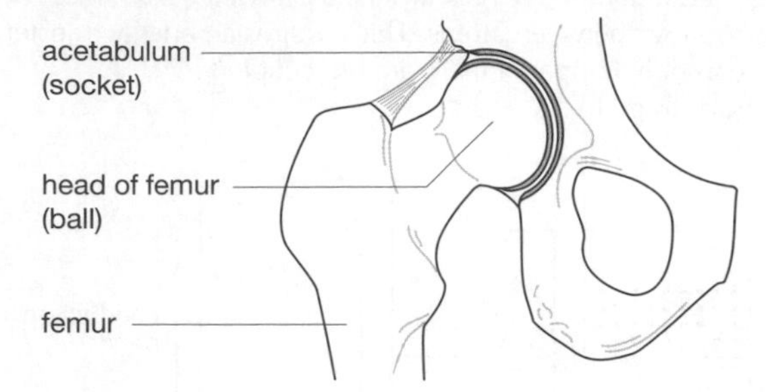

A worn or damaged hip joint can restrict a person's activities and cause a lot of pain. Hip replacements are performed when the lining or tissue in the hip joint is worn or otherwise damaged so that the bones rub together painfully, or if the bone is damaged or fractured and cannot repair itself. At least 50 000 hip replacements are carried out in the UK each year.

The acetabulum and the femoral head are replaced with artificial parts. The materials chosen for these are specially developed.

Materials that are too weak may break, but if materials are used that are too stiff (see C1), the natural bone surrounding the replacement is not subjected to much stress. Bone contains living cells and grows when required to support a load but wastes away when it is not. The bone around a hip replacement that is too stiff will waste away. This means that the Young modulus of the material chosen must be about the same as that of bone (about 1×10^{10} Pa). If it is too low, the bone will take too much stress, but if it is too high the bone will waste away.

Materials must be sterile and must not react with the body. They must not be rejected by the body's immune system.

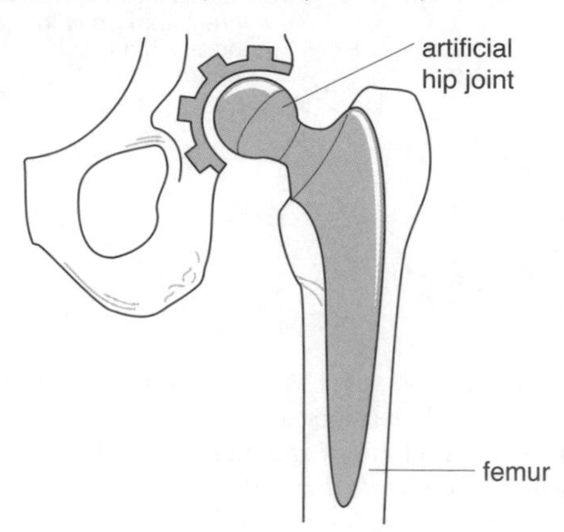

One metal used for hip implants is OXINIUM™ (oxidized zirconium) It is used because of its hardness, smoothness, and resistance to scratching and abrasion. The cup has a plastic liner.

The BIRMINGHAM HIP™ replacement system involves reshaping the femoral head and fitting a metal cap over it. Less bone is removed, so other operations are possible in the future. A metal cup fits into the acetabulum. Both parts of the joint are made of durable high-carbide chrome. This avoids the problems of wear of the polyethylene liner of some replacement joints. It is particularly suitable for young, active patients.

How Science Works

Ethical issues

The cost of a hip replacement varies, but in the UK it is likely to be between £6000 and £9000. With over 50 000 operations a year, it is clear that a large amount of money is required. How much funding should the NHS receive and how it should be spent? As the population ages, more people are likely to need replacement hips, but there are other treatments that also require funding.

Should the budget be increased? This would have to be paid for by increasing taxes. Should people pay more tax?

Whatever budget is set, there are likely to be more demands on it for different treatments. How do you decide between different people needing hip replacements, or between hip replacements and, for example, heart operations?

If new materials and designs of hip joint are produced that are more suitable, but more expensive, should they be used, or should we continue to use old, cheaper designs?

These are difficult questions and doctors and scientists try to make ethically sound decisions.

One approach is to make a decision that would benefit the most people. This would encourage the use of many small-scale cheap treatments rather than expensive hip replacements. Another approach would be to take into account how much the person is affected, how much pain they have, whether it stops them from moving about or working, and whether it could become more serious if not treated.

Newer designs for replacement body parts may be more expensive initially but will save in the long term if there are fewer complications and they last longer.

How Science Works

Kevlar™

Kevlar™ was developed at DuPont during the 1960s. It is a polymer with a high Young modulus and low density. This is because it is made of carbon, hydrogen, oxygen, and nitrogen. Take a sample of Kevlar with the same mass as a sample of steel and it will be five times as strong. It is lightweight and flexible, which makes it a good choice for body armour.

It is also used for lightweight ropes and for kayak construction. The properties of Kevlar have led to it being used for protective clothing for athletes and motorcyclists. It is also used in a variety of sports equipment, from racket strings to snowboards.

D1 Waves and rays

Types of wave motion

Waves transfer energy from one place to another. Where ever there is wave motion, there must be:

- a source of oscillation
- a material or field which can transmit oscillations.

Wave motion can be demonstrated using a 'slinky' spring, as shown below. The moving waves are called ***progressive waves***. There are two main types.

Transverse waves The oscillations are at right-angles to the direction of travel:

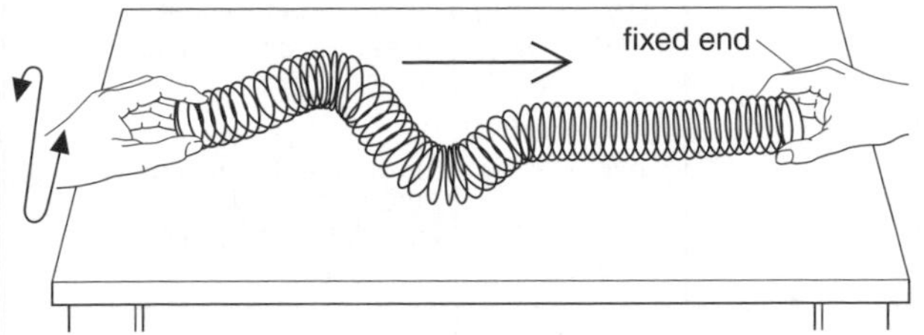

Longitudinal waves The oscillations are in line with the direction of travel, so that a compression ('squash') is followed by a rarefaction ('stretch'), and so on.

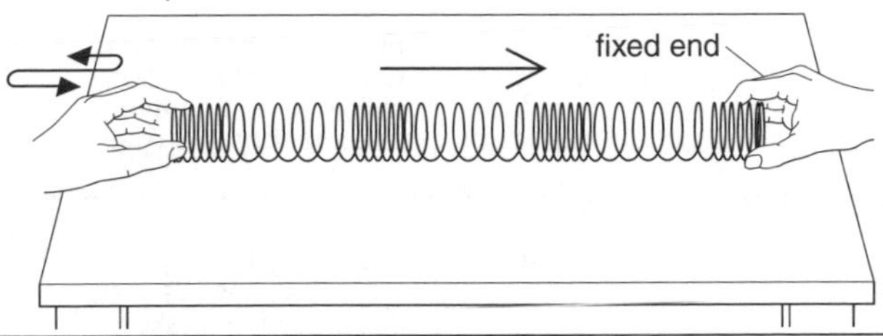

Wave features

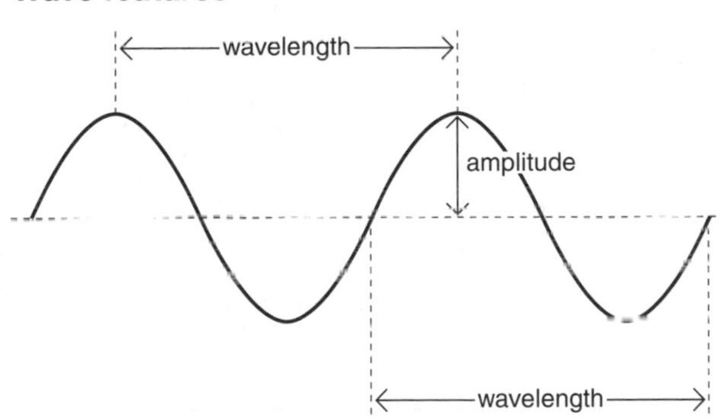

The waves above are tiny ripples moving across the surface of some water.

Amplitude This is the magnitude (size) of the oscillation.

Frequency This is the number of waves emitted per second. The SI unit is the ***hertz*** (**Hz**). A frequency of, say, 5 Hz means that 5 waves are being emitted per second.

Period This is the time for one complete wave to pass.

Wavelength In the example above, this is the distance between one wave crest and the next.

Speed The speed, frequency, and wavelength of a wave are linked by this equation:

wave speed = frequency × wavelength

For example, if the frequency is 5 Hz and the wavelength is 2 m, then the wave speed is 10 m s^{-1}. You could predict this result without the equation. If there are 5 waves per second and each occupies a length of 2 m, then each wave will travel 2 m in $\frac{1}{5}$ s, or 10 m in a second.

Waves in a ripple tank

Wave effects can be investigated using a **ripple tank** in which ripples travel across the surface of shallow water.

Reflection Waves striking an obstacle are reflected. The angle of incidence is equal to the angle of reflection.

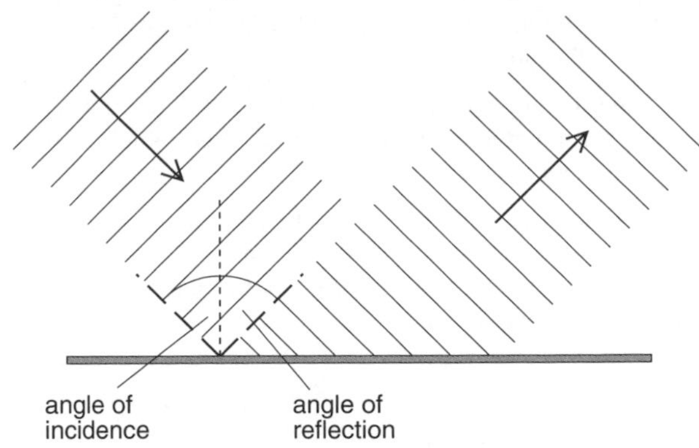

Refraction When waves are slowed down, they are ***refracted*** (bent), provided the angle of incidence is not zero. In a ripple tank, the waves can be slowed by using a flat piece of plastic to make the water shallower.

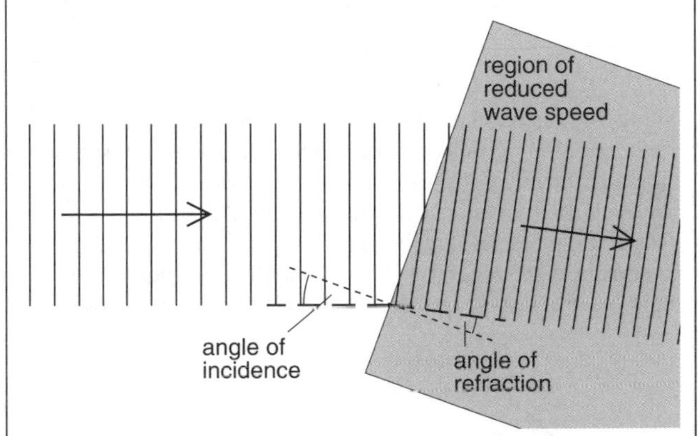

Diffraction Waves bend round the edges of a narrow gap. This is called ***diffraction***. It is significant if the gap size is about a wavelength. Wider gaps cause less diffraction.

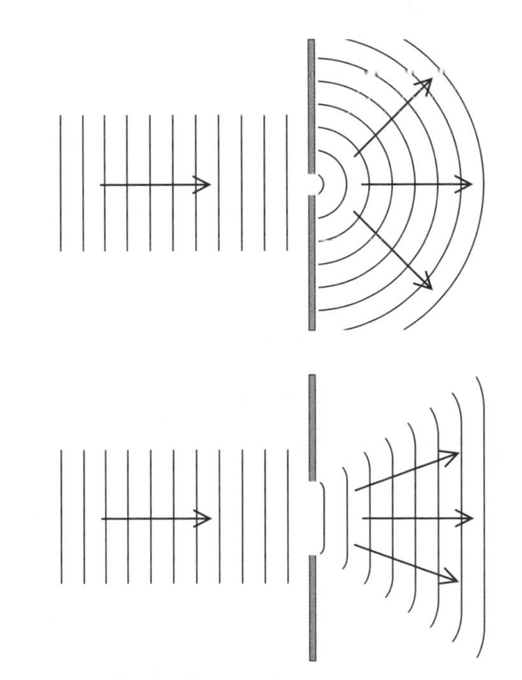

Interference If two identical sets of waves overlap, they may either reinforce or cancel each other, depending on whether they are in phase ('in step') or out of phase.

Light waves and rays

Light waves are transverse electric and magnetic field ripples. They can travel through empty space (a vacuum).

Light waves come from atoms. A burst of wave energy is given off whenever an electron loses energy by dropping to a lower orbit. Sometimes, this burst of wave energy acts like a particle, a **photon**.

The speed of light in empty space is 300 000 km s^{-1}. It is less in transparent materials such as air and water, which is why these materials refract light.

Our eyes experience different wavelengths as different colours. These range from red (0.000 7 mm) down to violet (0.000 4 mm). As light waves are so short, they are not noticeably diffracted by everyday objects.

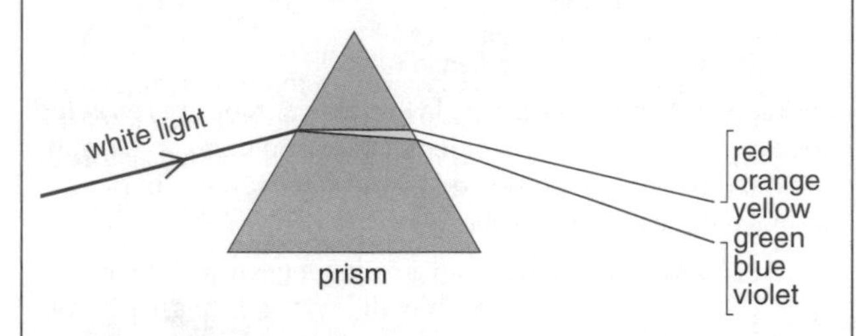

White light is not a single colour, but a mixture of all the colours of the rainbow. When white light enters a prism, the different wavelengths (i.e. colours) are slowed by different amounts, so they are refracted by different amounts. As a result, the white light splits into a range of colours called a ***spectrum***. The spreading effect is known as ***dispersion***. Violet light has the lowest speed in glass, so it is refracted most.

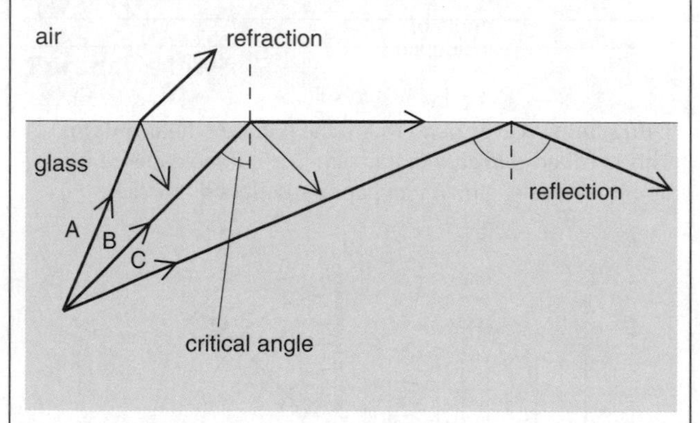

In diagrams, a ***ray*** is a line showing the direction in which light is travelling. In the diagram above:

- Ray A is mainly refracted when it passes from glass to air, although some light is also reflected.
- Ray B is also refracted, but only just. Beyond the ***critical angle***, no refraction can occur.
- Ray C strikes the surface at too great an angle for any refraction to occur. So all the light is reflected. This is called ***total internal reflection***.

In an optical fibre, light entering one end of the glass or plastic fibre is totally internally reflected until it comes out of the other.

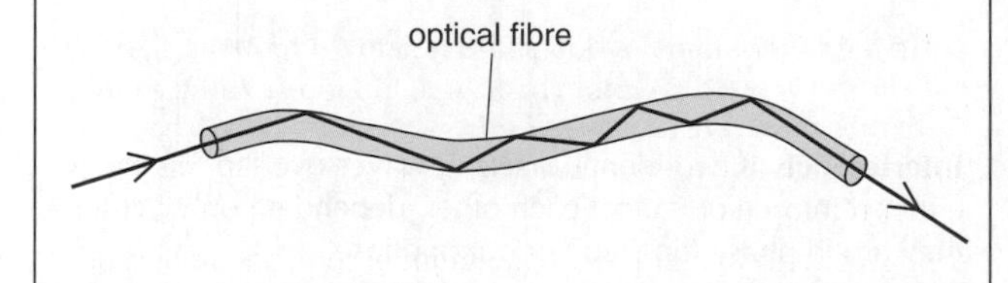

The electromagnetic spectrum

Light is one member of a whole family of transverse waves called the ***electromagnetic spectrum***. In empty space, these waves all travel at the same speed: 300 000 km s^{-1}.

Electromagnetic waves are emitted whenever electrons or other charged particles oscillate or lose energy. The greater the energy change, the lower the wavelength.

Wavelength in metres	Wave type	Typical sources, uses, and effects
10^5		
	radio waves: LW, MW, SW, VHF, UHF	From electrons oscillating in aerial. Used for communication.
3×10^{-2}		
	microwaves	From electrons oscillating in magnetron. Used for radar, communication, cooking.
10^{-3}		
	infrared	From hot objects. Used for heating.
7×10^{-7}		
	light	From very hot objects. Only form of radiation visible to human eye. Used for communication.
4×10^{-7}		
	ultraviolet	From very hot objects. Ionizes atoms. Causes fluorescence (makes some chemicals glow). Kills germs. Causes suntan.
10^{-9}		
	X-rays	From electrons stopped rapidly in X-ray tube. Causes ionization and fluorescence. Used for X-ray photography.
	gamma rays	From radioactive materials. Uses and effects as for X-rays.

Sound waves

Sound waves are longitudinal. When, say, a loudspeaker cone vibrates, it sends compressions and rarefactions through the air which the ear can detect.

Speed The speed of sound waves in dry air at room temperature is about 340 m s^{-1}. This rises if the air is warmer or damper. Sound travels faster through liquids than through gases, and faster still through solids.

Loudness The greater the amplitude of the sound waves, the louder the sound.

Wavelength This can vary from about 15 mm to 15 m, so sound waves will diffract round everyday objects.

Pitch The higher the frequency, the higher the pitch:

pitch: low — high

range of human hearing | ultrasound

frequency: 20 Hz — 20 000 Hz

D2 Moving waves

Progressive wave motion

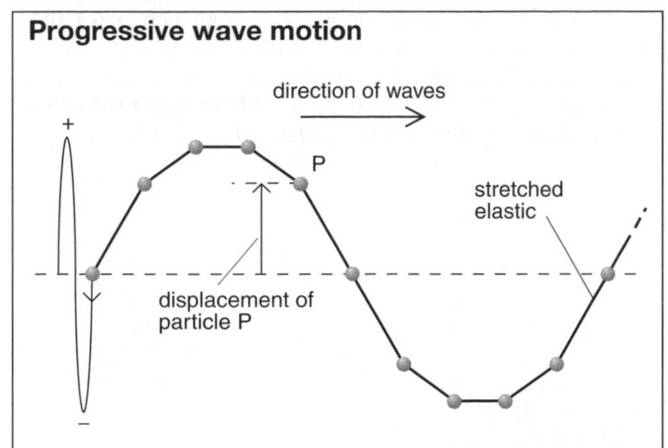

Above is one model of how waves travel. The first particle is oscillated up and down. This pulls on the next particle, making it oscillate up and down slightly later, and so on, along the line. As a result, ***progressive*** (moving) waves are seen travelling from left to right. The waves in this case are ***transverse*** (see D1) because the oscillations are at right-angles to the direction of travel.

The ***displacement*** of a particle at any instant is measured from the centre line, with a + or – to indicate an *upward* or *downward* direction.

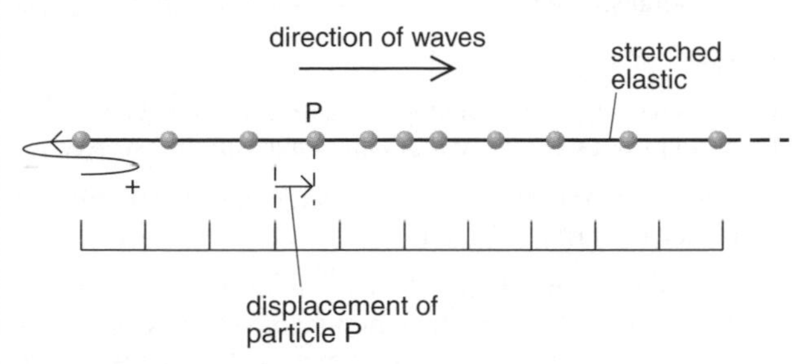

With ***longitudinal*** waves as above, the oscillations are in the direction of travel (see D1). So a particle can have a displacement to the *right* (+) or *left* (–).

The speed of waves depends on the properties of the ***medium*** (material) through which they are travelling. For example, if the particles above are lighter, or the elastic tighter, then each particle is affected more rapidly by the one before, so the wave speed is greater.

Speed, frequency, and wavelength

The waves on the right have a frequency f. So in time $1/f$, they move forward one wavelength, λ. As their speed c equals distance (λ) divided by time ($1/f$),

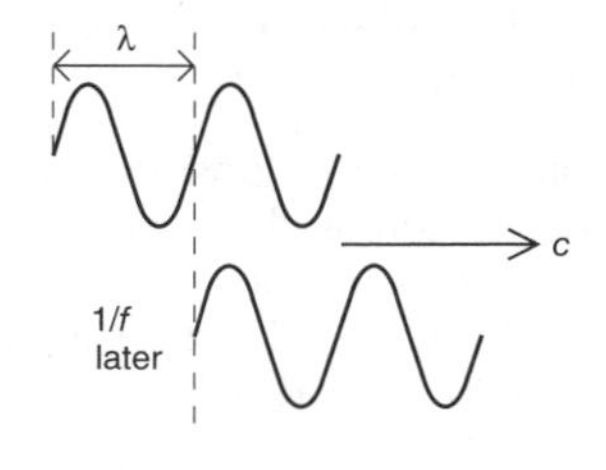

$$c = f\lambda$$

When refraction occurs, the frequency of the waves is unchanged. However, if the speed decreases, it follows from the above equation that the wavelength must also decrease.

Refraction of light waves

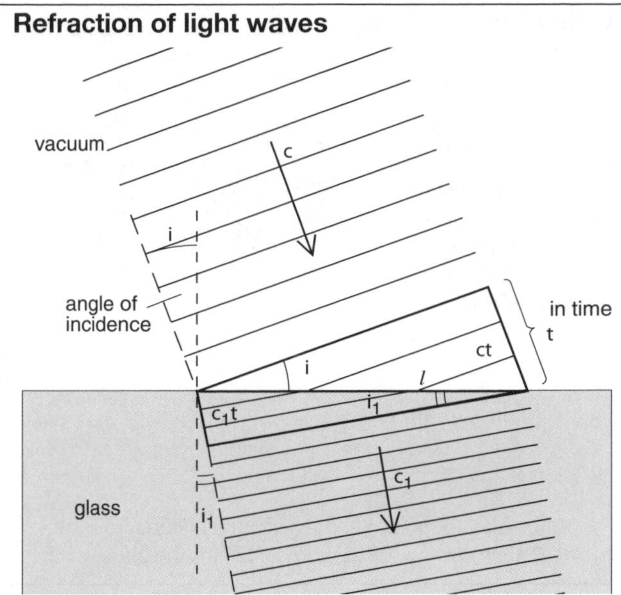

Above, light waves are shown as a series of lines called ***wavefronts***. All points on a wavefront are in phase. As the waves enter the glass, their speed slows from c to c_1. As a result, ***refraction*** (bending) occurs. In time t, one side of the beam travels a shorter distance in the glass (c_1t) than the other side does in the vacuum (ct). From the triangles in the diagram:

$$\sin i = \frac{ct}{l} \quad \text{and} \quad \sin i_1 = \frac{c_1t}{l} \quad \therefore \quad \frac{c}{c_1} = \frac{\sin i}{\sin i_1} = n_1$$

n is a constant called the ***refractive index*** of the medium (glass).

When light enters a typical glass, its speed slows from 3.0×10^8 m s^{-1} to 2.0×10^8 m s^{-1}. So, from the above equation, the refractive index of the glass is 1.5. Water does not slow light so much. Its refractive index is 1.3.

Note:
- $\sin i_1 \propto \sin i$.
- The refraction at an air–glass boundary is effectively the same as at a vacuum–glass boundary.
- The refractive index is slightly different for different wavelengths, which is why dispersion occurs (see D1).

On the right, light passes from one medium into another (of greater refractive index in this case). The wave direction is indicated by a single ray. The following equation applies:

$$\frac{c_1}{c_2} = \frac{\sin i_1}{\sin i_2} = {}_1n_2 \qquad (1)$$

${}_1n_2$ is the ***relative refractive index*** for light passing from medium 1 to medium 2. It can be shown that

$${}_1n_2 = n_2/n_1$$

From this and equation (1), it follows that

$$n_1 \sin i_1 = n_2 \sin i_2 \quad \text{and} \quad n_1c_1 = n_2c_2$$

Note:
- The equation above left is known as **Snell's Law**.
- n for a vacuum (or air) = 1.
- By the ***principle of reversibility***, a ray from B to A has the same path as one from A to B.

  So $\;{}_2n_1 = \dfrac{1}{{}_1n_2}$

Critical angle

On the right, light travels towards a boundary with a medium of lower refractive index. i_c is the ***critical angle*** (see D1). For angles greater than this, all the light is reflected by the surface and none is refracted. There is ***total internal reflection***.

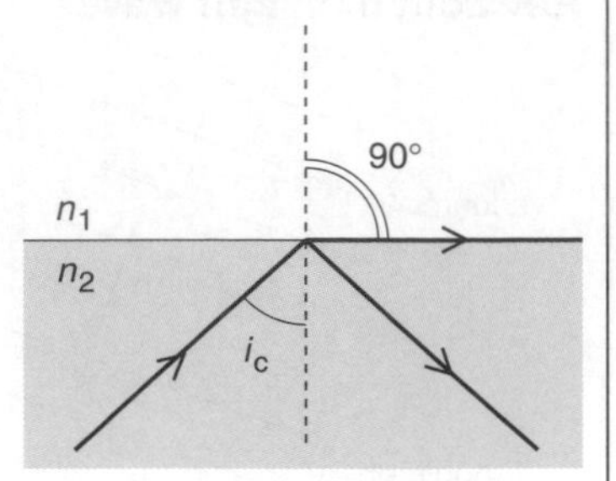

From equation (1)

$${}_1n_2 = \frac{\sin 90°}{\sin i_c} = \frac{1}{\sin i_c}$$

So $$\sin i_c = \frac{1}{{}_1n_2}$$

For an air–glass boundary, ${}_1n_2$ is effectively equal to n_2. If $n_2 = 1.5$, then $\sin i_c = 1/1.5$. So $i_c = 42°$ for the glass.

Optical fibres

Optical fibres (see also D1) can carry data, in the form of infrared pulses. They make use of total internal reflection.

Step-index multimode fibre This has a core surrounded by a cladding of lower refractive index. In the core, zig-zag paths (modes) of many different lengths are possible, so different pulses may overlap by the time they reach the end.

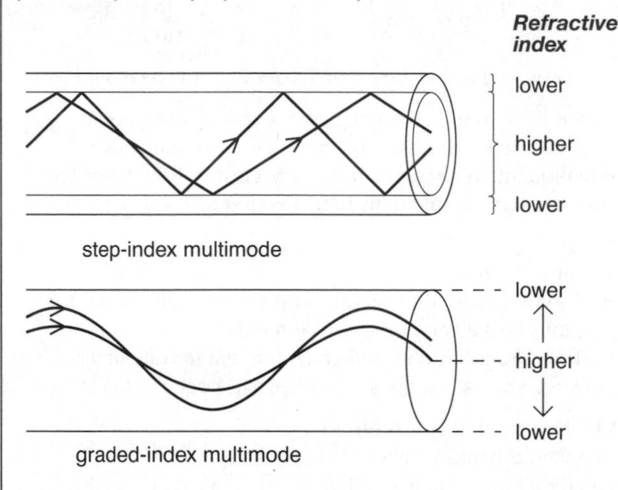

Graded-index multimode fibre The refractive index gradually reduces from the centre out. This means that the pulses take curved paths. But the longer paths are faster, so the travel times are about the same for all of them.

Intensity

Waves transmit energy. If waves pass through a surface, their ***intensity*** (in W m^{-2}) is calculated like this:

$$\text{intensity} = \frac{\text{power crossing surface}}{\text{area of surface}}$$

Intensity is proportional to (amplitude)^2.

On the right, waves are radiating uniformly from a source of power output P. At a distance r from the source, the power is spread over an area $4\pi r^2$. So intensity $I = P/4\pi r^2$

Note that $I \propto 1/r^2$. This is an example of an inverse square law.

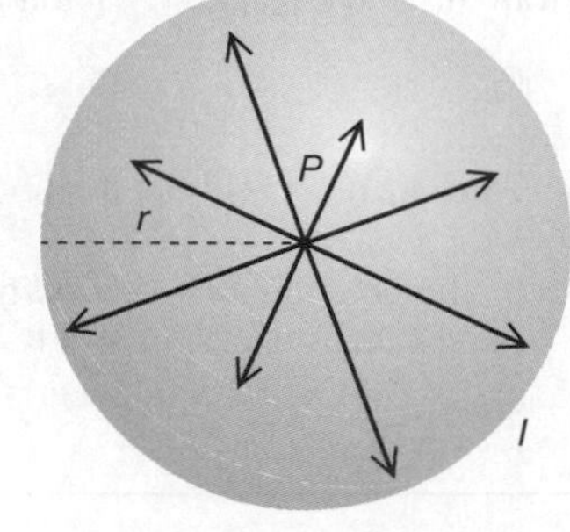

Polarization

In the top diagram on the opposite page, the particles oscillate in a vertical ***plane of vibration***. For a light wave, the plane of vibration is taken as that of the E vector. Light is usually a mixture of waves with different planes of vibration. It is ***unpolarized***. Polaroid transmits light in one plane of vibration only. Light like this is ***polarized***.

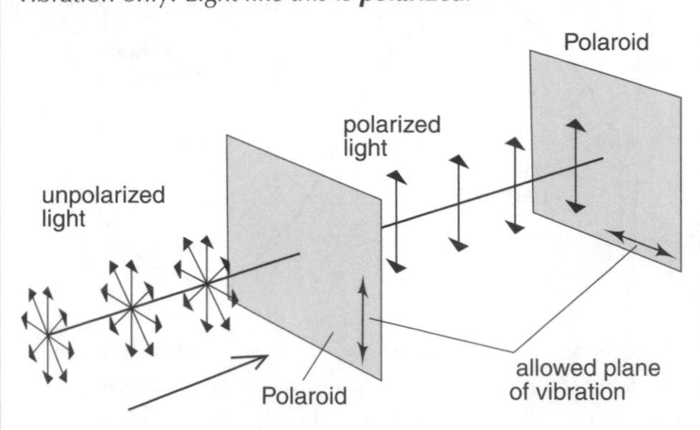

Above, polarized light from one Polaroid strikes a second. The light is blocked because its plane of vibration has no component in the allowed direction.

Malus' law

If polarized light with intensity I_0 is directed at a Polaroid at an angle θ to the allowed plane of vibration, the intensity θ of the light transmitted is

$I = I_0 \cos^2 \theta$

Only tranverse waves can be polarized. Experiments with Polaroid provide evidence that light waves are transverse.

Microwaves and radio waves are polarized, so the transmitter and receiver must be aligned.

Polarization by reflection When an unpolarized light ray strikes the surface of a transparent medium such as water, the refracted ray is partly polarized. At most angles, the reflected ray is also partly polarized.

But if the reflected ray is at 90° to the refracted ray, it is *totally* polarized.

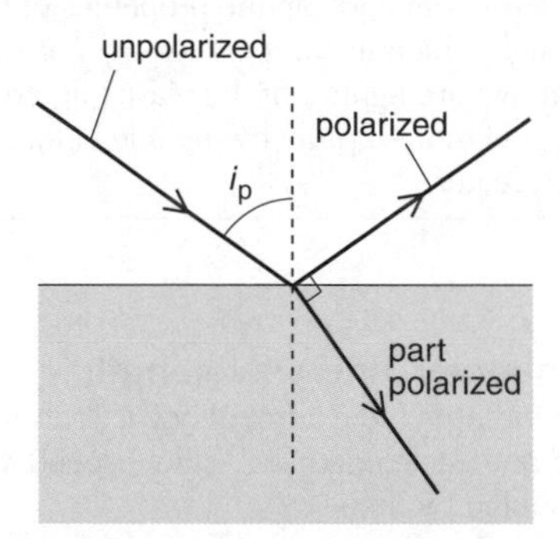

Polaroid sunglasses reduce the glare from wet surfaces by blocking the reflected, polarized light.

Optically active materials rotate the plane of polarization. If they are placed between crossed Polaroids, the second Polaroid must be rotated to produce extinction. Sugar solution is optically active, and the angle through which it rotates the plane of polarization can be used to measure the concentration of the solution.

Stress analysis

Perspex models, for example of a bridge, are viewed through crossed Polaroids. The colour and intensity of the light changes as the model is loaded, showing the areas where the stress is greatest.

D3 Combining waves

Superposition and interference

Two sets of waves can pass through the same point without affecting each other. However, they have a combined effect, found by adding their displacements (as vectors). This is known as the ***principle of superposition***.

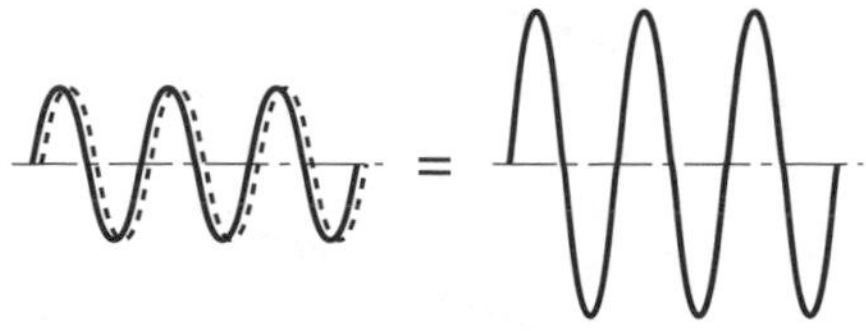

The waves above are *in phase* and reinforce each other. This is called ***constructive interference***.

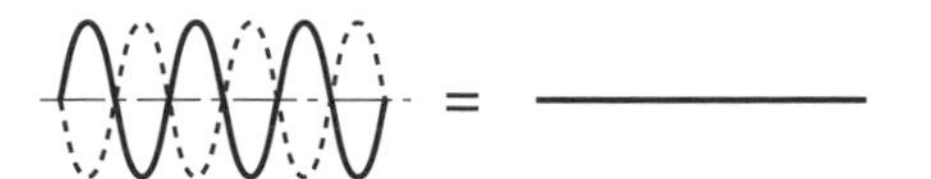

The waves above have a *phase difference* of $\frac{1}{2}$ cycle (180°) and cancel each other. This is called ***destructive interference***.

For interference to be observed:

- The sets of waves must be ***coherent***: there must be a constant phase difference between them. For this, they must have the same frequency.
- The sets of waves must have approximately the same amplitude and plane of vibration.

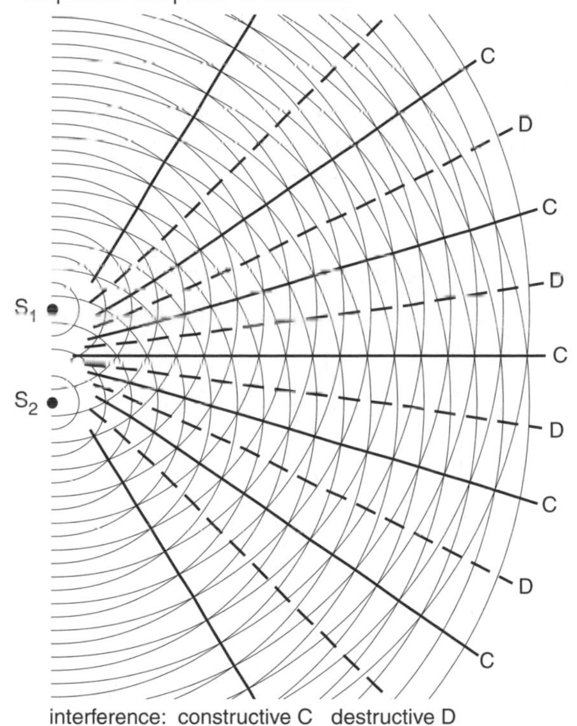

interference: constructive C destructive D

Above, waves from two coherent sources, S_1 and S_2, produce regions of reinforcing and cancelling called an ***interference pattern***. At each point of constructive interference, the path from one source is an exact number of wavelengths longer than from the other source (or the same length). The ***path difference*** is 0 or λ or 2λ, and so on.

Light waves will produce an interference pattern. However, waves from separate sources are not normally coherent, so the two sets of waves must originate from the same source. Light of one frequency (and therefore of one wavelength and colour) is called ***monochromatic light***. A laser emits monochromatic light which is coherent across its beam.

Double-slit experiment

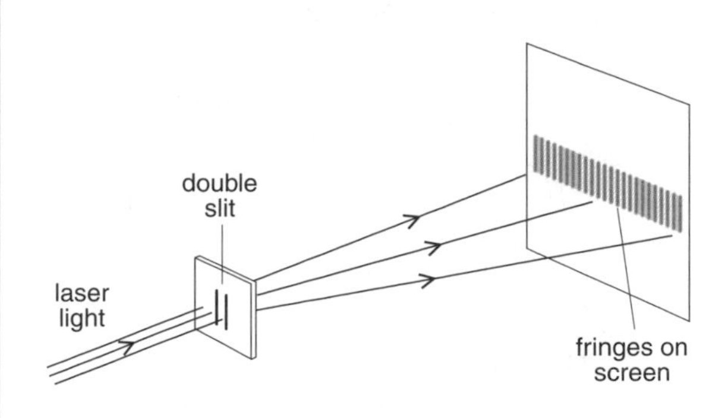

Above, light waves from a laser spread out from two slits (typically less than $\frac{1}{2}$mm apart). The interference pattern produces a series of bright and dark ***fringes*** on the screen. The bright fringes are regions of constructive interference.

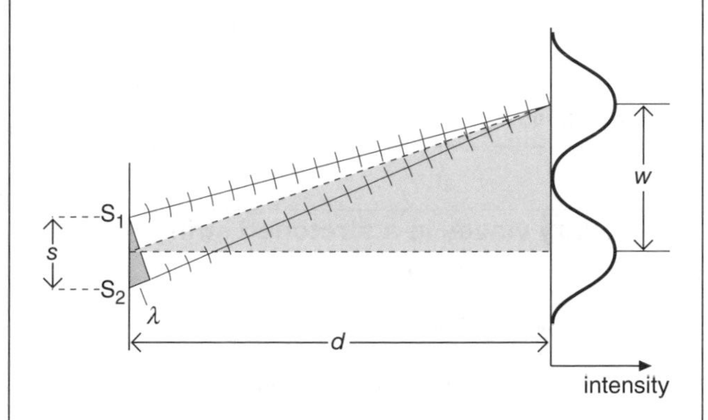

The first bright fringe occurs where the path difference is λ. For small angles, the shaded triangles above are similar, and the following equation applies:

$$\frac{\lambda}{s} = \frac{w}{D}$$

So $w = \frac{\lambda D}{s}$ — w is the ***fringe spacing***

Note:

- The fringe spacing is increased if the slits are closer together or light of longer wavelength is used.
- By measuring w, D, and s, the wavelength of light can be found using the above equation. Light wavelengths range from 7×10^{-7} m (red) down to 4×10^{-7} m (violet).

Single-slit diffraction

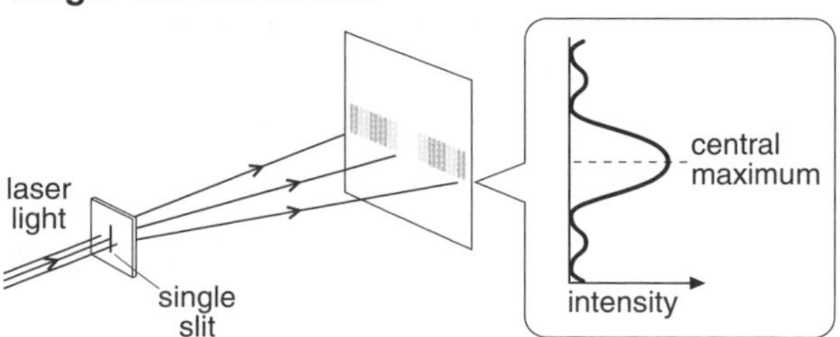

The spreading of light from a slit is an example of ***diffraction*** (see D1). Interference occurs between the different waves diffracted by the slit. The result is a pattern as above. The pattern becomes wider if the slit is made narrower or light of longer wavelength is used.

In optical instruments, diffraction limits the amount of detail in the image.

Diffraction grating

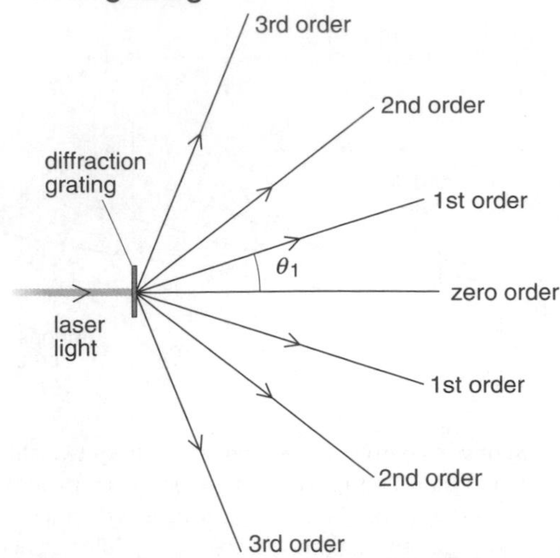

A ***diffraction grating,*** as above, has many slits (typically, 500 per mm). Constructive interference produces sharp lines of maximum intensity at set angles either side of a sharp, central maximum. In between, destructive interference gives zero or near-zero intensity. To identify the lines, they are each given an ***order number*** (0, 1, 2 etc.).

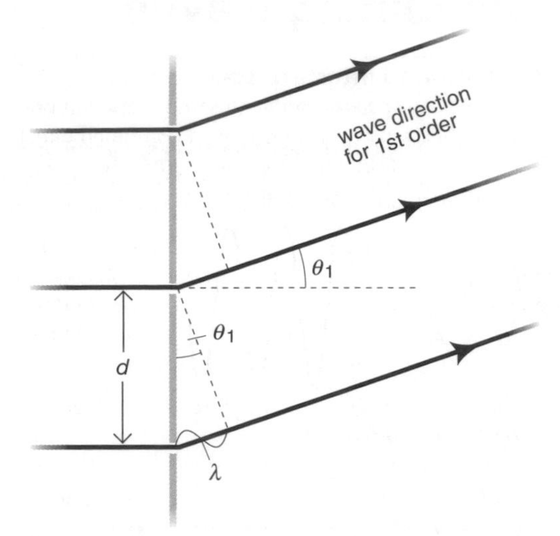

A close-up of part of the grating is shown above. d is the ***grating spacing***. θ_1 is the angle of the first order maximum. In this case, the path difference for any two adjacent slits is one wavelength, λ. From the triangles,

$$\sin \theta_1 = \frac{\lambda}{d} \qquad \text{So } d \sin \theta_1 = \lambda$$

For higher orders, the path differences are 2λ, 3λ etc. The following equation gives values of θ for all orders:

$$d \sin \theta = n\lambda$$

where n is the order number (0, 1, 2 etc.).

Note:

- If d, θ_1, and n are known, λ can be calculated.
- A longer wavelength gives a larger angle for each order.
- If the incoming light is a mixture of wavelengths (e.g. white), each order above zero becomes a spectrum.

Stationary waves in a stretched string

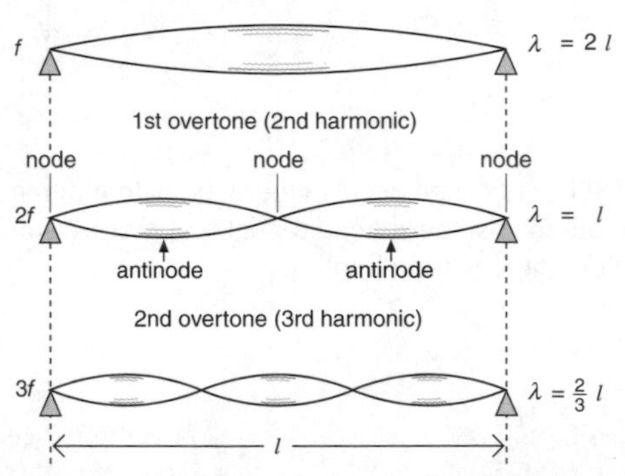

A stretched string can vibrate in various ***modes***, some of which are shown above. Each has a different frequency. The waves produced are known as ***stationary waves***. At ***nodes***, the amplitude of the oscillation is always zero. At ***antinodes***, it is always a maximum.

Stationary waves are produced by the superposition of two sets of progressive waves (of equal amplitude and frequency) travelling in opposite directions. For example, when a stretched string is vibrated, waves travel along the string, reflect from the ends, and are superimposed on waves travelling the other way.

On the right, you can see how a node is formed. As one wave moves to the right and the other to the left, the + and – displacements always cancel, so the resultant displacement is zero.

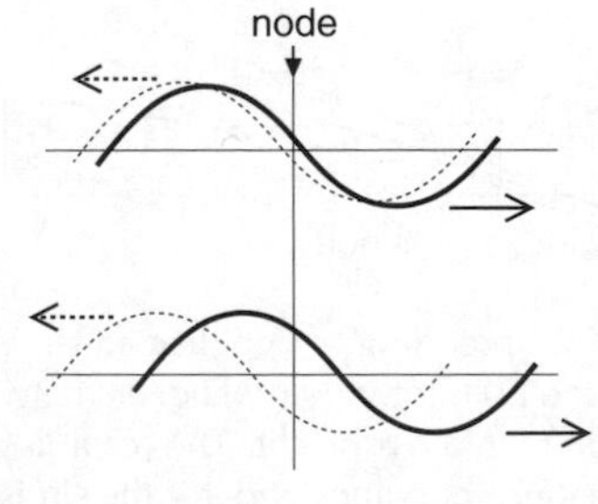

Frquency depends on tension T and the mass per unit length μ of the string.

$$f = \frac{1}{2l}\sqrt{\frac{T}{\mu}} \quad \text{for the fundamental frequency.}$$

Stationary waves in an air column

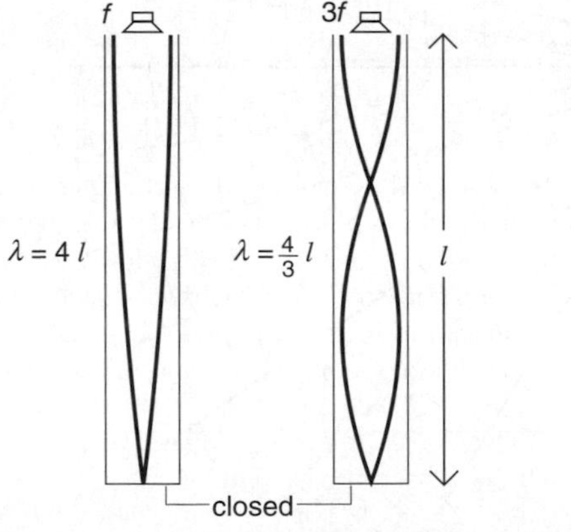

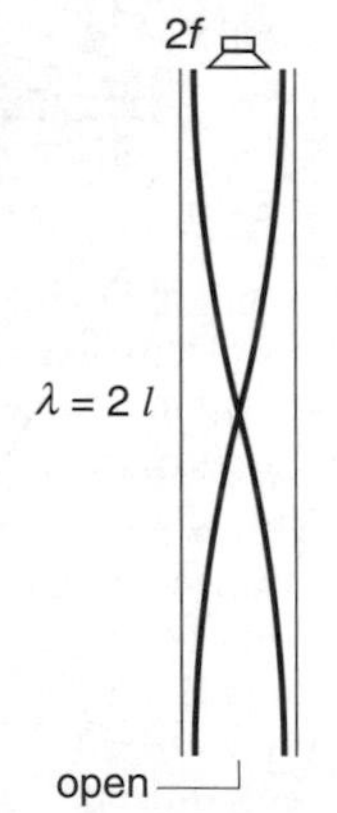

If a sound source is placed near the open end of a pipe, there are certain frequencies at which stationary waves are set up in the air column and the sound intensity reaches a maximum. This is another form of ***resonance*** (see also B16). Three examples are shown above.

Note:

- Sound waves are *longitudinal*. The 'waves' in each pipe are a *graphical representation* of the amplitude.
- Where the end of a pipe is open, there is an antinode. Where the end of a pipe is closed, there is a node.
- Knowing the frequency needed to produce resonance, and the wavelength from the pipe length, the speed of sound c can be calculated using $c = f\lambda$.

D4 Lenses

Convex and concave lenses

The diagrams on the right show how a convex and a concave lens each refract incoming rays parallel to the axis. Outgoing rays either converge towards or diverge from a ***principal focus***, F. Rays can come from either side, so there is another principal focus, F′, in an equivalent position on the opposite side of each lens. P is the ***optical centre***.

Note:

- In the rest of this unit, lenses are assumed to be thin, with no distorting effects. Diagrams will show refraction occuring at a line through the optical centre.

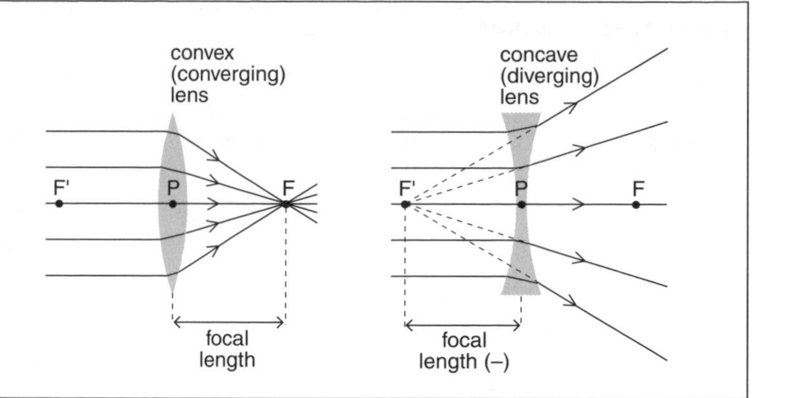

Convex lens equations

On the right, three rays have been used to show a convex lens forming a real, inverted image of an object:

1 A ray parallel to the axis is refracted through F.
2 A ray though P is undeviated (straight).
3 A ray through F′ is refracted parallel to the axis.

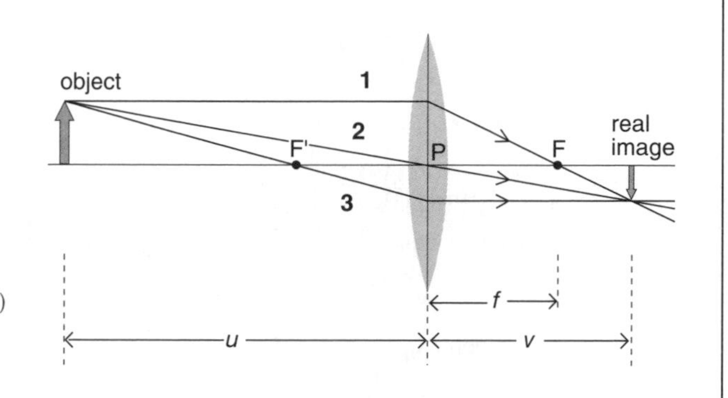

From pairs of similar triangles in the diagram, it is possible to link the object distance u, the image distance v and the focal length f with an equation:

$$\frac{1}{u} + \frac{1}{v} = \frac{1}{f} \qquad (1)$$

For example, for a lens of focal length 20 cm, the equation predicts that an object 60 cm from the lens will produce an image 30 cm from it, because

$$\frac{1}{60} + \frac{1}{30} = \frac{1}{20}$$

Power $P = \frac{1}{F}$ the unit of power is the ***dioptre (D)*** 1 D = 1 m^{-1}

Linear magnification *m* This is defined as follows:

$$\text{linear magnification} = \frac{\text{height of image}}{\text{height of object}}$$

By comparing similar triangles in the diagram above right,

$$m = \frac{v}{u} \qquad (2)$$

Convex lens as a magnifier

If u is less than f, a convex lens produces a virtual, upright, magnified image as on the right. Equation (1) applies, but the *virtual* image gives a *negative* value for v. For example, if f is 20 cm, and u is 15 cm, solving the equation gives $v = -60$ cm. This tells you that there is a virtual image 60 cm from the lens. Also, as $v/u = -4$, the image is four times the height of the object.

If the object above is moved towards F′, the image distance increases and the image gets larger. When u is $2f$, v is also $2f$, and the image and object are the same size ($m = 1$).

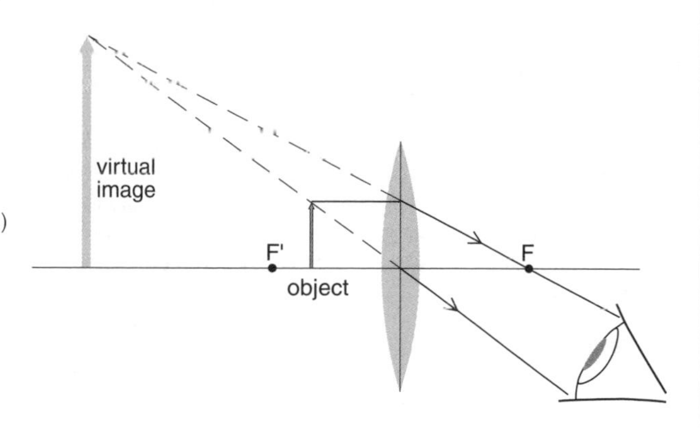

D5 The nature of light

Properties of waves and particles

This table shows the processes that occur for waves and for particles.

behaviour	waves	particles
reflection	✓	✓
refraction	✓	✓
diffraction	✓	
interference	✓	

Interference of other waves

Interference of sound waves: the speakers are connected to the same signal so the sound is coherent.

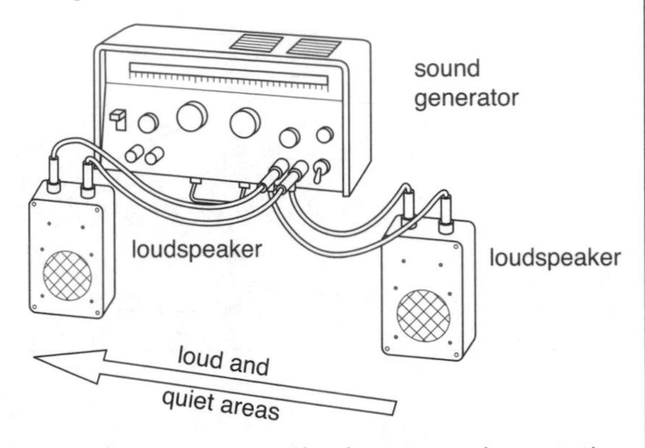

Interference of microwaves can be demonstrated in a similar way, with two transmitters and a movable detector.

How Science Works

The wave–particle debate

For centuries scientists debated whether a wave model or a particle model described the behaviour of light most accurately.

- The ancient Greeks thought that light was a stream of particles.
- Christiaan Huygens developed a wave theory that also explained reflection and refraction.
- Isaac Newton knew that he should be able to observe diffraction if light was a wave. For diffraction effects to be a maximum, the wavelength should be the same size as the gap. Newton did not realize that the wavelength of light was so small (about 5×10^{-7} m), so he saw no diffraction and said that light must be a particle.
- Thomas Young reported seeing interference fringes from his double-slit experiment (see D3). Young did not have a laser but used light from a single slit to illuminate the double slit). He said that light was a wave but was not believed at first because Newton was so respected.

The wave model works well, but then physicists discovered the photoelectric effect (see E1). Light with a frequency above a specific threshold frequency caused electrons to be emitted from the surface of some metals. It made sense that low-intensity light produced few electrons, and high-intensity light produced more electrons, but physicsists could not explain why there was a threshold frequency. Why did low-intensity ultraviolet light cause photoelectrons to be emitted, but very high-intensity red light did not?

Einstein used the idea of the photon, a quantum of light energy, to explain the effect (see E1).

Today we use a model that says that sometimes light has particle properties and sometimes it has wave properties. It is interesting to wonder whether this model will need changing again in the future.

How Science Works

Lasers

Lasers are a source of intense coherent radiation so they are ideal for demonstrating interference effects (see D3).

Because of their intense localized energy great care has to be taken to avoid injury when using lasers. Even with low-energy beams the eyes are particularly vulnerable. It is necessary to avoid inadvertently looking at a beam reflected from a mirror or other good reflector. The use of safety goggles to reduce the intensity is advisable.

Uses of lasers

The laser beam is a very narrow beam of intense energy, and as a result it has many and varied uses. Here are just some of the uses of lasers.

Tracking CDs

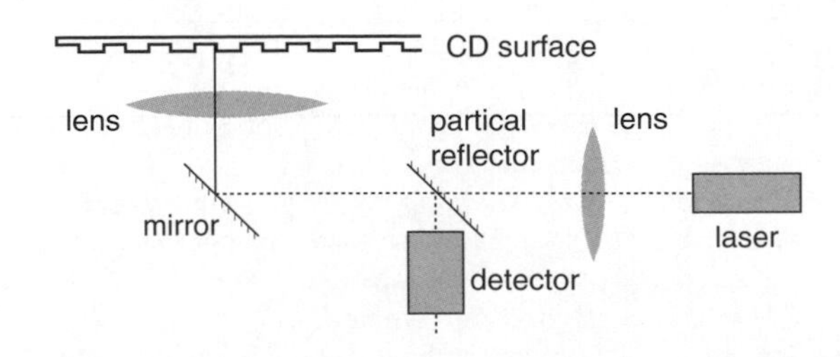

Within a CD, there is a metal layer with a spiral track of tiny steps (bumps) on it. These and the spaces between represent the 0s and 1s of the digitized signal.

Light from a laser is focused onto the track and reflected. But where there is a step, the reflected light is cancelled because of interference effects (see D3). The result is a series of light pulses, which the detector converts into electrical signals for processing.

Other uses of lasers

In medicine

- to destroy tissue in a localized area
- to break up kidney stones
- to repair broken tissue (e.g. detached retina)
- to restore sight impaired by a cateract
- to remove decay in teeth (white teeth reflect the laser energy but darker decayed areas do not)

In industry

- to drill fine holes in hard material
- to produce very accurate surveys
- to produce holograms

In communications

- to produce the light beam in fibre optic transmission

E1 Quantum theory

Quantum energy

To explain certain features of thermal radiation (see I7), Planck (in 1900) put forward the theory that energy cannot be divided into smaller and smaller amounts. It is only emitted in discrete 'packets', each called a ***quantum***. The energy E of a quantum depends on the frequency f of the radiating source, as given by this equation:

$$E = hf$$

where h is known as the ***Planck constant***. Its value, found by experiment, is 6.63×10^{-34} J s.

For electromagnetic radiation, $c = f\lambda$ (see D2), so the equation on the left can be rewritten as $E = hc/\lambda$,
where c is the speed of light and λ the wavelength.

Note:
- The shorter the wavelength (and therefore the higher the frequency), the greater the energy of each quantum.
- A quantum is an extremely small amount of energy.

quantum of red light: energy = 2 eV
quantum of violet light: energy = 4 eV
(1 eV = 1.60×10^{-19} J – see E2)

Photons

Some effects indicate that light is a wave motion. Examples include interference and diffraction (see D3). But there are others which suggest that light has particle-like properties. These include the ***photoelectric effect***, described below. Einstein (in 1905) was able to explain this by assuming that light (or other electromagnetic radiation) is made up of 'packets' of wave energy, called ***photons***. Each photon is one quantum of energy.

The photoelectric effect

When some substances are illuminated by light (or shorter wavelengths), electrons are emitted from their surface. This is called the ***photoelectric effect***. The electrons are emitted instantaneously with a range of kinetic energies, up to a maximum.

Experiments show that:
- Increasing the intensity of the light increases the number of electrons emitted per second.
- For light beneath a certain ***threshold frequency***, f_0, no electrons are emitted, even in very intense light.
- Above f_0, the maximum KE of the electrons increases with frequency, but is not affected by intensity. Even very dim light gives some electrons with high KE.

The wave theory cannot explain the threshold frequency, or how low-amplitude waves can cause high-KE electrons.

Einstein's quantum explanation Each photon delivers a quantum of energy, hf, which is absorbed by an electron. Energy Φ is needed to free the electron from the surface. If hf is more than this, the remainder is available to the electron as KE (though most electrons lose some KE before emission because they interact with other atoms). So

$$hf = \Phi + \tfrac{1}{2}m_e v_{max}^2 \qquad (1)$$

hf	Φ	$\frac{1}{2}m_e v_{max}^2$
energy delivered by photon	energy needed to free electron from surface	KE of electron (with no further energy losses)

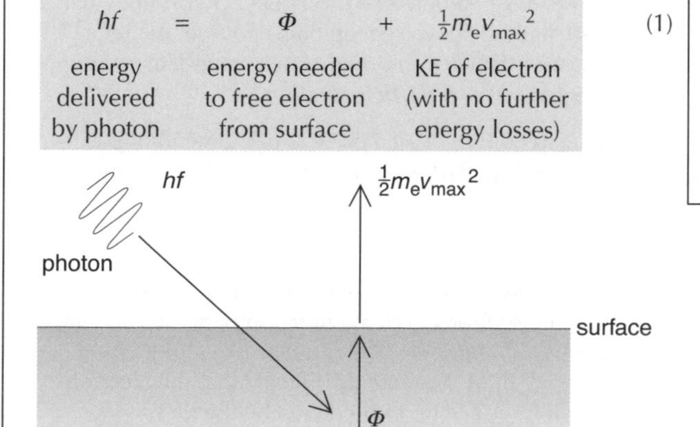

Investigating the photoelectric effect

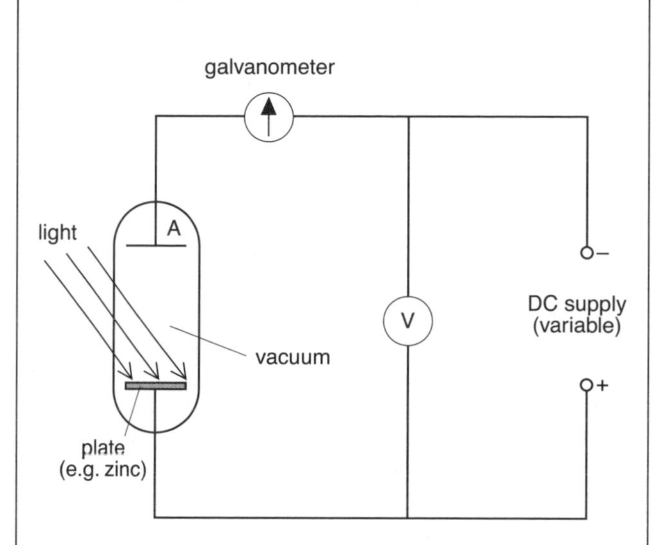

The principle of an experiment to investigate the photoelectric effect is shown above. The material being investigated (e.g. zinc) is illuminated with light of known frequency, f. Emitted electrons reach plate A, so the galvanometer detects a current in the circuit. The maximum KE of the emitted electrons is found by applying just enough *opposing* voltage, V_s, to *stop* them reaching A, so that the galvanometer reading falls to zero.

V_s is called the ***stopping voltage***. At this voltage

$$eV_s = \tfrac{1}{2}m_e v_{max}^2 \quad \text{(see E2)}$$

So, if equation (2) below is correct,

$$eV_s = hf - hf_0$$

Therefore, if V_s is measured for light of different frequencies, a graph of V_s against f should be of the form shown above.

Note:
- The number of electrons emitted is proportional to the number of photons absorbed.
- Φ is called the ***work function***. Materials with a low Φ emit electrons in visible light. Those with a higher Φ require the higher-energy photons of ultraviolet.
- If $hf < \Phi$, no electrons are emitted.
- The energy of a photon at the threshold frequency = hf_0 = Φ. So, equation (1) can be rearranged and rewritten:

$$\tfrac{1}{2}m_e v_{max}^2 = hf - hf_0 \qquad (2)$$

Spectral lines

A ***spectrum*** (see D1) contains a mixture of wavelengths, but not always in a continuous range. For example, if there is an electric discharge through hydrogen at low pressure, the gas emits particular wavelengths only, so the spectrum is made up of lines (visible colours, ultraviolet and infrared), some of which are shown below:

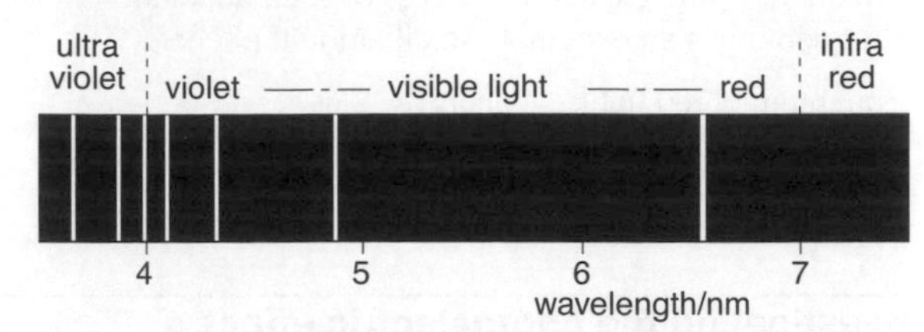

Bohr's quantum explanation (1913) In an atom, the electrons can move around the nucleus in certain allowed orbits only (top right). An electron has a different amount of energy (KE + PE) in each orbit. It may be raised to a higher orbit, for example, by colliding with an electron from another atom. When it jumps back to a lower orbit, it loses energy ($E_2 - E_1$), which is emitted as a photon. So

$$hf = E_2 - E_1$$

An electron jump is called a ***transition***.

- The greater the energy change ($E_2 - E_1$) of the transition, the higher the frequency f of the photon.
- Each possible transition gives a different spectral line.

Bohr's allowed-orbit analysis only works for the simplest atom, hydrogen. It has now been replaced by a mathematical, ***wave mechanics*** model of the atom in which allowed orbits are replaced by allowed ***energy levels***. However, the above equation still applies.

A line spectrum is a feature of any gas in which individual atoms do not interact. If atoms exert forces on each other, many more energy levels are created. Tightly-packed atoms or molecules which are vibrating, rotating, or colliding with each other have so many possible energy states that the spectrum is a continuous range of colours.

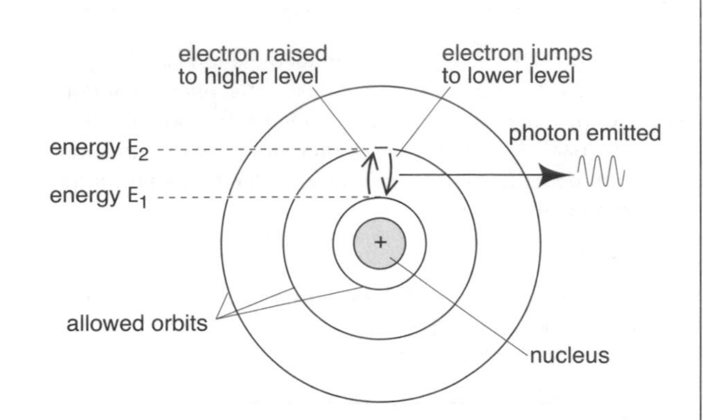

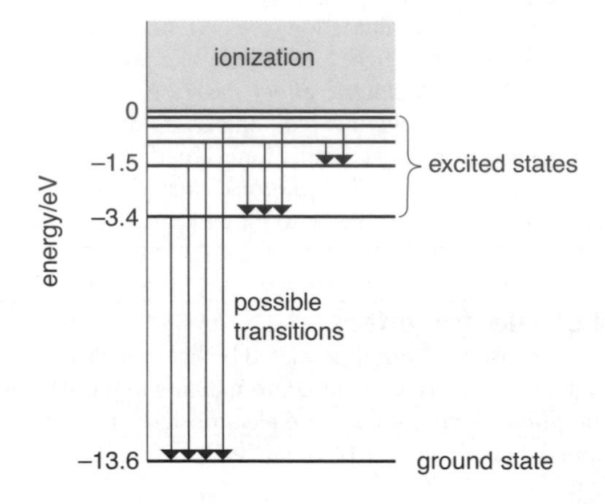

The main energy levels and transitions for hydrogen (with isolated atoms) are shown above.

Note:

- If an atom is in its ***ground state***, no electron has an unoccupied energy level beneath it.
- If an atom is in an ***excited state***, an electron has been raised to a higher energy level, so there is an unoccupied level beneath it.
- If an atom is in an ***ionized state***, an electron has been raised above the highest energy level (i.e. it has escaped). From the energy scale on the above chart, the minimum energy required to ionize a hydrogen atom is 13.6 eV.

Emission and absorption spectra

If light is radiated directly from its source, its spectrum is called an ***emission spectrum***. Examples include the line spectrum above and the continuous spectrum of the Sun.

The Sun's emission spectrum is crossed by many faint, dark lines. These are an ***absorption spectrum***. They occur because some wavelengths emitted by the Sun's core are absorbed by cooler gases (e.g. hydrogen) in its outer layers.

Some of the lines in the absorption spectrum of hydrogen are shown below. When the Sun's radiation passes through the gas, the atoms *absorb* photons whose energies match those in their emission spectrum. They then re-emit photons of these energies, but in all directions, so the intensity in the forward direction is reduced for those wavelengths.

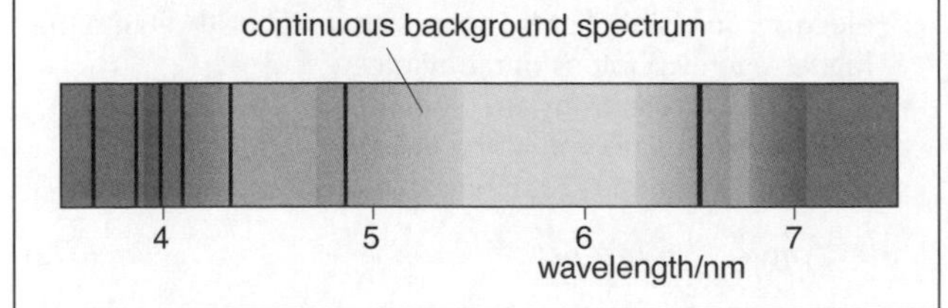

Wave–particle duality

Light waves have particle-like properties. De Broglie (in 1922) suggested that the converse might also be true: matter particles, such as electrons, might have wave-like properties. There might be ***wave–particle duality***.

According to de Broglie, if a particle of momentum p is associated with a ***matter wave*** of wavelength λ, then

$$\lambda = \frac{h}{p}$$

If a beam of electrons is passed through a thin layer of graphite, the electrons form a diffraction pattern. This suggests that the rows of atoms are acting rather like a diffraction grating (see D3). Measurements indicate that the electron wavelength is ~10^{-10} m, as predicted by the de Broglie equation. This is much shorter than light wavelengths. For more on ***electron diffraction***, see E2.

E2 Applications of quantum theory

Producing an electron beam

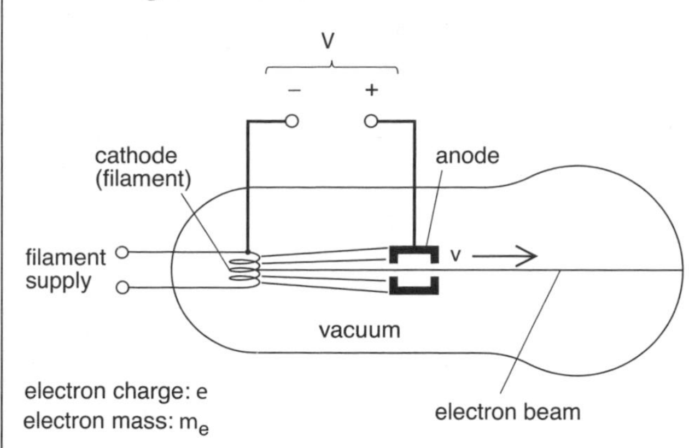

In the vacuum tube above, electrons are given off by a hot tungsten filament. The effect is called ***thermionic emission***. The electrons gain kinetic energy (KE) as they are pulled from the ***cathode*** (–) to the ***anode*** (+).

Some pass through the hole in the anode and emerge as a narrow beam, at speed v. Electrons in a beam are sometimes called ***cathode rays*** because they come from the cathode.

As an electron (charge e) moves from cathode to anode,

KE gained = work done = charge × PD = eV (see H1)

So $\frac{1}{2} m_e v^2 = eV$

Electron gun This is a device that produces a narrow beam of electrons. It uses the principle described above. In many electron guns, the cathode is an oxide-coated plate, heated by a separate filament.

Electronvolt (eV) This is a unit often used for measuring particle energies. 1 eV is the energy gained by an electron when moving through a PD of 1 V.

If an electron (1.60×10^{-19} C) moves through 1 V,
KE gained = charge × PD = $1.60 \times 10^{-19} \times 1$ J

So $1 \text{ eV} = 1.60 \times 10^{-19}$ J

Electron diffraction

Electrons have wave-like properties. This is an example of ***wave–particle duality***.

According to de Broglie's equation, electrons with a momentum p (= mv) have a wavelength λ associated with them given by

$$\lambda = \frac{h}{p}$$ where h is the Planck constant (6.63×10^{-34} J s)

The interpretation of the ***wave amplitude*** at any point is that it is related to the ***probability*** of finding the particle at that point. The greater the amplitude the more chance there is of detecting the particle at the point.

In an electron diffraction tube electrons are accelerated and hit a thin film of polycrystalline graphite. The graphite behaves in a similar way to a diffraction grating. The waves are diffracted by the graphite and produce an interference pattern. The pattern observed with an electron diffraction tube is two bright concentric rings. These are produced by two different spacings of atomic layers in the graphite structure.

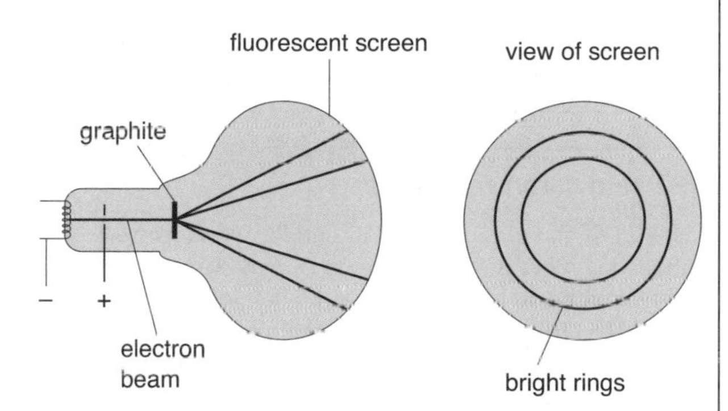

At a point where the electron waves interfere constructively the amplitude is a maximum. This means that there is a high probability of an electron arriving at that point. Lots of electrons arrive there and the screen coated with a phosphor (such as zinc sulphide) glows brightly.
At a point where the amplitude is zero, there is no chance of an electron arriving there so no light is emitted.

Typical de Broglie wavelengths of electrons

An electron accelerated through 2000 V has an associated wavelength of 2.7×10^{-11} m and a kinetic energy of 3.2×10^{-16} J.

The short wavelength is the reason why electron diffraction cannot be observed using ordinary diffraction gratings. The spacing of the 'slits' needs to be very small. Because atoms have diameters of the order of 10^{-10} m the layers of atoms in the graphite behave like a diffraction grating with a small enough spacing.

Electron microscope

The short wavelength of matter waves associated with electrons is put to use in electron microscopes.

Resolution

When two objects can be seen as two separate images, we say they can be resolved. A perfect human eye can resolve lines 0.1 mm apart at a distance of 25 cm.

Objects, or features, that are the same size or smaller than a wavelength cause diffraction. A clear image cannot be obtained. For visible light, diffraction affects images of objects that are about 750 nm and smaller, and an optical microscope cannot produce images of features smaller than 200 nm. No matter how high the magnification, using an optical microscope the resolution is limited to about 200 nm.

Using the most advanced electron microscopes, features as small as 0.1 nm can be seen.

The electron microscope has

- an electron gun to produce a beam of electrons
- electromagnetic lenses to focus the beam
- a fluorescent screen, where the image can be seen.

(Compare with the optical microscope, which uses a lamp, glass lenses, and the eye or a camera to see the image.)

The whole microscope is enclosed in a vacuum, so the electron beam can be accelerated. The specimen must be very thin and specially prepared (often coated with gold) to be conducting.

Electrons pass though the specimen and an aperture, or they are scattered. Only those that pass through the aperture reach the screen and produce fluorescence, while other areas remain dark. False colour can be added to stored images to show any difference in intensity.

The transmission electron microscope (TEM)

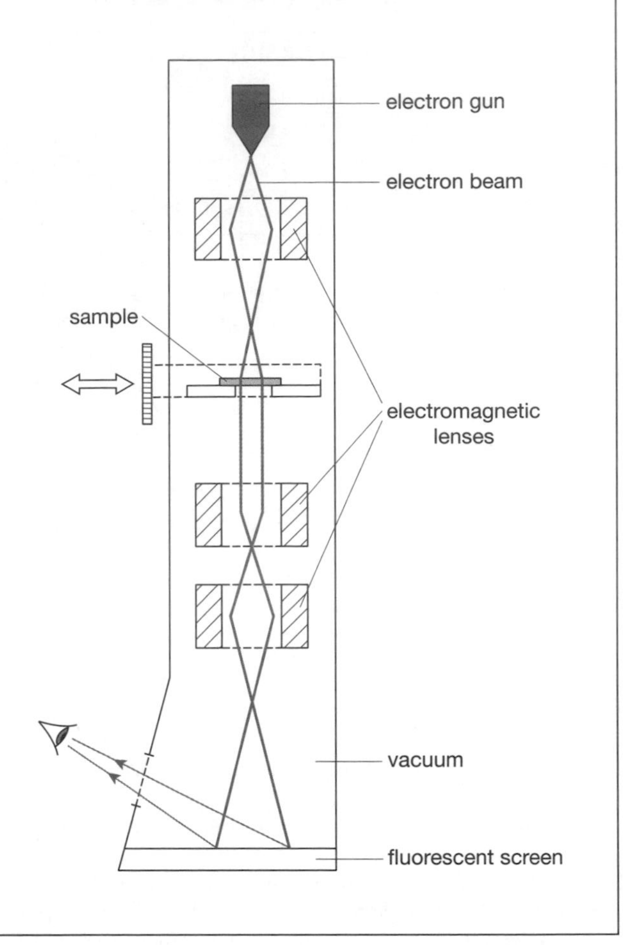

How Science Works

CRT television receivers

Before LCD and plasma TVs, TVs and PC monitors were cathode ray tubes (CRTs). These contained electron guns, and the electron beam was scanned across a fluorescent screen using an electromagnet to produce the picture.

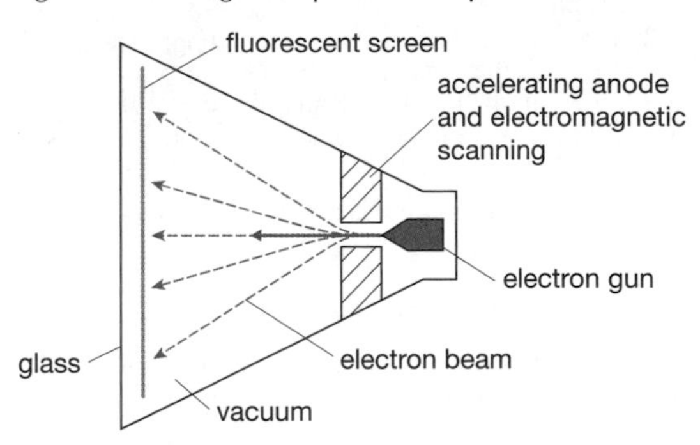

Television is an application of science and technology with the following benefits and risks.

Benefits

- provides education
- provides information (e.g. news)
- provides entertainment
- improves understanding between nations/ethnic groups.

Risks

- encourages inactivity
- encourages imitation of undesirable behaviour (e.g. violence)
- may cause health risks from viewing screens (e.g. epilepsy).

Note: care is taken in design to ensure that X-rays are not produced when the electrons are decelerated as they hit the screen.

How Science Works

Fluorescent lamps

Fluorescent lamps use the same principle as the electron beam tube. The glass tube contains a small amount of mercury vapour and an inert gas (usually argon) at low pressure. When the fast electrons collide with atoms, excited atoms and positive ions are produced. The excited mercury atoms return to the ground state by emitting ultraviolet photons.

The inside of the glass is coated with phosphor, which fluoresces in ultraviolet radiation. Fluorescence occurs when an atom absorbs an ultraviolet photon, which has higher energy than visible light, and returns to the ground state by emitting lower energy photons in the visible range.

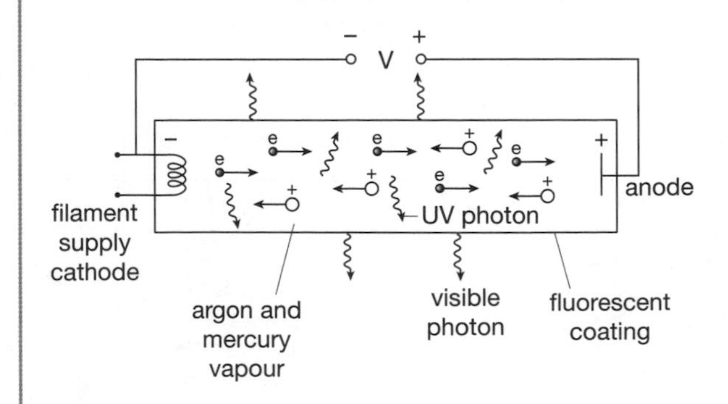

Benefits

Fluorescent lamps are about five times more efficient than incandescent lamps.

Risks

Fluorescent lamps contain mercury, which is toxic. Some landfill sites collect them separately to reduce pollution.

F1 Charges and circuits

Static electricity

If two materials are rubbed together, electrons may be transferred from one to another. As a result, one gains negative charge, while the other is left with an equal positive charge. If the materials are ***insulators*** (see right), the transferred charge does not readily flow away. It is sometimes called ***static electricity***.

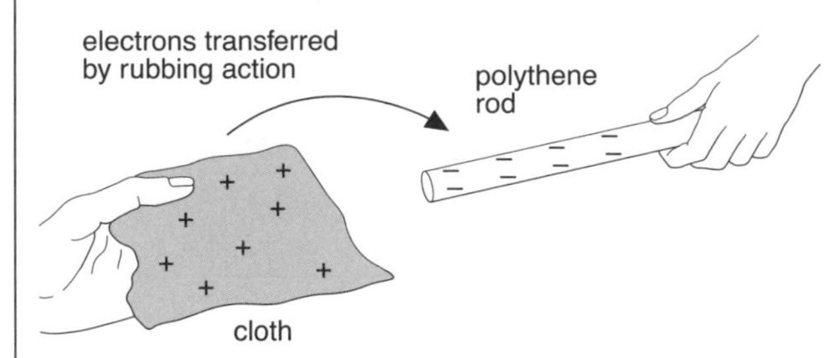

A charged object will attract an uncharged one. On the right, the charged rod has extra electrons. Being uncharged, the foil has equal amounts of – and + charge. The – charges are repelled by the rod and tend to move away, while the + charges are attracted. However, the force of attraction is greater because of the shorter distance.

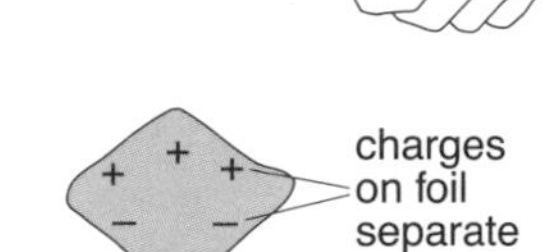

Charge that collects in one region because of the presence of charge on another object is called ***induced*** charge.

Current

cell

electron flow

conventional direction

ammeter

bulb

In the circuit above, chemical reactions in the cell push electrons out of the negative (–) terminal, round the circuit, to the positive (+) terminal. This flow of electrons is called a ***current***.

An arrow in the circuit indicates the direction from the + terminal round to the –. Called the ***conventional direction***, it is the *opposite* direction to the actual electron flow.

The SI unit of current is the ***ampere*** (**A**).

A current of 1 A is equivalent to a flow of 6×10^{18} electrons per second. However, the ampere is not defined in this way, but in terms of its magnetic effect (see H6).

Current may be measured using an ***ammeter*** as above.

Conductors and insulators

Current flows easily through metals and carbon. These materials are good ***conductors*** because they have free electrons which can drift between their atoms (see I1).

Most non-metals are ***insulators***. They do not conduct because all their electrons are tightly held to atoms and not easily moved. Although liquids and gases are usually insulators, they do conduct if they contain ions.

Semiconductors, such as silicon and germanium, are insulators when cold but conductors when warm.

Charge

Charge can be calculated using this equation:

charge = current × time

The SI unit of charge is the ***coulomb*** (**C**).

For example, if a current of 1 A flows for 1 s, the charge passing is 1 C. (This is how the coulomb is defined.) Similarly, if a current of 2 A flows for 3 s, the charge passing is 6 C.

Voltage (PD and EMF)

In the circuit below, several cells have been linked in a line to form a ***battery***. The ***potential difference*** **(PD)** across the battery terminals is 12 volts (V). This means that each coulomb (C) of charge will 'spend' 12 joules (J) of energy in moving round the circuit from one terminal to the other.

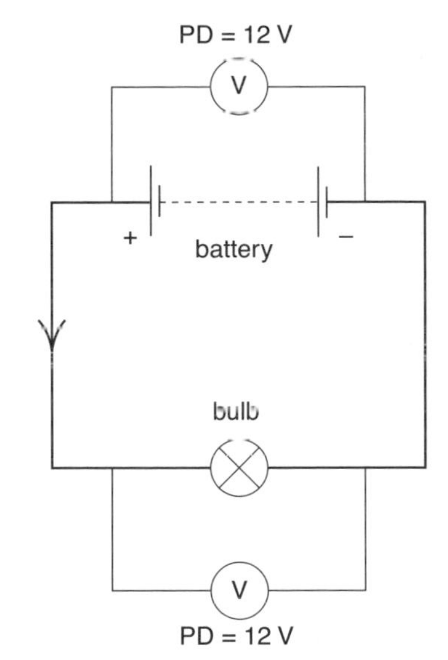

The PD across the bulb is also 12 V. This means that, for each coulomb pushed through it, 12 J of electrical energy is changed into other forms (heat and light energy).

PD may be measured using a ***voltmeter*** as shown above.

PD, energy, and charge are linked by this equation:

energy transformed = charge × PD

For example, if a charge of 2 C moves through a PD of 3 V, the energy transformed is 6 J.

The voltage produced by the chemical reactions inside a battery is called the ***electromotive force*** (**EMF**). When a battery is supplying current, some energy is wasted inside it, which reduces the PD across its terminals. For example, when a torch battery of EMF 3.0 V is supplying current, the PD across its terminals might be only 2.5 V.

Ohm's law and resistance

If a conductor obeys ***Ohm's law***, then the current *I* through it is directly proportional to the PD *V* across it, provided the temperature is constant.

Metals obey Ohm's law. If a graph of *I* against *V* is plotted for a metal conductor at constant temperature, the result is as on the right. Expressed mathematically this is

$$\frac{V}{I} = \text{constant}$$

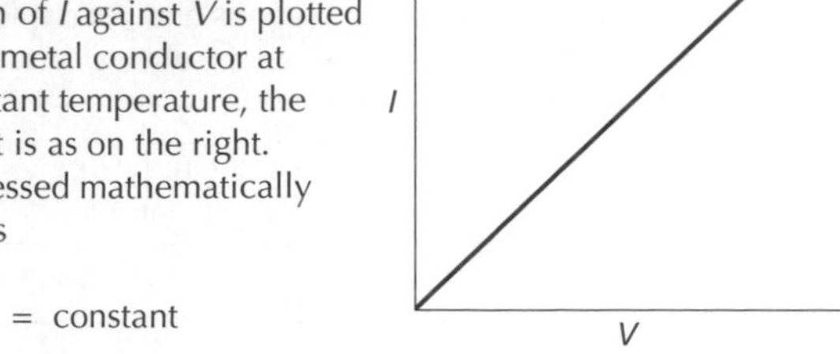

The ***resistance*** *R* of a conductor is calculated like this:

$$\text{resistance} = \frac{\text{PD}}{\text{current}} \qquad \text{In symbols} \quad R = \frac{V}{I}$$

The SI unit of resistance is the ***ohm*** (Ω).

For example, if a PD of 1 V causes a current of 1 A, then the resistance is 1 Ω. (This is how the ohm is defined.)

Similarly, if a PD of 12 V causes a current of 4 A, then the resistance is 3 Ω.

The resistance of a metal conductor (such as a wire) depends on various factors:

- **Length** A long wire has more resistance than a short one.
- **Cross-sectional area** A thin wire has more resistance than a thick one.
- **Temperature** A hot wire has more resistance than a cold one.
- **Type of material** A nichrome wire has more resistance than a copper wire of the same dimensions.

Note:
- While the resistance of a metal *increases* with temperature, that of a semiconductor *decreases*.

Resistance components

Heating elements If a conductor (such as a wire) has resistance, then electrical energy is changed into heat when a current passes. This effect is used in heating elements.

Resistors These are components specially designed to provide resistance. In electronic circuits, they are needed so that other components are supplied with the correct current.

Variable resistors These have a control for varying the length of resistance material through which the current passes.

Thermistors These components have a resistance which changes considerably with temperature (e.g. high when cold, low when hot). They contain semiconducting materials.

Light-dependent resistors (LDRs) These have a high resistance in the dark but a low resistance in the light.

Diodes These have an extremely high resistance in one direction but a low resistance in the other. In effect, they allow current to flow in one direction only.

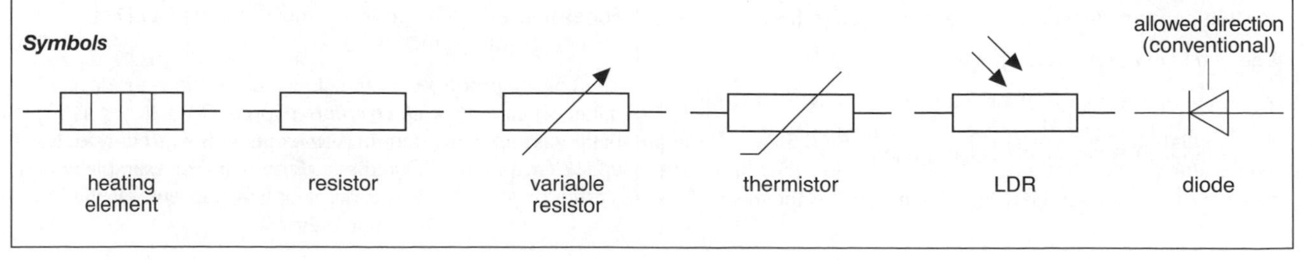

Circuit rules

Resistors in series The current through the battery and each resistor is the same. However, the voltage arcoss the battery is shared by the resistors.

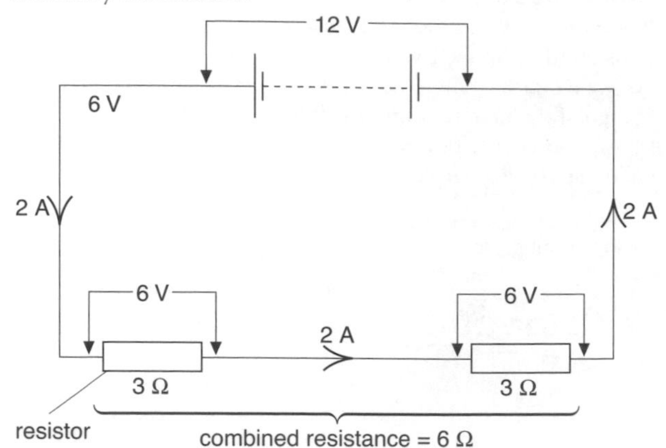

Resistors in parallel The voltage across the battery and each resistor is the same. However, the current from the battery is shared by the resistors.

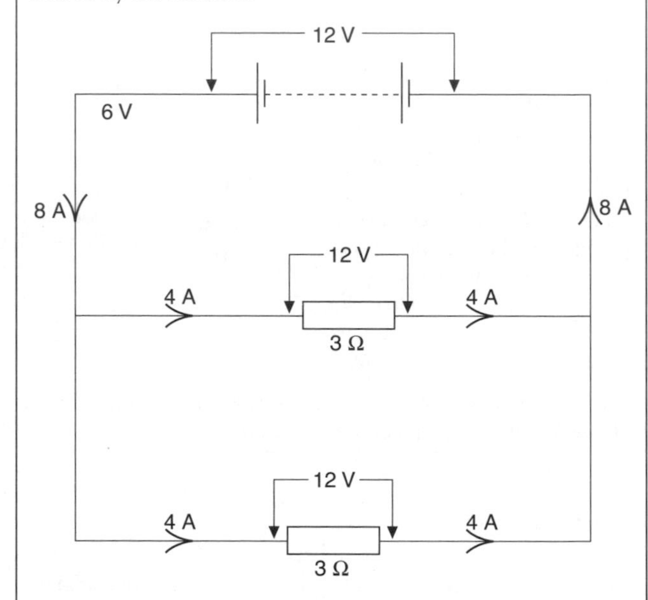

Power

Power *P* is calculated with this equation:

$$\text{power} = \text{PD} \times \text{current} \qquad \text{In symbols} \quad P = VI$$

For example, if there is a PD of 12 V across a resistor and a current of 3 A through it, then the power = 12 × 3 = 36 W (watts). In other words, the resistor is changing 36 joules of electrical energy into heat every second.

Alternative forms of the power equation are

$$P = I^2R \qquad P = \frac{V^2}{R}$$

F2 Current and resistance

Current, charge, and electrons

In a wire, the current is a flow of electrons. If I is the current and Q the charge passing any point in time t, then

$$Q = It$$

From the above, a current of 1 ampere (A) means that charge is flowing at the rate of 1 coulomb (C) per second. The charge on an electron is $e = -1.60 \times 10^{-19}$ C.

Therefore, 1 C is the charge carried by $1/e$ electrons, i.e. 6.24×10^{18} electrons. So a flow of 6.24×10^{18} electrons per second gives a current of 1 A. However, as e is negative, an electron flow to the *right* is a current to the *left*.

Resistance

If a conductor has resistance, then energy is *dissipated* (changed to internal energy) when a current passes through.

The PD V (in V) across a conductor, current I (in A) through it, and its resistance R (in Ω) are linked by this equation:

$$R = \frac{V}{I}$$

If a conductor obeys ***Ohm's law***, its resistance is constant for any given temperature (i.e. R is independent of V).

The link between I and V can be investigated using the circuit below. Graphs for three different components are shown. (A negative V means that the DC supply connections have been reversed.)

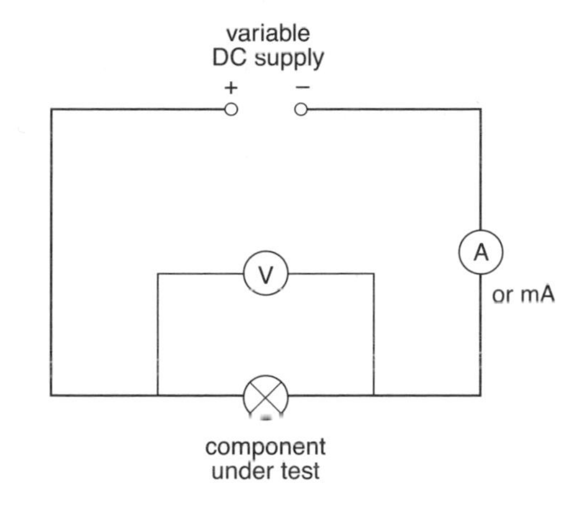

Current and drift speed

Most electrons are bound to their atoms. However, in a metal, some are ***free electrons*** which can move between atoms. When a PD is applied, and a current flows, the free electrons are the ***charge carriers***.

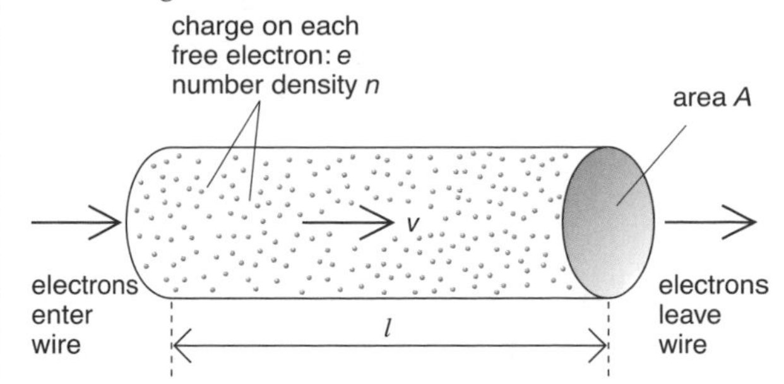

In the wire above, free electrons (each of charge e) are moving with an average speed v. n is the ***number density*** of free electrons: the number per unit volume (per m^3).

In the wire the number of free electrons = nAl
So total charge carried by free electrons = $nAle$

As time = distance/speed:
time taken for all the free electrons to pass through A $= l/v$

As current I = charge/time $\quad I = \dfrac{nAle \times v}{l}$

$$\therefore \quad I = nAev$$

v is called the ***drift speed***. Typically, it can be less than a millimetre per second for the current in a wire.

The ***current density*** J is the current per unit cross-sectional area (per m^2).

$$J = \frac{I}{A} \quad \text{so} \quad J = nev$$

Note:

- The number density of free electrons is different for different metals. For copper, it is 8×10^{28} m^{-3}.
- When liquids conduct, ions are the charge carriers. The above equations apply, except that e and n must be replaced by the charge and number density of the ions.

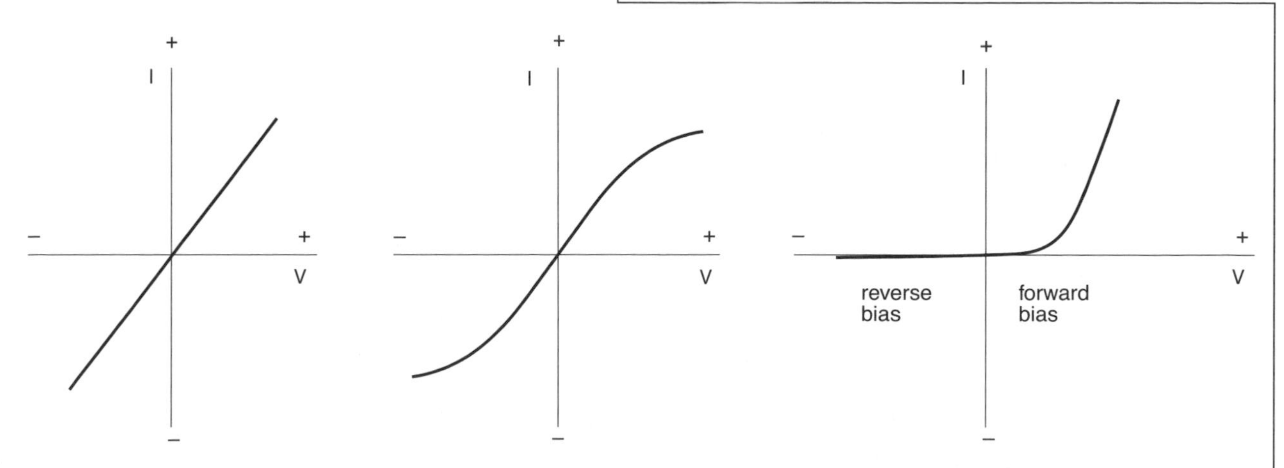

Wire (metal) kept at constant temperature V/I is constant, so R is constant.

Bulb filament (metal) As the current rises the filament heats up. V/I increases, so R increases.

Diode (semiconductor) R is very high in one direction. It is much lower in the other direction and decreases as the current rises.

Resistance and temperature

A conducting solid is made up of a ***lattice*** of atoms. When a current flows, electrons move through this lattice.

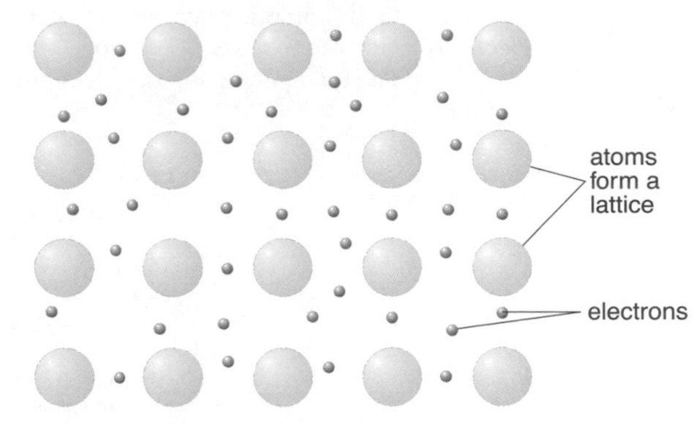

Metals When free electrons drift through a metal, they make occasional collisions with the lattice. These collisions are inelastic and transfer energy to the lattice as internal energy. That is why a metal has resistance. If the temperature of a metal rises, the atoms of the lattice vibrate more vigorously. Free electrons collide with the lattice more frequently, which increases the resistance.

Semiconductors (e.g. silicon) At low temperature, the electrons are tightly bound to their atoms. But as the temperature rises, more and more electrons break free and can take part in conduction. This easily outweighs the effects of more vigorous lattice vibrations, so the resistance decreases. At around 100–150 °C, ***breakdown*** occurs. There is a sudden fall in resistance – and a huge increase in current. That is why semiconductor devices are easily damaged if they start to overheat.

The conduction properties of a semiconductor can be changed by ***doping*** it with tiny amounts of impurities. For example, a diode can be made by doping a piece of silicon so that a current in one direction increases its resistance while a current in the opposite direction decreases it.

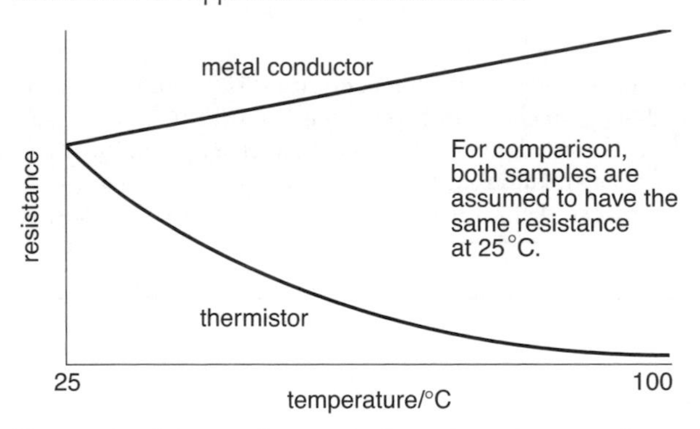

The graphs above are for a typical metal conductor and one type of thermistor. The thermistor contains semiconducting materials.

Superconductivity

When some metals are cooled towards absolute zero, a ***transition temperature*** is reached at which the resistance suddenly falls to zero. This effect is called ***superconductivity***. It occurs when there is no interaction between the free electrons and the lattice, and is explained by the quantum theory. Some specially developed metal compounds have transition temperatures above 100 K.

If an electromagnet has a superconducting coil, a huge current can be maintained in it with no loss of energy. This enables a very strong magnetic field to be produced.

Energy transfer

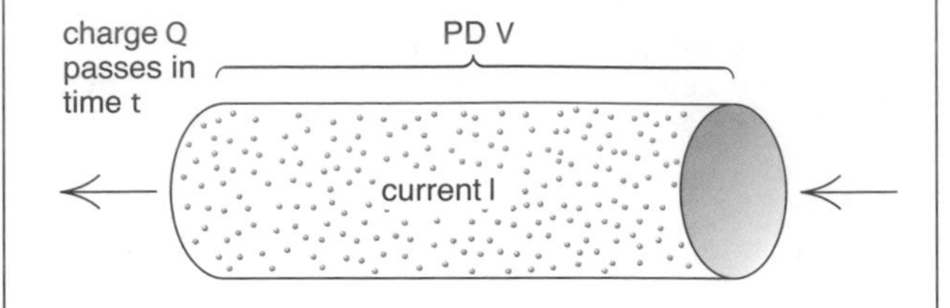

Above, charge Q passes through a resistor in time t. Work W is done by the charge, so energy W is transformed – the electrons lose electrical potential energy and the lattice gains internal energy (it heats up).

W, Q, and V are linked by this equation (see also E1):

$$W = QV$$

But $Q = It$, so $W = VIt$ (1)

Applying $V = IR$ to the above equation gives

$W = I^2Rt$ and $W = V^2t/R$ (2)

For example, if a current of 2 A flows through a 3 Ω resistor for 5 s, W = $2^2 \times 3 \times 5 = 60$ J. So the energy dissipated is 60 J. *Double* the current gives *four* times the energy dissipation.

Note:

- Equation (1) can be used to calculate the total energy transformation whenever electrical potential energy is changed into other forms (e.g. KE and internal energy in an electric motor). Equations (2) are only valid where *all* the energy is changed into internal energy. Similar comments apply to the power equations which follow.

As power $P = W/t$, it follows from (1) and (2) that

$P = VI$ $P = I^2R$ $P = V^2/R$

Resistivity

The resistance R of a conductor depends on its length l and cross-sectional area A:

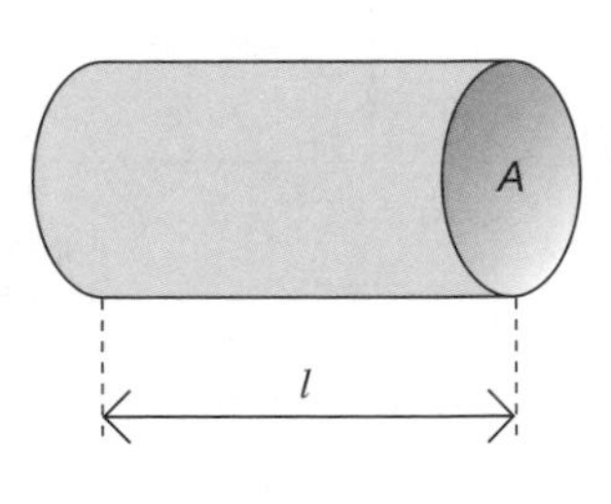

$$R \propto \frac{l}{A}$$

This can be changed into an equation by means of a constant, ρ, known as the ***resistivity*** of the material:

$$R = \frac{\rho l}{A}$$

With this equation, the resistance of a wire can be calculated if its dimensions and resistivity are known.

Resistivities, in Ω m

copper: 1.55×10^{-8} aluminium: 2.50×10^{-8}

Conductance and conductivity

If a PD V is applied across a conductor, and a current I flows, then $V = IR$. However, as V is the cause of the current and I is the effect, it is more logical to write this as:

$$I = \frac{1}{R} \times V$$

$1/R$ is called the ***conductance***. $1/\rho$ is the ***conductivity***.

F3 Analysing circuits

Note: in this unit, the symbol E stands for EMF and not electric field strength.

Kirchhoff's first law

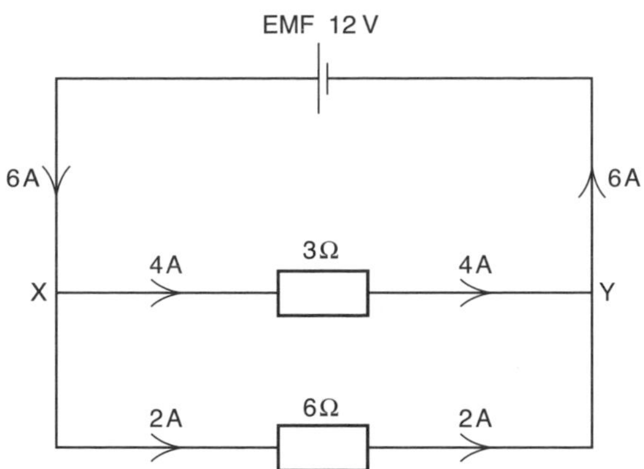

The currents at junctions X and Y above illustrate a law which applies to all circuits:

total current out of junction = total current into junction

This is known as ***Kirchhoff's first law***. It arises because, in a complete circuit, charge is never gained or lost. It is conserved. So the total rate of flow of charge is constant.

Kirchhoff's second law

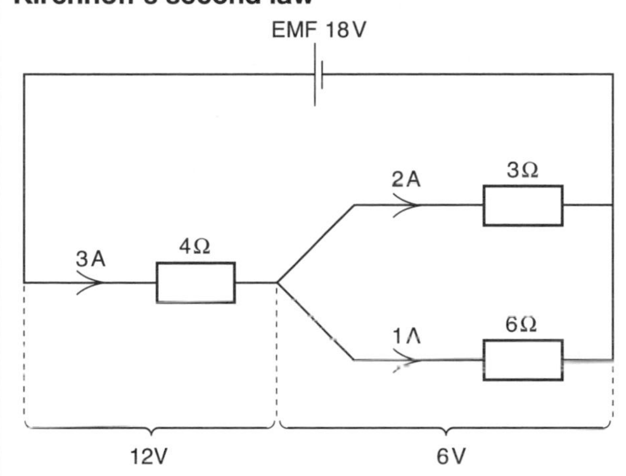

The arrangement above is called 'a circuit'. But, really, there are *two* complete circuits through the battery. To avoid confusion, these will be called *loops*.

In the circuit above, charge leaves the battery with electrical potential energy. As the charge flows round a loop, its energy is 'spent' – in stages – as heat. The principle that the total energy supplied is equal to the total energy spent is expressed by ***Kirchhoff's second law***.

Round any closed loop of a circuit, the algebraic sum of the EMFs is equal to the algebraic sum of the PDs (i.e. the algebraic sum of all the *IR*s).

Note:

- From the law, it follows that if sections of a circuit are in parallel, they have the same PD across them.
- 'Algebraic' implies that the direction of the voltage must be considered. For example, in the circuit on the right, the EMF of the right-hand battery is taken as *negative* (–4 V) because it is opposing the current. Therefore:

  algebraic sum of EMFs = 18 + (–4) = +14 V
  algebraic sum of *IR*s = (2 × 3) + (2 × 4) = +14 V

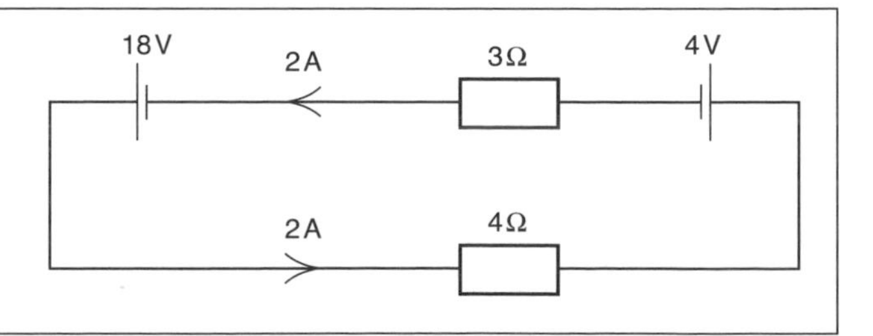

Resistors in series

If R_1 and R_2 below have a total resistance of R, then R is the single resistance which could replace them.

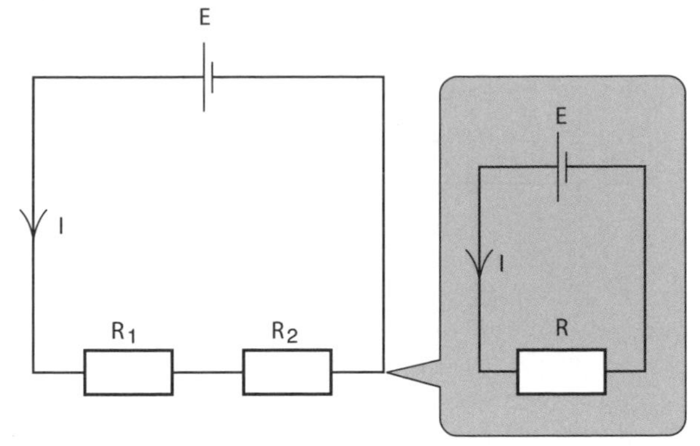

From Kirchhoff's first law, all parts of the circuit have the same current I through them.

From Kirchhoff's second law $E = IR$ and $E = IR_1 + IR_2$.

So $IR = IR_1 + IR_2$

$\therefore$ $R = R_1 + R_2$

For example, if $R_1 = 3\ \Omega$ and $R_2 = 6\ \Omega$, then $R = 9\ \Omega$.

Resistors in parallel

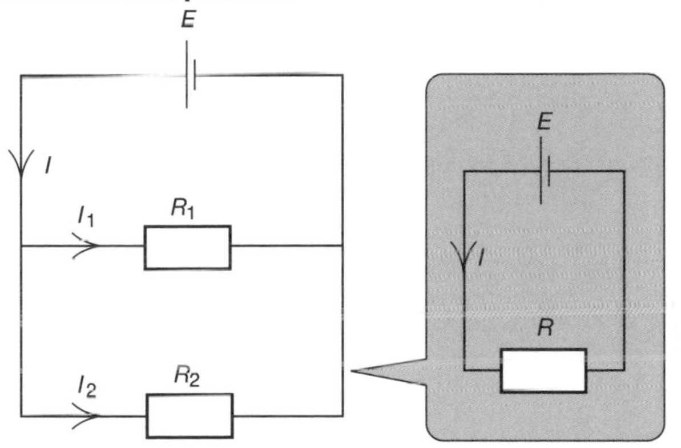

From Kirchhoff's second law (applied to the various loops):

$E = IR$ and $E = I_1R_1$ and $E = I_2R_2$

From Kirchhoff's first law $I = I_1 + I_2$.

So $\frac{E}{R} = \frac{E}{R_1} + \frac{E}{R_2}$

$\therefore$ $\frac{1}{R} = \frac{1}{R_1} + \frac{1}{R_2}$

For example, if $R_1 = 3\ \Omega$ and $R_2 = 6\ \Omega$, $1/R = 1/3 + 1/6 = 1/2$. So $R = 2\ \Omega$.

Internal resistance

On the opposite page, it was assumed that each battery's output PD (the PD across its terminals) was equal to its EMF. In reality, when a battery is supplying current, its output PD is *less* than its EMF. The greater the current, the lower the output PD. This reduced voltage is due to energy dissipation in the battery. In effect, the battery has ***internal resistance***. Mathematically, this can be treated as an additional resistor in the circuit.

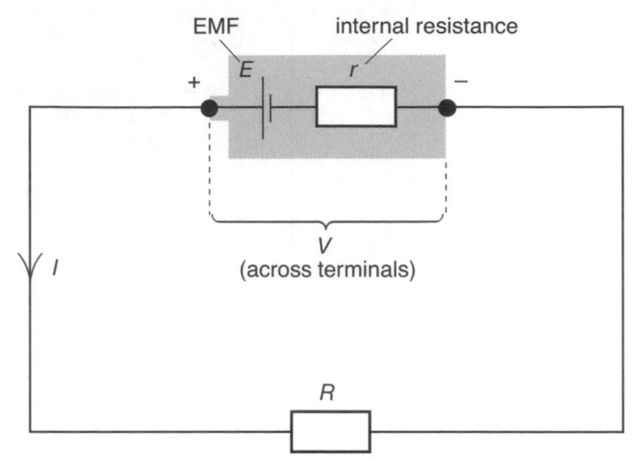

The battery above is supplying a current I to an external circuit. The battery has a constant internal resistance r.

From Kirchhoff's second law $\quad E = IR + Ir$

But $\quad V = IR$, so $\quad E = V + Ir$

So $\quad V = E - Ir \quad (1)$

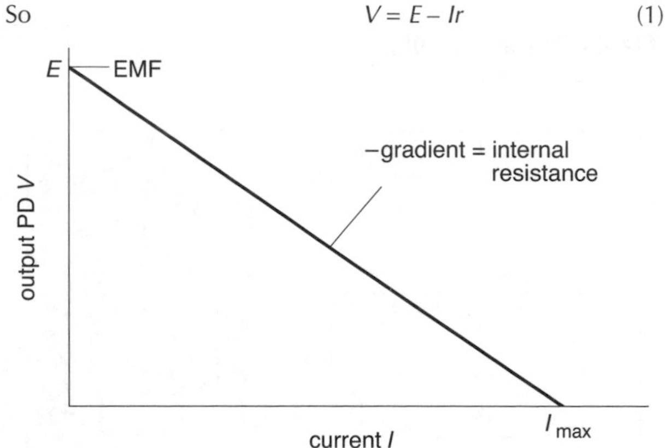

The graph above shows how V varies with I. Unlike earlier graphs, V is on the vertical axis.

Note:

- When I is zero, $V = E$. In other words, when a battery is in ***open circuit*** (no external circuit), the PD across its terminals is equal to its EMF.
- When R is zero, V is zero. In other words, when the battery is in ***short circuit*** (its terminals directly connected), its output PD is zero. In this situation, the battery is delivering the maximum possible current, I_{max}, which is equal to E/r. Also, the battery's entire energy output is being wasted internally as heat.
- As $I_{max} = E/r$, it follows that $r = E/I_{max}$. So the gradient of the graph is numerically equal to the internal resistance of the battery.

If both sides of equation (1) are multiplied by I, the result is $VI = EI - I^2r$. Rearranged, this gives the following:

EI	=	VI	+	I^2r
power released by chemical action		power delivered to external circuit		power dissipated inside battery

Potential divider

A ***potential divider*** or ***potentiometer*** like the one below passes on a fraction of the PD supplied to it.

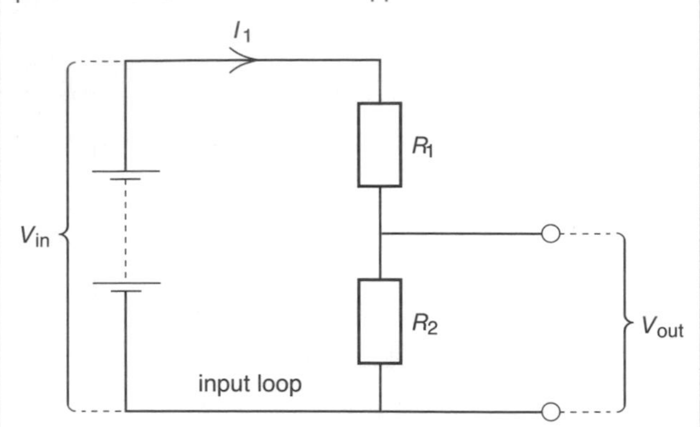

In the input loop above, the total resistance = $R_1 + R_2$.

So $\quad I = V_{in}/(R_1 + R_2)$

But $V_{out} = IR_2$, $\quad$ so $V_{out} = \left(\dfrac{R_2}{R_1 + R_2}\right)V_{in}$

For example, if R_1 and R_2 are both 2 kΩ, then $R_2/(R_1 + R_2)$ works out at 1/2, so V_{out} is a half of V_{in}.

Note:

- The above analysis assumes that no external circuit is connected across R_2. If such a circuit is connected, then the output PD is reduced.

In electronics, a potential divider can change the signals from a sensor (such as a heat or light detector) into voltage changes which can be processed electrically. For example, if R_2 is a thermistor, then a rise in temperature will cause a fall in R_2, and therefore a fall in V_{out}. Similarly, if R_2 is a ***light-dependent resistor*** (**LDR**), then a rise in light level will cause a fall in R_2, and therefore a fall in V_{out}.

Potential dividers are not really suitable for high-power applications because of energy dissipation in the resistors.

Balanced PDs

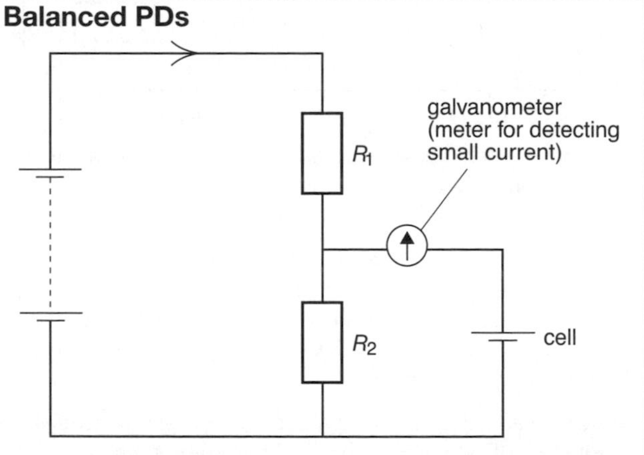

Above, a cell has been connected across the output of a potential divider, and the values of R_1 and R_2 adjusted so that the galvanometer reads zero. This happens when the PD across R_2 exactly balances the cell's output PD.

The above method can be used to compare the EMFs of different cells. It has several advantages.

- The cell is effectively in open circuit, so the PD across the cell's terminals is equal to its EMF.
- As the meter is only being used to test for zero current, it does not need an accurately calibrated scale.
- A very sensitive meter can be used.

F4 AC and DC

Direct current (DC)

Direct current may be a constant current as supplied by a battery, or varying in size but still moving in the same direction.

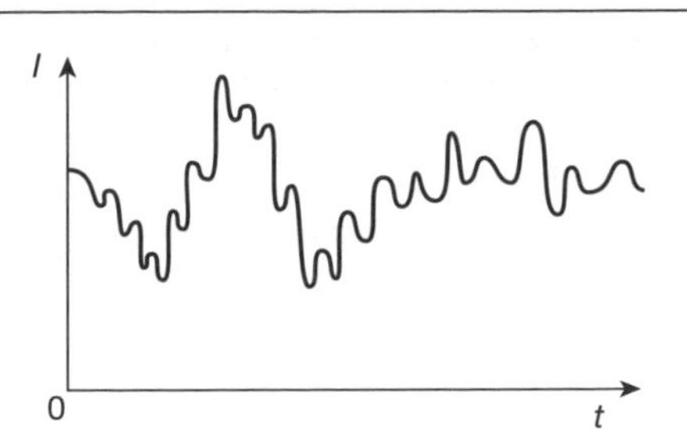

Alternating current (AC)

The graph below shows how the current from an AC supply varies with time. Here are some of the terms used to describe AC.

Frequency f This is the number of cycles per second. The unit is the hertz (Hz). For example, in the UK, the frequency of the AC mains is 50 Hz (50 cycles per second).

Peak current I_0 This is the maximum current during the cycle. It is the amplitude of the waveform in the graph below.

The current I at any instant is related to the peak current by the following equation:

$I = I_0 \sin \theta$ where $\theta = 360 \times f \times t$

The graph is an example of a ***sinusoidal*** waveform.

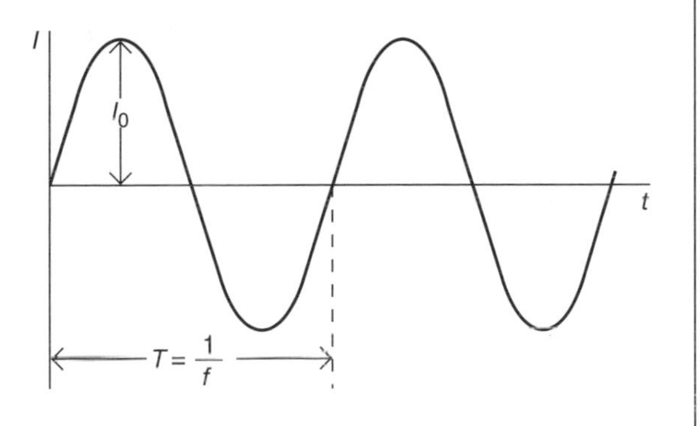

Supplying AC for the mains

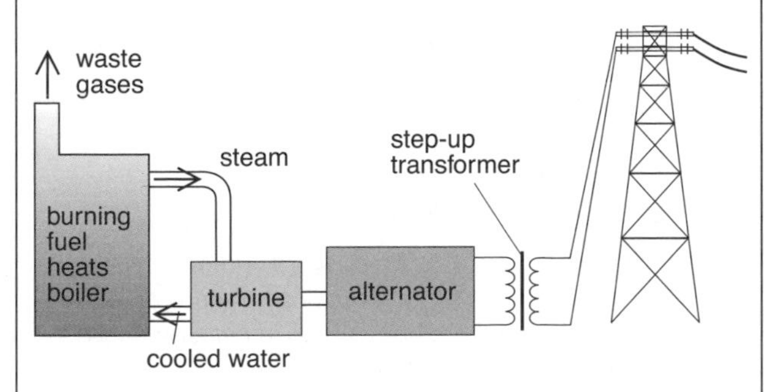

Mains AC is generated in power stations. The layout of a typical fuel-burning station is shown above. The power is fed into a distribution network called the ***Grid***.

Power is sent across country through overhead lines at very high voltage (typically 400 000 V). The voltage is increased to this level by transformers and then reduced again at the far end. As power = voltage × current, transmitting at a higher voltage means a lower current and, therefore, less power wasted as heat because of line resistance.

Transformers

Transformers will work only with alternating voltage. They increase or decrease the voltage (for details see H10).

Step-up transformers

The input coil has fewer turns of wire than the output coil. The output voltage is higher than the input voltage. (The current will be less.)

Step-down transformers

The input coil has more turns than the output coil. The output voltage is less than the input voltage. (The current will be greater.)

For an ideal transformer, input power = output power, for a practical transformer:

$V_2 I_2 = e V_1 I_1$

where e is the efficiency (typically over 0.95).

The oscilloscope

An oscilloscope is an instrument used to draw a graph of a voltage as it varies in time. The vertical (Y) direction is the voltage axis. The horizontal (X direction) is the time axis.

Displaying an alternating voltage The line below has two components. In the Y direction, an alternating PD, coming from an external source, moves the spot up and down. In the X direction, a ***time base*** circuit in the oscilloscope changes the PD so that the spot moves from left to right at a steady rate – until it reaches the edge of the screen and flicks back. The result is a graph of PD against time, drawn over and over again.

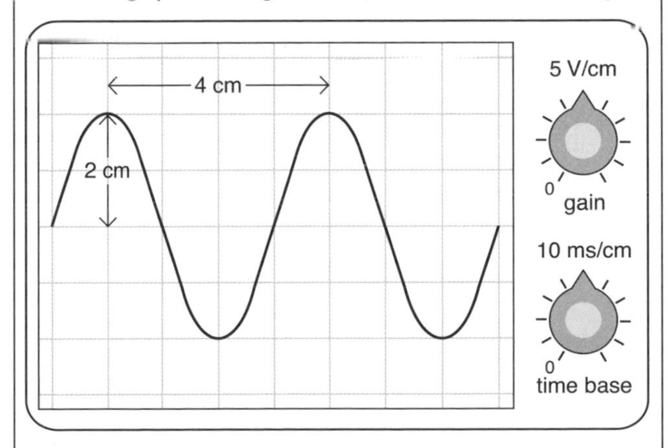

Measuring peak alternating PD The setting on the gain control (above) means that 5 V across the Y-input terminals cause the spot to move 1 cm vertically. On the screen, the amplitude of the wave trace is 2.0 cm. So in this case, the peak voltage is 2.0 × 5 = 10 V.

Measuring time and frequency The setting on the time base control (above) means that the spot takes 10 ms to move 1 cm horizontally. On the screen, the wave peaks are 4.0 cm apart. So the time between the peaks = 4.0 × 10 = 40 ms (0.04 s). This is the period of the wave cycle. So the frequency = 1/period = 1/0.04 = 25 Hz.

F5 Alternating current

Power dissipated in a resistor

Below, an alternating current I flows through a resistance R. The power P dissipated in the resistor varies through the cycle. At any instant $P = I^2R$.

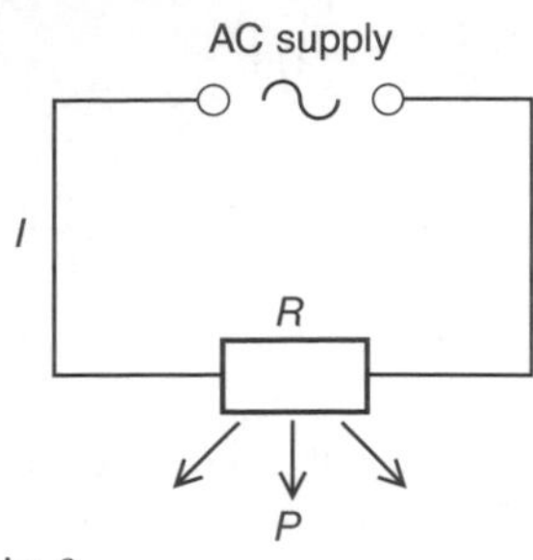

But $I = I_0 \sin\theta$

So $P = I_0^2 R \sin^2\theta$

where $\theta = 360\,ft$

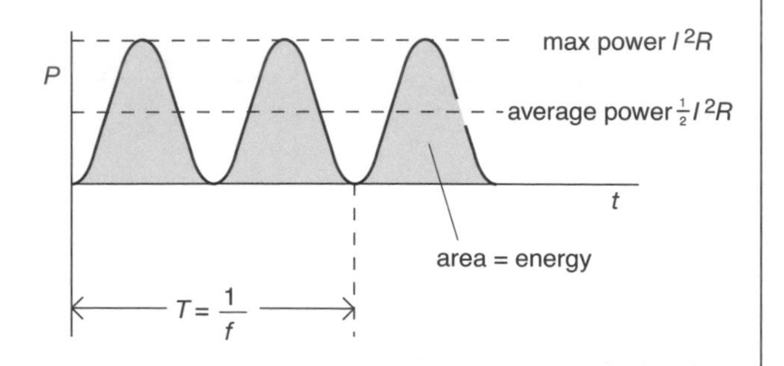

The graph shows how P varies with t. P is always positive because all values of $\sin^2 \omega t$ lie between 0 and 1. Note:

average power $= \frac{1}{2}$maximum power $= \frac{1}{2}I_0^2R$

RMS voltage and current

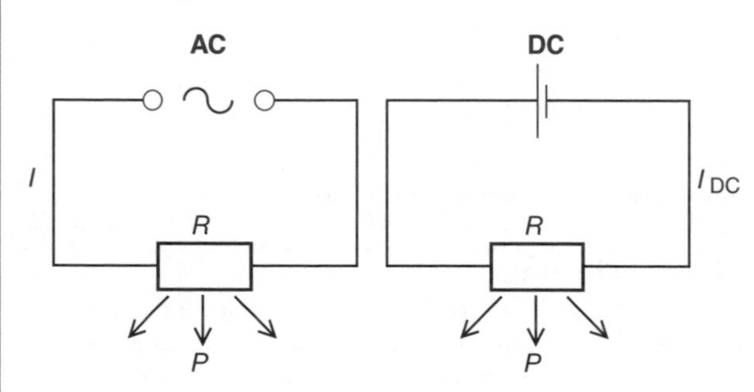

The current from a battery is one-way ***direct current*** **(DC)**. In the DC circuit above, there is a steady current I_{DC} such that power is dissipated in the resistor at exactly the same rate as in the AC circuit. So

$$I_{DC}^2 R = \frac{1}{2} I_0^2 R$$

From which it follows that $I_{DC} = I_0/\sqrt{2}$.

In the AC circuit, the ***root mean square (RMS) current*** is equal to I_{DC}. So it is equal to the steady current which gives the same power dissipation as the alternating current. (It is so named because it is the square root of the mean of I^2 throughout the sinusoidal cycle.) So:

$$I_{RMS} = \frac{I_0}{\sqrt{2}}$$

Similarly, for the alternating PD of peak value V_0,

$$V_{RMS} = \frac{V_0}{\sqrt{2}}$$

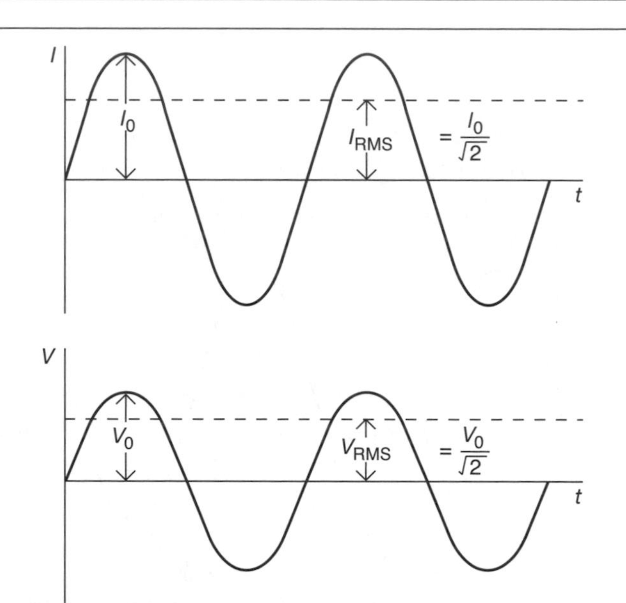

The graphs above show how the alternating V across a resistor, and I through it, vary with time. The resistor is assumed to obey Ohm's law.

Note:

- When the PD is zero, the current is zero. And when the PD is at its peak, the current is at its peak. In other words, the PD and current are ***in phase***.
- $R = \frac{V}{I} = \frac{V_0}{I_0} = \frac{V_{RMS}}{I_{RMS}}$
- The average power dissipated $= \frac{1}{2}V_0 I_0 = V_{RMS} I_{RMS}$

The UK AC mains voltage of 230 V is a RMS value. The peak voltage = 230 × $\sqrt{2}$ = 325 V. In doing calculations on power dissipation, RMS values are normally used because the factor $\frac{1}{2}$ does not need to be included in the working.

G1 Information and telecommunication – 1

Telecommunications systems

Telecommunications systems (e.g. radio, TV, and telephone) send information from one place to another using electric currents or electromagnetic radiation. The information being sent may be sounds, pictures, or computer data.

- The ***channel of communication*** may be radio waves, microwave beam, metal cable, or optical fibre.
- Signals for sounds (e.g. speech) are called ***audio*** signals.

Simple sound information: frequency spectrum

Sound information may be very simple. For example, the sinusoidal variation of the displacement of air against time, shown in the figure below, represents a pure a single note

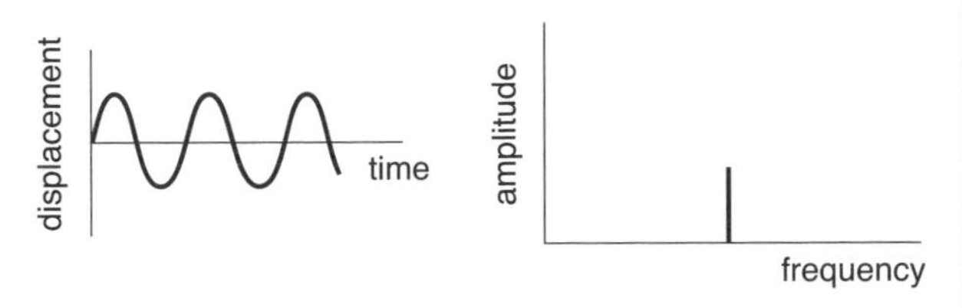

The sound can be represented by a ***frequency spectrum***. This shows the frequencies that are present, plotted on the *x*-axis, and their relative amplitudes, shown by the height of the vertical line.

More complex sounds: speech and music transmission

Speech consists of a succession of complex waveforms such as this:

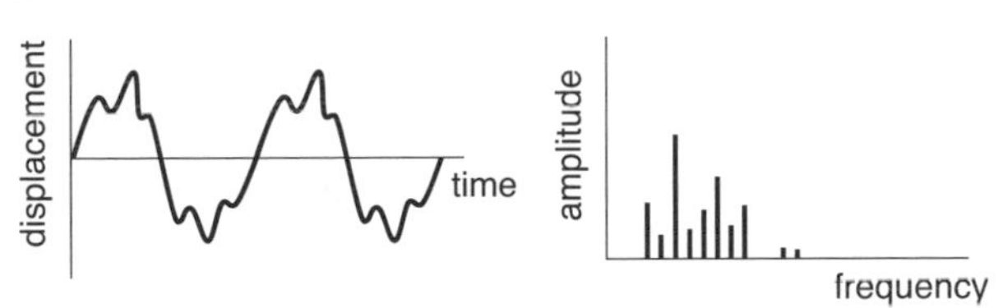

The frequency spectrum of the transmitted information is continually changing but the frequencies always lie in the same range. Music contains a wider range of frequencies.

Base bandwidth

The base bandwidth of a communication channel

- is the range of frequencies present in the original information that is transmitted (note that the actual channel bandwidth is twice the base bandwidth (see below))
- is the range of frequencies that are transmitted for effective transmission
- affects the quality of the sound that is received.

Although all frequencies to which the ear is sensitive are present in speech most of the energy lies in the frequency range below about 3.4 kHz. Adequate speech is transmitted for a base bandwidth of 300 Hz to 3000 Hz. This range is used in telephone systems.

For high-quality transmission of music, audible frequencies from 15 Hz to 15 kHz have to be transmitted.

- For reasonable-quality music and good-quality speech the base bandwidth range has to be 50 Hz to 4500 Hz.
- This range is used when transmitting signals by medium- and long-waveband radio. AM (amplitude modulated) transmissions are used in this case. This is used in long-wave and medium-waveband radio.

Video information

To transmit black and white video (TV) information with sound a much wider base bandwidth, of about 8 MHz, is needed. For colour TV a channel is needed for each of the three colours, red, blue, and green.

Analogue information

The displacement of a sound wave varies continually with time. When transmitting analogue information a voltage is transmitted that is proportional to the displacement. The transmitted voltage variation is a replica of the displacement time variation. This voltage variation is converted back to the original signal at the receiver. The radio system below is transmitting analogue signals.

Digital signals

These are pulses, produced because a circuit's output voltage is either HIGH or LOW. They can be represented by the by the logical numbers 1 and 0. The advantages of using them are described in G2.

Encoding information

All types of information can be encoded into a digital form. Video information is converted into digital information in a similar way to audio.

Letters and numbers can be transmitted by means of a code to represent each individual character

Image production

Each image on a video display unit is made up of pixels that are arranged in an orderly way on the tube. In colour display units there are three types of pixel. One type glows red when electrons are incident on it, another blue, and another green. The brightness of each pixel depends on how many electrons hit the pixel each second. The full range of colours is obtained by suitably illuminating pixels that are close together. The eye interprets the mixture of red, blue, and green as a different colour or hue.

Resolution

If you sit close to a TV screen you will be able to see the individual pixels. In this case the eye is able to resolve (see E2) the pixels that are close together so that the effect of a full range of colours is not seen. The minimum angle for resolution is about 1×10^{-4} rad (about 6×10^{-3}°).

To view a screen at 0.5 m (which is typical when working at a computer) the pixels need to be closer together than 50 μm if they are not to be seen as individual pixels.

Bandwidth Mathematically, a carrier of frequency f_c, amplitude modulated at a frequency f_m, is equivalent to a pure sine wave of frequency f_c, with two ***side frequencies*** of $f_c + f_m$ and $f_c - f_m$ (i.e. at the carrier frequency $\pm f_m$).

Speech and music contains a variety of frequencies, mainly between 50 Hz and 4.5 kHz, so a modulated carrier normally has two bands of side frequencies, called ***sidebands***. The frequency range occupied by the signal is the ***bandwidth***.

- A radio station needs a bandwidth of about *twice* its AF range. In the UK, the bandwidth used for AM is 9 kHz.

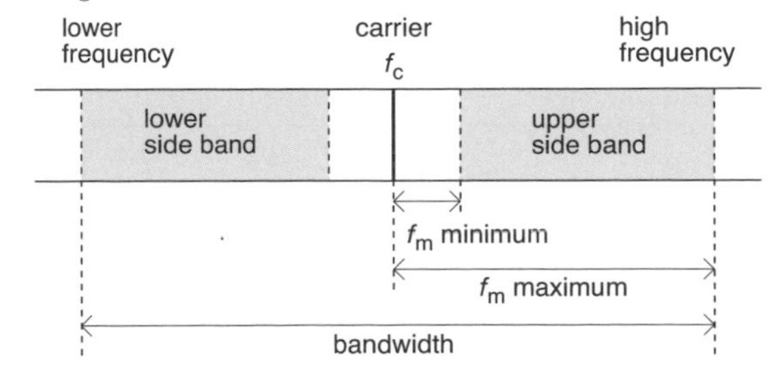

Digital transmission

Most telephone systems use digital transmission between exchanges. Audio signals from one telephone are converted from analogue to digital form, transmitted, and then changed back into analogue signals for the other telephone.

To convert the analogue signal on the right into a digital signal, it is ***sampled*** by measuring the voltage level at regular intervals of time. The levels are then changed into ***binary*** codes, consisting only of 0s and 1s, and these are transmitted as a sequence of pulses. The process is called ***pulse code modulation (PCM)***. Reversing the process produces the analogue signal again.

- Telephone systems normally use 256 voltage levels. Each requires an 8-bit binary code, e.g. 10011001.
- The ***sampling rate*** needs to be at least *twice* the highest frequency in the signal being sampled. Telephone systems normally use a sampling rate of 8000 Hz, i.e. the signal is being sampled every 1/8000 s.

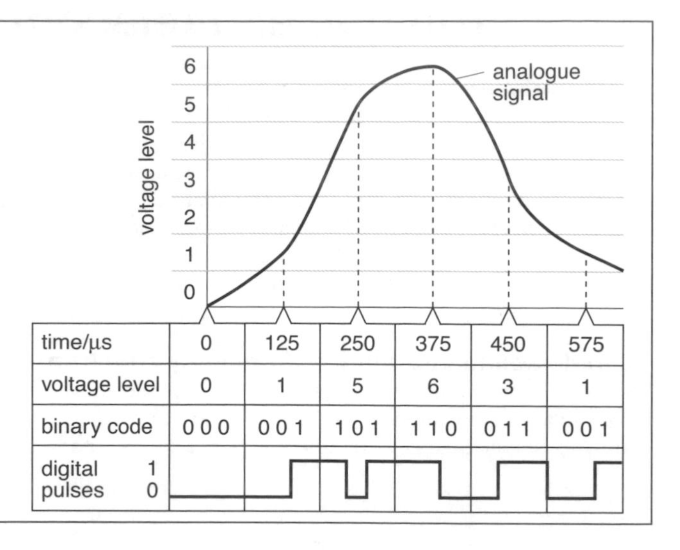

time/μs	0	125	250	375	450	575
voltage level	0	1	5	6	3	1
binary code	0 0 0	0 0 1	1 0 1	1 1 0	0 1 1	0 0 1
digital pulses 1 / 0						

Advantages of digital transmission

Regeneration Signals lose power as they travel along a cable. They are also affected by ***noise*** (additions caused by electrical interference and thermal activity). To restore their power and quality, digital signals can be amplified and 'cleaned up' at intervals by ***regenerators***. The digital signal must be a 0 or a 1, so the signal can be regenerated with just these values. (Analogue signals can be amplified by ***repeaters***, but unfortunately, the noise is amplified as well.)

Data handling Computers operate digitally, so digital transmission is ideal for long-distance computer links.

Disadvantages

If the signal has so much noise that it is not possible to identify a value as a 0 or a 1, then the signal is lost entirely, whereas a very noisy analogue signal can sometimes still be picked up.

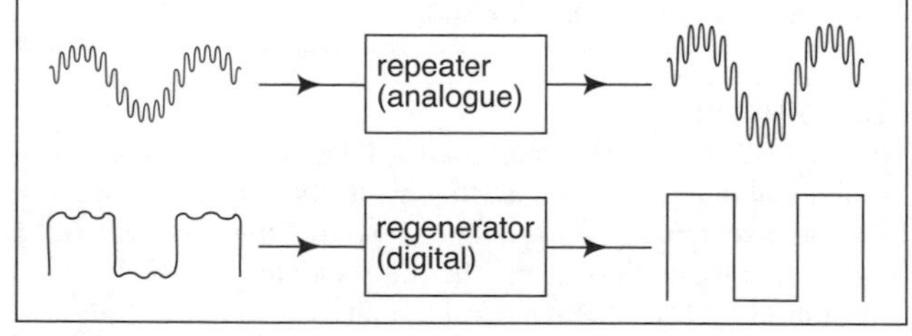

Transmitting more channels

Using higher frequency ranges to carry the information allows more channels to be transmitted. The range available for UHF (ultra high frequency) transmission as is used for analogue TV transmission is 300 MHz to 3000 MHz, a total of 2700 MHz. This range could carry 2700/24, which is approximately 110 colour TV channels.

$$\text{number of channels} = \frac{\text{total bandwidth available}}{\text{bandwidth required for each channel}}$$

Using optical frequency ranges and transmitting information down optical fibres increases the available frequency range. The range of visible light is about 5×10^{14} Hz so theoretically $(5 \times 10^{14})/(24 \times 10^{6}) = 2.1 \times 10^{7}$ analogue TV channels could be transmitted down one optical fibre. Fewer channels are available in practice.

Optical fibres

Read D2 first.

Optical fibres carry infrared pulses from a laser or LED. At the far end, these are detected by a sensor (a photodiode).

For transmitting signals, optical fibre cables have many advantages over metal cables and radio waves:

- They are ideal for digital transmission,
- They have a much higher signal-carrying capacity,
- They are free of noise and crosstalk (signals crossing over from one fibre or wire to the next),
- They offer better security, e.g. they cannot be 'tapped',
- They are thinner and lighter than metal cables,
- Attenuation is low.

Attenuation Infrared is absorbed as it passes along an optical fibre. If P_0 is the power input, P is the power output, and x is the length of the fibre:

$$P = P_0 e^{-\alpha x}$$

where α is the ***linear attenuation coefficient***.

Power losses and gains

Signals lose power as they travel along a cable. The effect is called ***attenuation***. Repeaters or regenerators compensate by increasing the power.

In telecommunication, the unit used for measuring power change is the ***decibel (dB)*** (see also H8). If P_0 is the power input into a cable and P is the power output,

$$\text{power increase in dB} = 10 \log_{10} (P/P_0)$$

So, if the power input is 200 mW and the power output is 2 mW, the power increase = $10 \log_{10}(2/200) = -20$ dB. In this case, there is a power *loss* of 20 dB.

Note:

- The attenuation caused by a cable is often expressed in dB per km.
- dB changes can be added algebraically. If there is a power loss of 20 dB in a cable, followed by a power gain of 15 dB in a repeater, the overall power loss is 5 dB.

G2 Information and telecommunication – 2

Radio communication

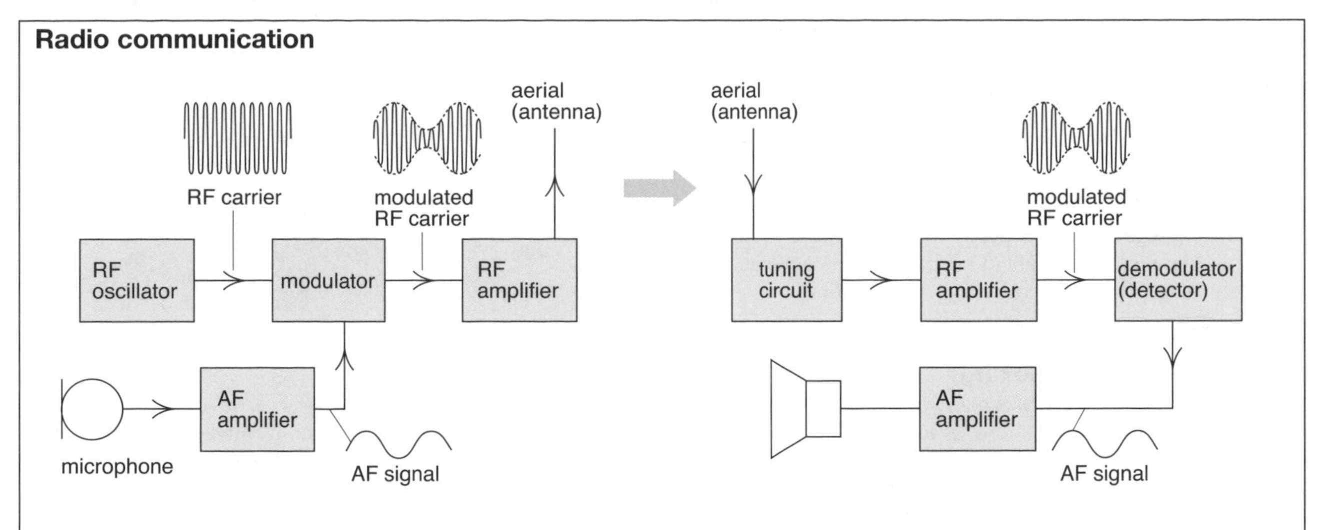

Transmitter

Receiver

The principle of simple radio system is shown below:

Carrier Radio frequencies (e.g. 1 MHz) are much higher than audio frequencies (e.g. 1 kHz). The principle of radio transmission is to produce a steady radio signal called the carrier, and then ***modulate*** (vary) it with the audio signal.

Amplitude modulation The instantaneous amplitude of the carrier frequency depends on the displacement of the signal. This is used for medium- and long-wavelength transmissions.

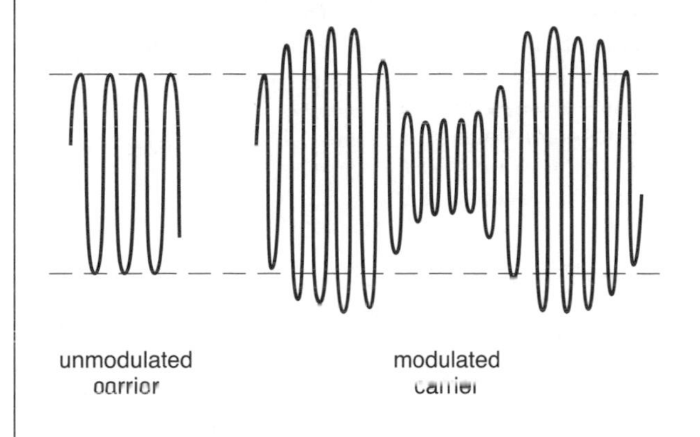

Frequency modulation The instantaneous frequency of the carrier frequency depends on the displacement of the signal. The maximum frequency change of the carrier depends on the amplitude of the signal.

The diagrams below right show how simple signals – in this case a positive signal followed by a negative – can be carried by modulating the carrier frequency. The positive signal increases the frequency, the negative signal decreases it.

Note: In practice the frequency is very high.

Frequency modulation is less affected by noise than amplitude modulation but needs a much greater bandwidth. It is used for transmission at high frequencies (UHF and VHF).

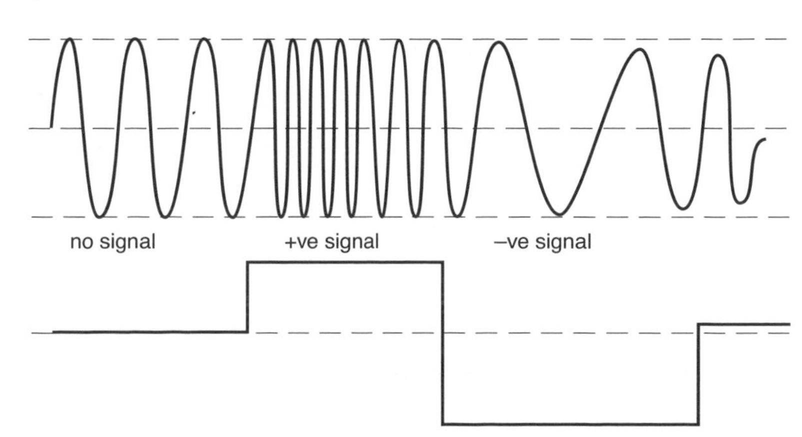

Tuning circuit Each radio station broadcasts at its own carrier frequency. The tuning circuit uses a resonant circuit (see B16) to select the incoming frequency required.

Demodulation The demodulator removes the ***RF*** (radio frequency) part of the signal, and passes on only the ***AF*** (audio frequency) part. The key component in demodulation is the diode. This rectifies the RF signal so that only its 'forward' parts are left, as pulses of varying amplitude. The AF signal is recreated from these.

Frequencies used for radio communication

Frequency band		Examples of uses
30 kHz – 300 kHz	low (LF)	long-wave radio
300 kHz – 3 MHz	medium (MF)	medium-wave radio
3 MHz – 30 MHz	high (HF)	short-wave radio
30 MHz – 300 MHz	very high (VHF)	FM radio
300 MHz – 3 GHz	ultra-high (UHF)	TV
1 GHz – 30 GHz	microwave	telephone and TV links, satellite links, radar

How radio waves travel

Despite the curvature of the Earth's surface, radio waves can travel between places which are a long way apart.

Ground (surface) waves These follow the Earth's surface. Low frequencies travel furthest (up to 1000 km).

Sky waves These are reflected by the ionosphere (see H15) and the Earth. They are mostly high frequency waves.

Space waves These are waves with frequencies above 30 MHz. They are not reflected by the ionosphere. They can only be used for 'line of sight' (i.e. straight) communication, but their range can be extended by a satellite link.

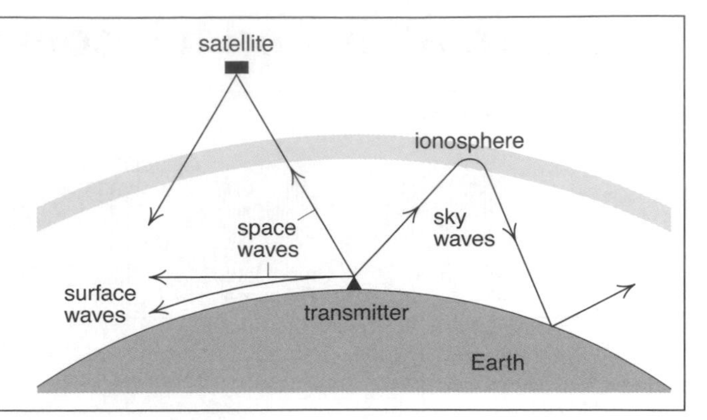

Time division multiplexing

This is the system that enables many digitized transmissions, such as telephone conversations, to use the same communications channel. The signal of each user is sampled as explained in the section on digital transmission opposite. To transmit a single conversation the voltage level of each signal is sampled 8000 times a second and for each sample 8 bits (0s or 1s) have to be transmitted. This is one ***byte*** of information.

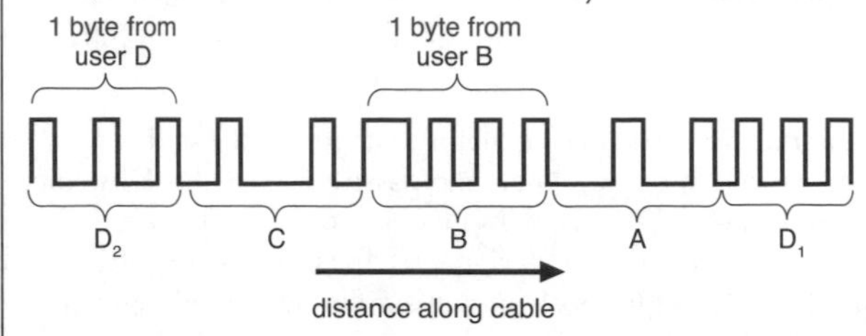

When transmitting alpha-numeric data (numbers or letters) each byte will represent a number or letter. A single transmission requires (8 × 8000) = 64 000 bits to be transmitted each second. The diagram shows a simplified version in which four signals from users **A**, **B**, **C**, and **D** are transmitted along a single channel. The byte D_1 from user **D** was sent first, then bytes from users **A**, **B**, and **C**. Then user **D**'s signal was sampled again to transmit byte D_2, and so on.

Bit rates

Bits can be sent at much higher rates than that required to transmit a single conversation. A coaxial cable can carry 140 Mbit s^{-1}. This is called the ***bit rate***. This means that a single coaxial cable can carry $140 \times 10^6/64 \times 10^3 = 2200$ telephone conversations. A single cable can be shared by 2200 users with each user's signal being transmitted in turn. Each signal only uses the cable for 1/2200 of the time hence the term 'time division'.

How Science Works

Compression

MP3 is a music compression format that can compress, for example, a CD track that usually takes up about 30 megabytes down to about 3 megabytes.

MP3 uses the fact that the human ear:

- cannot hear frequencies outside the range 10 Hz to 20 kHz
- cannot hear a quiet sound at the same time as a loud sound
- can hear some sounds better than others.

After removing the frequencies that cannot be heard, the file is much smaller. It can then be compressed further, for example by looking for repeating patterns, recording them once and referring to them each time.

Satellite links

For radio waves, the transmitter is an aerial called a dipole and the oscillating signal in the aerial causes electromagnetic waves to spread out through space. The receiving aerial picks up the electromagnetic waves, and an electrical signal is generated in the aerial.

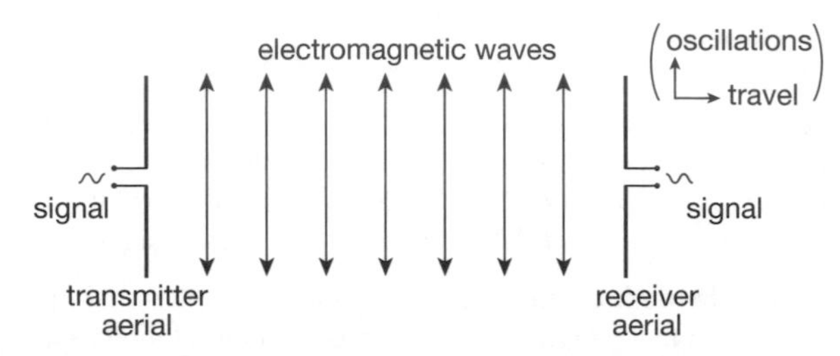

Satellites use microwaves in the range 3 to 30 GHz. The aerial is a horn at the centre of a metal dish. The dish focuses the waves onto the receiving horn. (In the transmitter the dish sends the microwaves out as a parallel beam.)

Diffraction effects

The beam from the transmitting satellite dish is spread by diffraction (see D1 and D3).

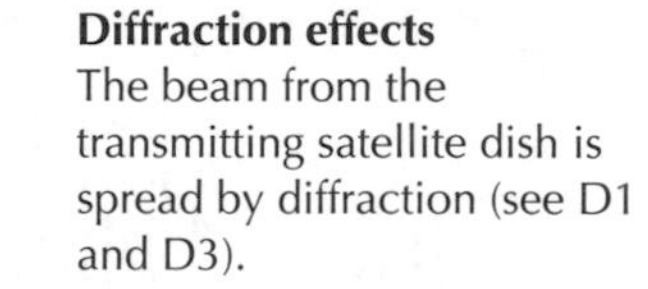

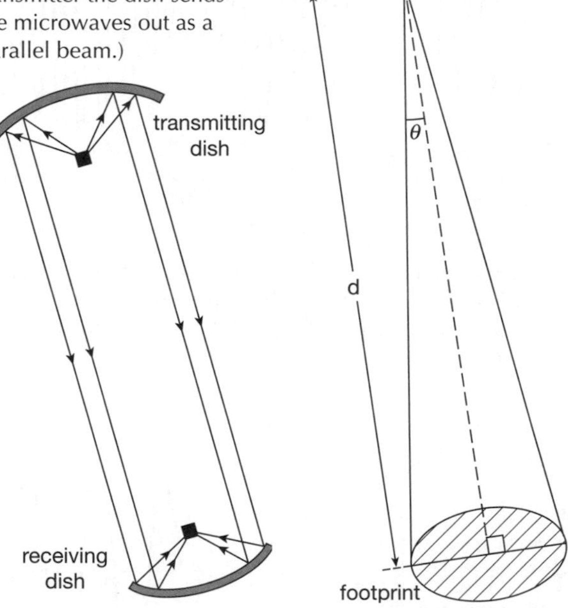

The half-beam width θ is given by $\sin \theta = \dfrac{\lambda}{b}$

where λ = the wavelength and b = the dish diameter.

Note: the same effect happens to all waves from apertures, for example sound waves from loudspeakers. The smaller the value of b the more the beam will spread out.

The area covered by a satellite is called the ***footprint***. If the orbit is a height d above the surface of the Earth, the footprint will have area $A = \pi (d \tan \theta)^2$.

H1 Electric charges and fields

Electric force

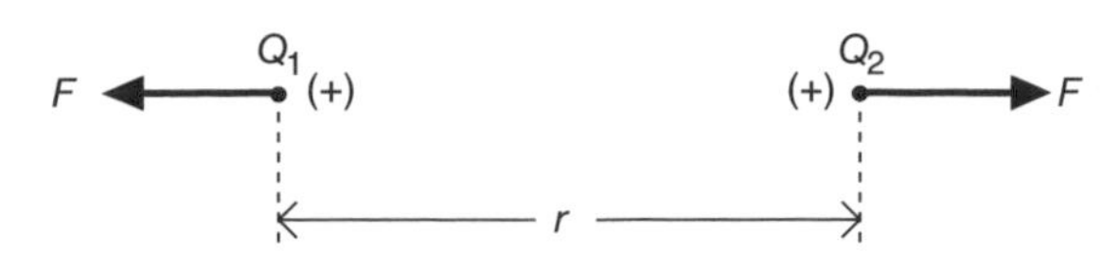

Charges attract or repel each other with an ***electric force***. If point charges Q_1 and Q_2, are a distance r apart, and F is the force on each, then according to ***Coulomb's law***:

$$F \propto \frac{Q_1 Q_2}{r^2}$$

This is an example of an ***inverse square law***. If r doubles, the force F drops to one quarter, and so on.

With a suitable constant, the above proportion can be turned into an equation:

$$F = \frac{kQ_1 Q_2}{r^2}$$

The unit of charge for Q_1 and Q_2 is the coulomb (C).

The value of k is found by experiment. It depends on the ***medium*** (material) between the charges. For a vacuum, k is 8.99×10^{-9} N m^2 C^{-2}, and is effectively the same for air.

In practice, it is more convenient to use another constant, ε_0, and rewrite the equation on the left in the following form:

In a vacuum $$F = \frac{Q_1 Q_2}{4\pi\varepsilon_0 r^2}$$

ε_0 is called the ***permittivity of free space***. Its value is 8.85×10^{-12} C^2 N^{-1} m^{-2}.

Note:
- Although '4π' complicates the above equation, it simplifies others derived from it.
- In the above equation, if, say, Q_1 and Q_2 are *like* charges (e.g. – and –), then F is *positive*. So a positive F is a force of *repulsion*. Similarly, it follows that a negative F is a force of *attraction*.

Electric field

If a charge feels an electric force, then it is in an ***electric field***. If a charge q feels a force F, then the ***electric field strength*** E is defined like this:

$$E = \frac{\text{electric force}}{\text{charge}} \qquad \text{In symbols} \quad E = \frac{F}{q} \qquad (1)$$

For example, if a charge of 2 C feels an electric force of 10 N, then E is 5 N C^{-1}.

Note:
- Electric field strength is a vector. Its direction is that of the force on a positive (+) charge.

The force acting on a charge in an electric field can be found by rearranging the equation above:

$$F = qE$$

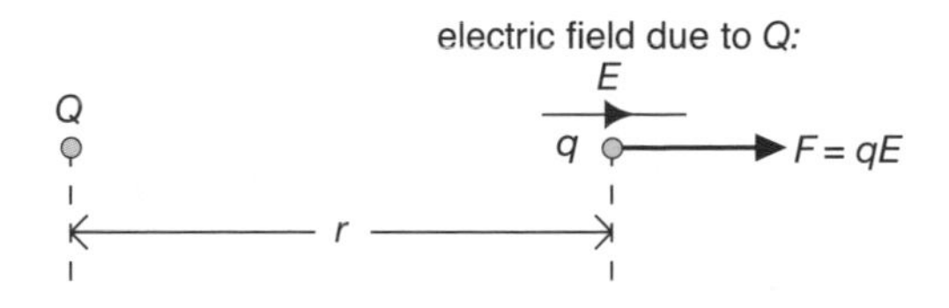

Above, charge Q produces an electric field which acts on a small charge q. As $F = qE$ and

$$F = \frac{Qq}{4\pi\varepsilon_0 r^2} \qquad \text{it follows that} \quad E = \frac{Q}{4\pi\varepsilon_0 r^2} \qquad (2)$$

The electric field round a charged, spherical conductor is shown on the right. It is a ***radial*** field.

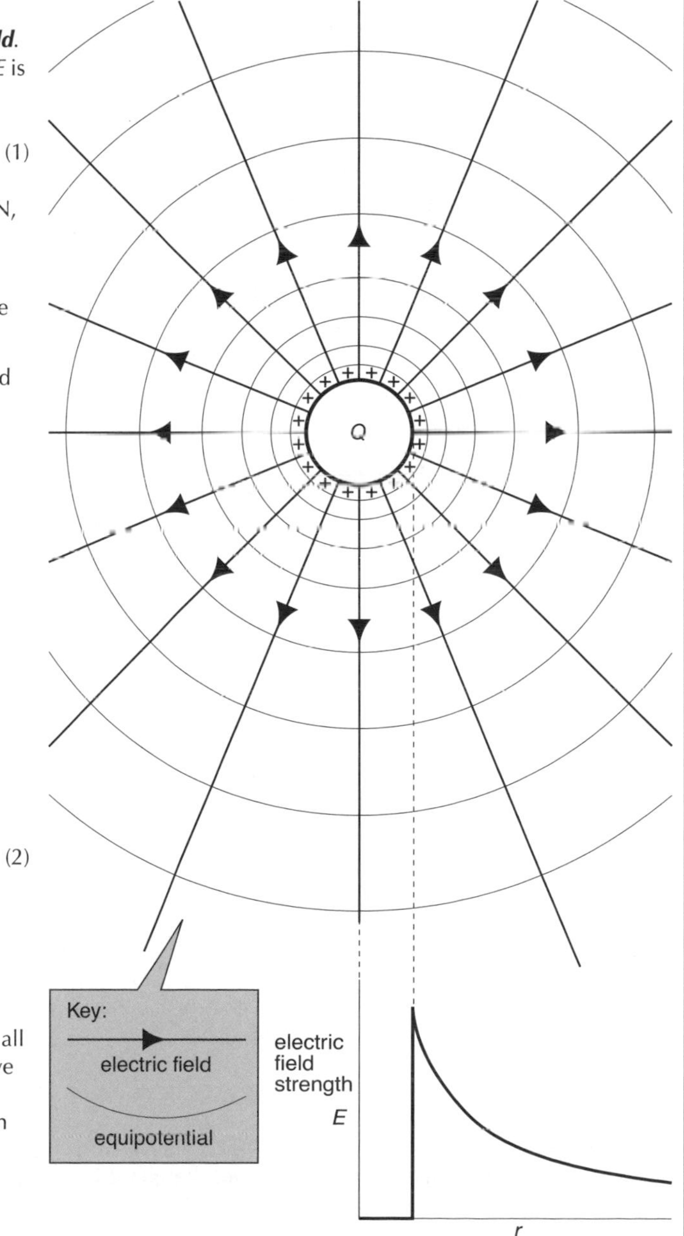

Note:
- The charge is on the surface of the conductor.
- Outside the conductor, the electric field is the same as if all the charge were concentrated at the centre, and the above equation applies.
- Inside the conductor, there is no electric field. The reason for this is given on the opposite page.
- The equipotential lines in the diagram on the right are explained on the opposite page.

Electric potential

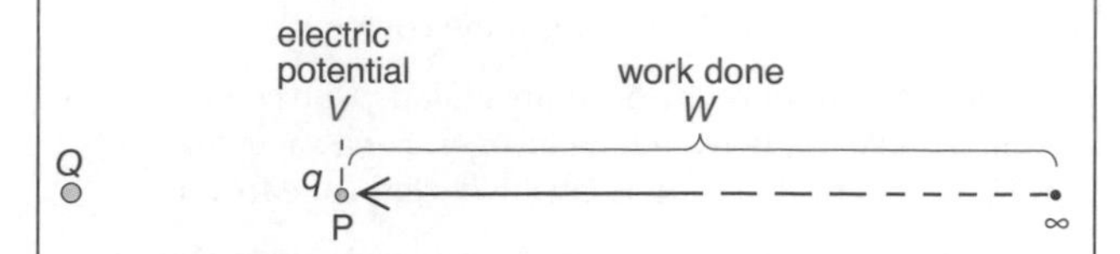

Above, charge Q causes an electric field. A small charge q has been moved through this field, from an infinite distance (where the electric force is zero), to point P.

The ***electric potential*** V (at point P) is defined as follows:

$$V = \frac{W}{q} \qquad (3)$$

where W is the work done in moving a charge q from infinity (∞) to point P.

The SI unit of electric potential is the ***volt*** (V). For example, if 1000 J of work is done in moving a charge of +2 C from ∞ to P, then the potential at P = 1000/2 = 500 V.

Note:

- Electric potential is a scalar.
- At infinity, the electric potential is *zero.*
- Elsewhere, the electric potential due to *positive* charge is *positive.* Similarly, the electric potential due to a *negative* charge is *negative.*
- Equation (3) can also be used to find the work done in moving a charge q between two points. In this case, V is the ***potential difference*** **(PD)** between the points.

Electric potential in a radial field

A radial field is shown on the opposite page. Provided r is not less than the radius of the sphere,

$$E = \frac{Q}{4\pi\varepsilon_0 r^2} \qquad \text{Also, } E = -\frac{dV}{dr} \qquad \text{(see bottom panel)}$$

Combining these and using calculus gives

$$V = \frac{Q}{4\pi\varepsilon_0 r} \qquad \text{(This also applies to a point charge.)}$$

Note:

- In the diagram on the opposite page, each ***equipotential line*** is a line joining points of equal potential.
- Inside the charged conductor, all points are at the same potential, so the potential gradient (see bottom panel) is zero. From this it follows that E is also zero, so there is no electric field inside the conductor.

Comparing electric and gravitational fields

For particles of similar size, electric forces are very much stronger than gravitational ones. For example, electric forces hold atoms together to form solids.

Electric and gravitational fields have similar features. That is why the equations is this unit have a similar form to those in H3. However, comparing equivalent equations, a minus sign may be present in one but absent from the other. This arises because of the differing force directions.

Gravity is always a force of attraction. Mass is always positive and it produces a gravitational field which is directed *towards* it.

Electric charges may attract or repel. However, if a charge is positive, then it produces an electric field which is directed *away* from it.

Linking potential and field strength

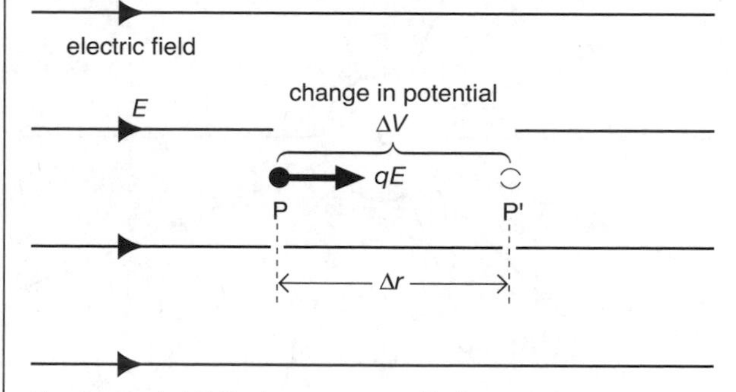

Above, work ΔW is done on a small charge q in moving it from P to P′ in a uniform electric field E. So, from (3):

$$\Delta W = q\Delta V \qquad (3)$$

This equation gives the work done *on* the charge.

So work done *by* charge = $-q\Delta V$

But work done by charge = force × distance moved = $qE\Delta r$

So $\qquad qE\Delta r = -q\Delta V$

Therefore $\qquad E = -\dfrac{\Delta V}{\Delta r} \qquad (4)$

In calculus notation, there is a more general version of this equation which also applies to non-uniform fields:

$$E = -\frac{dV}{dr}$$

Note:

- In the above equations, the minus sign indicates that E is in the direction of *decreasing* potential.

The metal plates on the right have a small test charge q between them. The charge feels a force F. So, from equation (1),

$$E = \frac{F}{q}$$

There is a potential difference V between the plates. From the graph on the right, and equation (4), the potential gradient is $-V/d$. So

$$E = \frac{V}{d} \qquad (5)$$

The constant potential gradient means that the electric field is uniform.

The above equations show different aspects of electric field strength. If, say, E is 10 N C^{-1}, you can think of this either as a force of 10 N per coulomb or a potential drop of 10 V per metre. (E can be expressed in N C^{-1} or in V m^{-1}.)

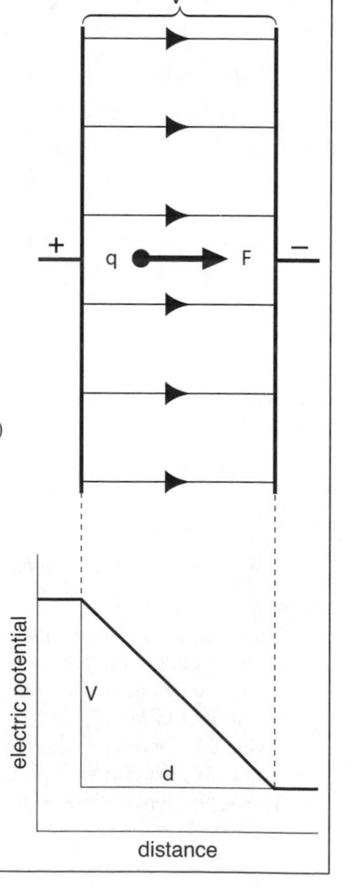

H2 Capacitors and fields

Capacitance

Capacitors store small amounts of energy by separating charge (they are sometimes said to 'store charge').

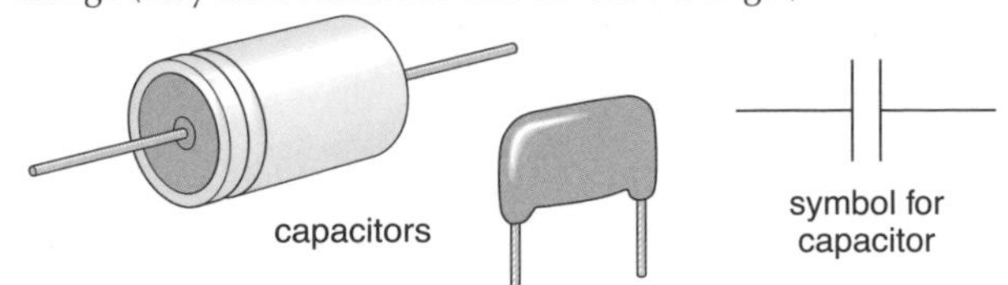

A capacitor can be charged by connecting a battery across it. The higher the PD V, the greater the charge Q stored. Experiments show that $Q \propto V$. Therefore, Q/V is a constant.

The ***capacitance*** C of a capacitor is defined as follows:

$$\text{capacitance} = \frac{\text{charge}}{\text{PD}} \qquad \text{In symbols} \quad C = \frac{Q}{V}$$

The higher the capacitance, the more charge is stored for any given PD.

Capacitance is measured in C V^{-1}, known as a ***farad*** (F). However, a farad is a very large unit, and the μF (10^{-6} F) is more commonly used for practical capacitors.

Energy stored by a capacitor

Work must be done to charge up a capacitor. Electrical potential energy is stored as a result.

If a charge of 2 C is moved through a *steady* PD of 10 V, then, using equation (3) in H1,

work done $W = QV = 2 \times 10 = 20$ J.

So the stored energy is 20 J. Numerically, this is the area under the graph below.

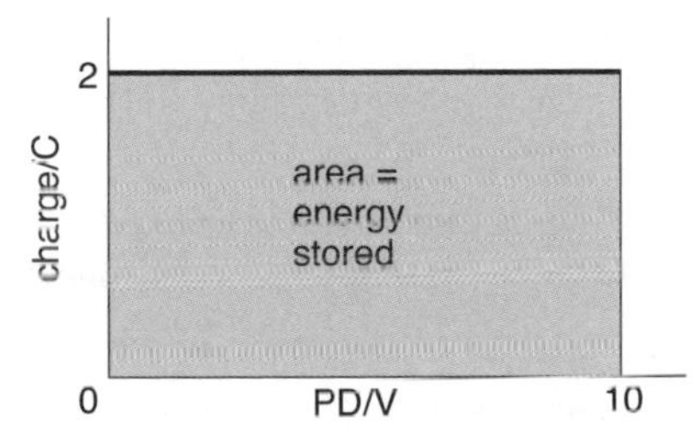

When a capacitor is being charged, Q and V are related as in the graph below. As before, the energy stored is numerically equal to the area under the graph, which is $\frac{1}{2}QV$. As $C = Q/V$, this can be expressed in three ways:

$$\text{energy stored} = \tfrac{1}{2}QV = \tfrac{1}{2}CV^2 = \tfrac{1}{2}Q^2/C$$

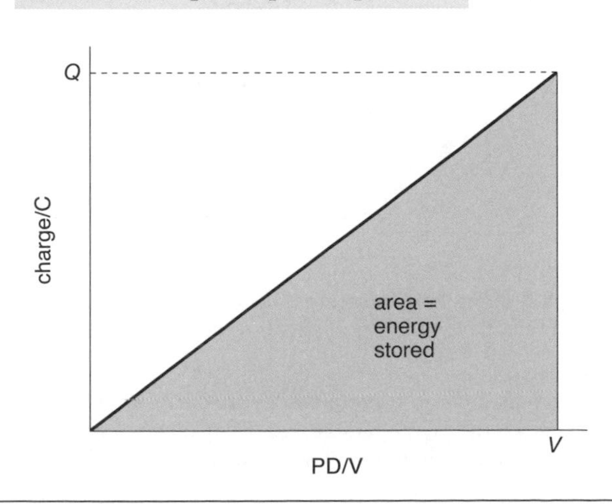

Electric field near a charged plate

On the right, a metal sphere has a charge Q uniformly distributed over its surface. The electric field E near the surface is given by equation (2) in H1:

$$E = \frac{Q}{4\pi\varepsilon_0 R^2}$$

But the surface area of the sphere $A = 4\pi R^2$. So

$$E = \frac{Q}{\varepsilon_0 A} \qquad (1)$$

charge Q
surface area A

This equation also applies to a flat, charged metal plate of surface area A.

Capacitors in series

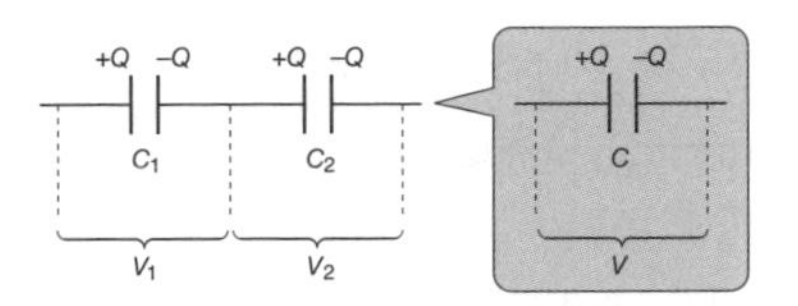

If C_1 and C_2 have a total capacitance of C, then C is the single capacitance which could replace them.

Two capacitors in series store only the same charge Q as a single capacitor. So $V = Q/C$, $V_1 = Q/C_1$, and $V_2 = Q/C_2$.

But $V = V_1 + V_2$ So $\frac{Q}{C} = \frac{Q}{C_1} + \frac{Q}{C_2}$

or $$\frac{1}{C} = \frac{1}{C_1} + \frac{1}{C_2}$$

For example, if $C_1 = 3$ μF and $C_2 = 6$ μF, then $1/C = 1/3 + 1/6 = 1/2$. So $C = 2$ μF.

Capacitors in parallel

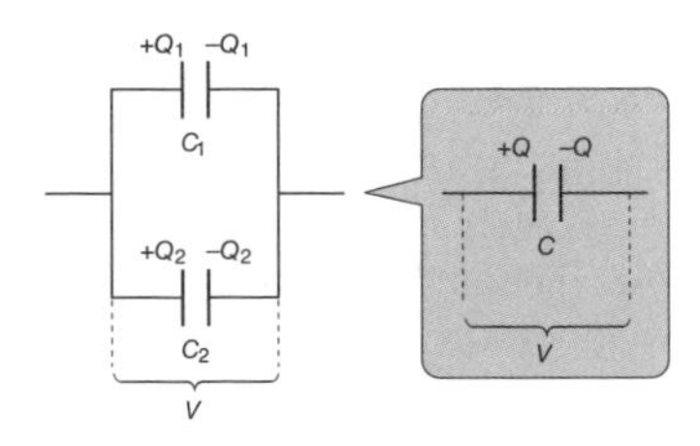

If C_1 and C_2 have a total capacitance of C, then C is the single capacitance which could replace them.

Capacitors in parallel each have the same PD across them. So $Q = CV$, $Q_1 = C_1V$, and $Q_2 = C_2V$.

Together, the capacitors act like a single capacitor with a larger plate area. So $Q = Q_1 + Q_2$

$$\therefore \quad CV = C_1V + C_2V$$

and $$C = C_1 + C_2$$

For example, if $C_1 = 3$ μF and $C_2 = 6$ μF, then $C = 9$ μF.

H3 Capacitors and exponential decay

Exponential relationships

Many changes, such as the discharge of a capacitor or radioactive decay (see J3), are exponential changes.
The rate of change at any time is proportional to the amount at that time.

For exponential decay:

$$\frac{dx}{dt} = -k\,t \text{ or } x = x_o\, e^{-kt}$$

For exponential growth:

$$\frac{dy}{dt} = k\,t \text{ or } y = y_o\, e^{kt}$$

It is difficult to see by eye if a curve is exponential. To test a relationship, either use log-linear graph paper or plot a graph of the natural logs against time.

Taking natural logs of the equation $Q = Q_0\, e^{-t/RC}$ gives

$$\ln Q = \ln Q_0 \frac{-t}{RC}$$

A graph of ln Q against t is a straight line with gradient $-1/RC$.

The diagram is a graph of ln(Q/C) against t/s.

Note: label the x-axis with the numerical value of the time t divided by the unit s. Label the y-axis with the natural log of the numerical value of the charge Q divided by the unit C.

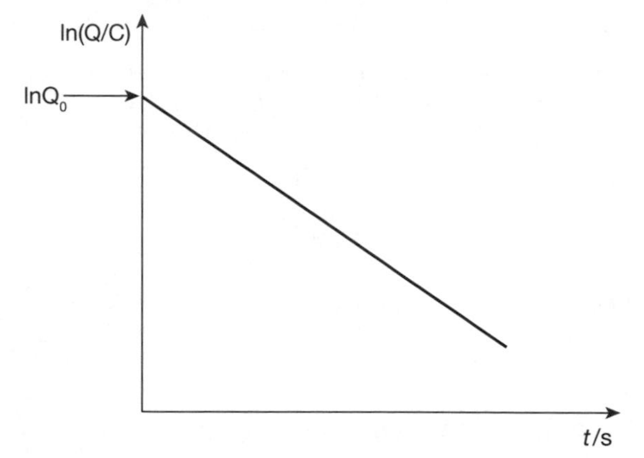

The gradient can be used to find the value of the time constant RC.

If you are given a graph like A (opposite) and asked to find RC, you can use the tangent at Q_0 as shown, but this may be difficult to draw. Another method is to use the time for Q to fall to 1/e of its original value Q_0.

$$\frac{Q_0}{e} = Q_0\, e^{-t/RC}$$

$$e^{-1} = e^{-t/RC}$$

$$-1 = \frac{-t}{RC}$$

$$t = RC$$

1/e is about 0.37 or 37%, but there is no need to remember this, as you can use any scientific calculator to calculate it when needed.

These methods can be used for radioactive decay graphs (see J4) and graphs of attenuation (see J4).

Charging a capacitor

The capacitor below is charged through a resistance R. The graph shows how the charge builds up.

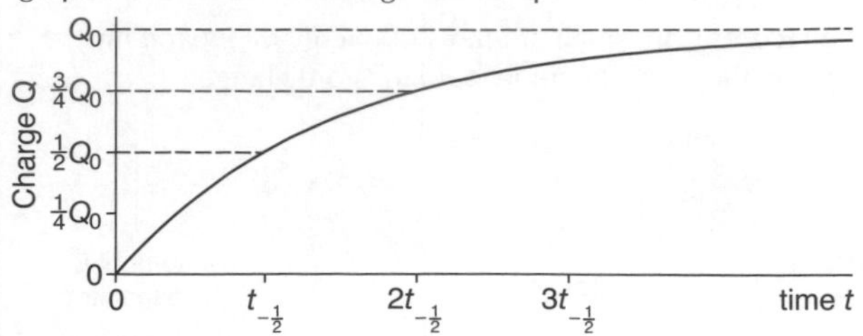

The equation for the charging process is

$$Q = Q_0(1 - e^{-t/CR})$$

The charge reaches a maximum value of Q_0 which is equal to VC.

The charging current starts at a maximum value of V/R and falls to a lower value in the same way as it does when the capacitor discharges.

Voltage–time graphs

Since the voltage across a capacitor is proportional to the charge on it the variations of voltage with time are the same shape as the charge–time graphs. The equations for calculating the voltage at any time are similar to those for charge, substituting V for Q.

Discharge of a capacitor

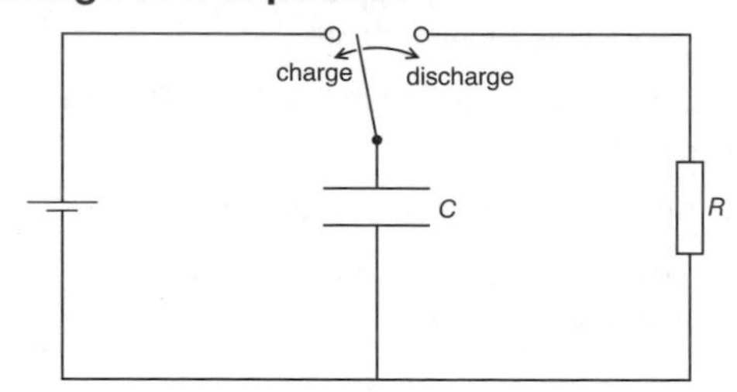

The capacitor above is charged from a battery and then discharged through a resistance R.

Graph A shows how, during discharge, the charge Q decreases with time t, according to the following equation:

$$Q = Q_0 e^{-t/RC} \qquad \text{where } e = 2.718$$

RC is called the ***time constant***. (It equals the time which the charge would take to fall to zero if the initial rate of loss of charge were maintained.) Increasing R or C gives a higher time constant, and therefore a slower discharge.

The gradient of the graph at any time t is equal to the current at that time.

Graph A **Graph B**

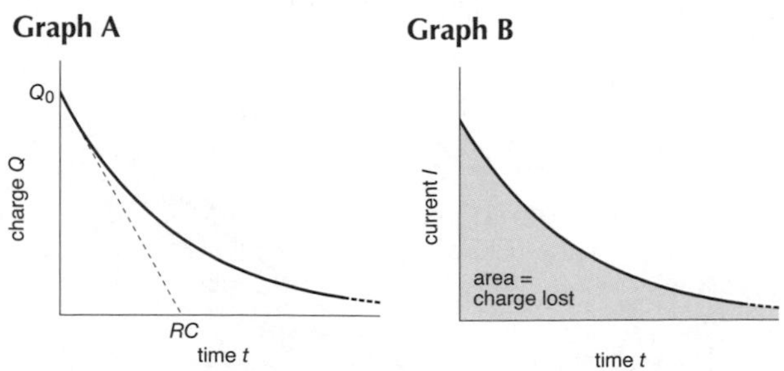

Graph B shows how the current decreases with time. The area under the graph is numerically equal to the charge lost.

Note:

- Each graph is an ***exponential decay curve***, with the same characteristics as a radioactive decay curve. A ***half-life*** can be calculated in the same way (see J4).

H4 Gravitation

Gravitational force

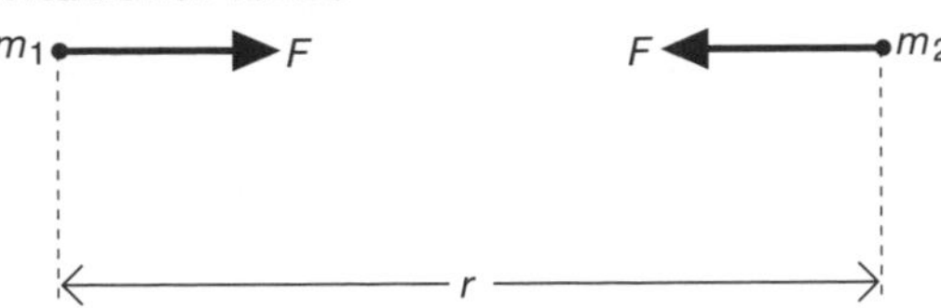

All masses attract each other with a ***gravitational force***. If point masses m_1 and m_2, are a distance r apart, and F is the force on each, then according to ***Newton's law of gravitation***

$$F \propto \frac{m_1 m_2}{r^2}$$

With a suitable constant, the above proportion can be turned into an equation:

$$F = \frac{Gm_1 m_2}{r^2}$$

G is called the ***gravitational constant***. It is found by experiment using large laboratory masses and an extremely sensitive force-measuring system. In SI units, the value of G is 6.67×10^{-11} N m^2 kg^{-2}.

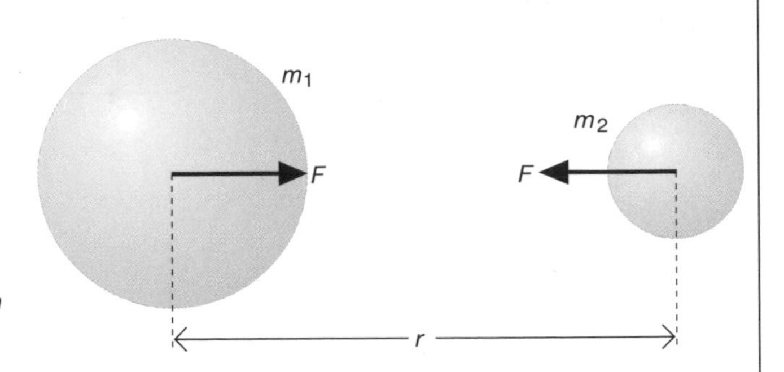

The equation on the left is also valid for spherical masses of uniform density, with centres r apart, as above.

Note:

- Newton's law of gravitation is an example of an ***inverse square law***. If the distance r doubles, the force F drops to one quarter, and so on.
- Gravitational forces are always forces of *attraction*.
- Gravitational forces are extremely weak, unless at least one of the objects is of planetary mass or more.

Gravitational field

If a mass feels a gravitational force, then it is in a ***gravitational field***. The ***gravitational field strength*** g is defined like this:

$$g = \frac{\text{gravitational force}}{\text{mass}} \qquad \text{In symbols } g = \frac{F}{m}$$

For example, if a mass of 2 kg feels a gravitational force of 10 N, then g is 5 N kg^{-1}.

Note:

- Gravitational field strength is a vector.
- g is a variable and can have different values. The symbol g above does not imply the particular value of 9.81 N kg^{-1} near the Earth's surface.
- The force acting on a mass in a gravitational field can be found by rearranging the equation above: $F = mg$.

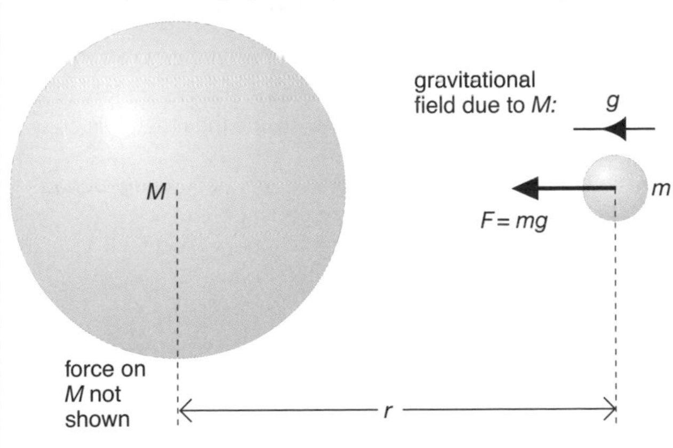

Above, mass M produces a gravitational field which acts on mass m.

As $F = \frac{GMm}{r^2}$ and $F = mg$

it follows that $$g = \frac{GM}{r^2}$$

It is equally true to say that m produces a gravitational field which acts on M. Either way, mass × gravitational field strength gives a force of the same magnitude, GMm/r^2.

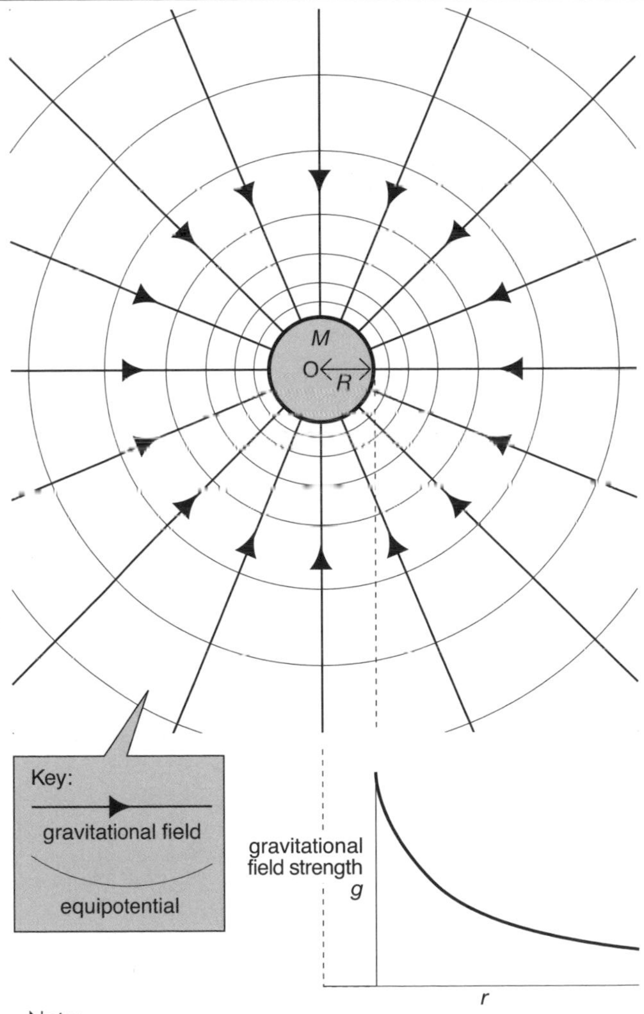

Note:

The gravitational field around a spherical mass is shown above. It is called a ***radial*** field because of its shape.

- Inside the mass, the equation on the left does not apply. g falls to zero at the centre.
- Equipotential lines are explained on the next page.

The Earth's gravitational field strength

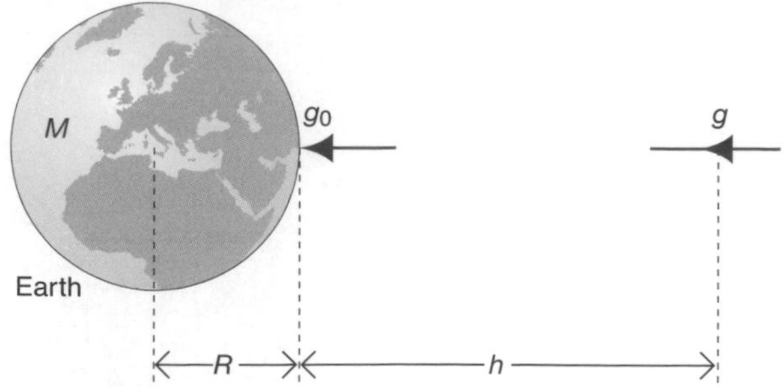

At the surface If M is the Earth's mass, R is its radius, and g_0 is the gravitational field strength at its surface, then

$$g_0 = \frac{GM}{R^2} \qquad (1)$$

Note:
- g_0 is 9.81 N kg^{-1}. It is more commonly known as g (without the $_0$). Here however, the $_0$ has been added to distinguish it from other possible values of g.
- Using measured values of g_0, R, and G in the above equation, the Earth's mass M can be calculated. With R known, the Earth's average density can also be found.

Above the surface In this case, $g = GM/r^2$. From this and equation (1), the following result is obtained:

$$g = \frac{g_0 R^2}{r^2}$$

So as the distance from the Earth increases, g decreases.

Gravitational potential

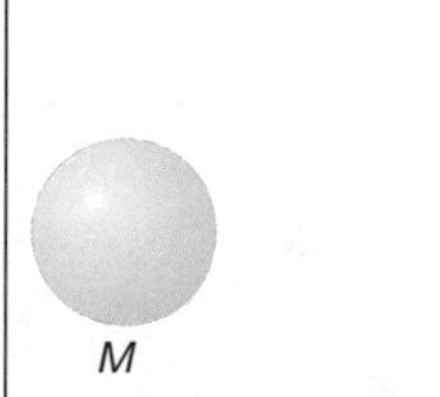

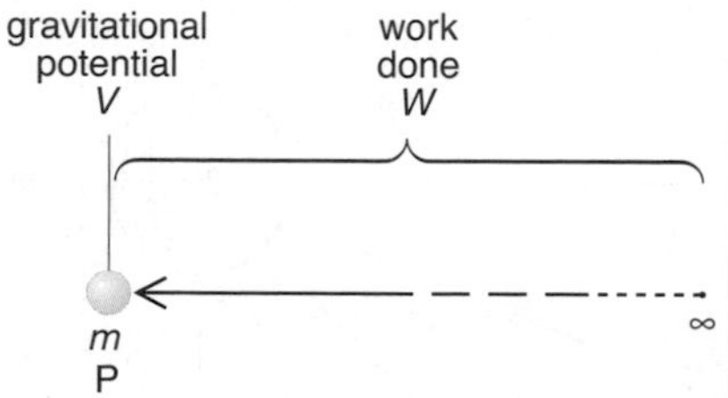

Work must be done to move a mass against a gravitational field. Above, mass M causes a gravitational field. Mass m has been moved through this field, from an infinite distance (where the gravitational force is zero) to point P.

The ***gravitational potential*** V (at point P) is defined as follows:

$$V = \frac{W}{m} \qquad (2)$$

where W is the work done in moving a mass m from infinity (∞) to point P.

Note:
- Like energy, gravitational potential is a scalar.
- At infinity, the gravitational potential is *zero*.
- Elsewhere, the gravitational potential is *negative*. This is because gravity is a force of attraction. Work is done *by* the mass as it is pulled from ∞ to P, so *negative* work is done *on* it. For example, if 1000 J of work is done by a 2 kg mass when it moves from ∞ to P, then -1000 J of work is done on it. So $V = -1000/2 = -500$ J kg^{-1}.

Linking potential and field strength

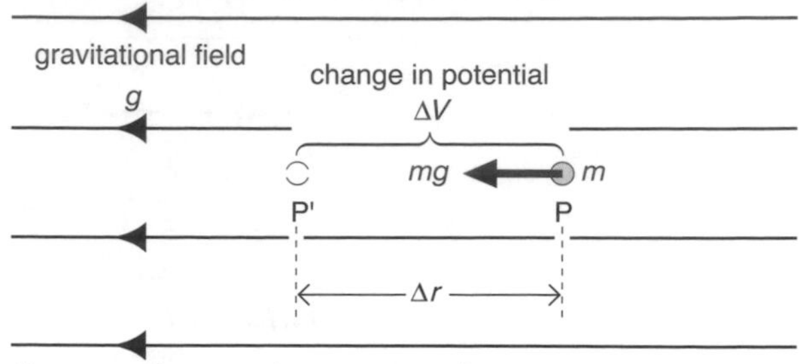

Above, work ΔW is done on a small mass m in moving it from P to P' in a uniform gravitational field g. So, from (2),

$$\Delta W = m\Delta V \qquad (3)$$

This equation gives the work done *on* the mass.

So work done *by* mass $= -m\Delta V$

But work done *by* mass = force × distance moved $= mg\Delta r$

So $\quad mg\Delta r = -m\Delta V$

Therefore $\quad g = -\dfrac{\Delta V}{\Delta r}$

In calculus notation, there is a more general version of this equation which also applies to non-uniform fields:

$$g = -\frac{dV}{dr}$$

Note:
- In the above equations, the minus sign indicates that g is in the direction of decreasing potential.

Gravitational potential in a radial field

A radial field is shown on the opposite page. Provided r is not less than the radius of the sphere:

$$g = \frac{GM}{r^2}$$

Also, $\quad g = -\dfrac{dV}{dr}$

Combining these, and using calculus, gives $V = -\dfrac{GM}{r}$

Note:
- $V \propto 1/r$. So if the distance r doubles, the gravitational potential V halves, and so on. (Inside the mass, this does not apply.)
- In the diagram on the opposite page, each ***equipotential line*** is a line joining points of equal potential.
- In the case of the Earth, the gravitational potential V_0 at the surface is $-GM/R$, where R is the radius.

  As $\quad g_0 = \dfrac{GM}{R^2}$

  it follows that $V_0 = -g_0 R$.

Escape speed

This is the speed, v_{esc}, at which an object must leave a planet's surface to completely escape its gravitational field (i.e. be 'thrown' to infinity). For this, the object must be given enough KE to do the work necessary to move it from the surface (where $V_0 = -g_0 R$) to infinity (where $V = 0$).
The required gain in potential is $g_0 R$. So, from (3), the amount of work to be done is $mg_0 R$. Therefore

$$\tfrac{1}{2} m v_{esc}^2 = mg_0 R \qquad \therefore \quad v_{esc} = \sqrt{2g_0 R}$$

The escape speed from the Earth is 11.2×10^3 m s^{-1}.

H5 Circular orbits and rotation

An orbit equation

Above, a satellite is in a circular orbit around the Earth. The gravitational force on the satellite provides the centripetal force needed for the circular motion. So

$$\frac{GMm}{r^2} = \frac{mv^2}{r} \qquad (1)$$

Note:
- The equation can be used to find the speed v needed for an orbit of any given radius r.
- As m cancels from each side of the equation, the speed needed does not depend on the mass of the satellite.

A value for *GM* To use the above equation, you do not need to know the Earth's mass M. Instead, a value for GM can be found using equation (1) in H4. Rearranged, this gives

$$GM = g_0R^2 \qquad (2)$$

where R is the Earth's radius (12.8×10^9 m), and g_0 is the gravitational field strength at its surface (9.81 N kg^{-1}).

Period of orbit

The period T is the time taken for one orbit.
Equation (1) can be rewritten using a different version of the equation for centripetal force:

$$\frac{GMm}{r^2} = mr\omega^2 \qquad (3)$$

Cancelling m, then substituting $\omega = 2\pi/T$ (see B14), and rearranging gives the following link between T and r:

$$\frac{T^2}{r^3} = \frac{4\pi^2}{GM} \qquad (4)$$

As $4\pi^2/GM$ is a constant, T^2/r^3 has the same value for all satellites. So as r increases, the period gets longer.

'Weightlessness'

An astronaut in a satellite is in a state of free fall. Her acceleration towards the Earth is exactly the same as that of the satellite, so the floor of the satellite exerts no forces on her. As a result, she experiences exactly the same sensation of weightlessness as she would in zero gravity. However, she is not really weightless. A few hundred kilometres above the Earth, the gravitational force on her is almost as strong as it is down on the surface.

Geostationary orbit

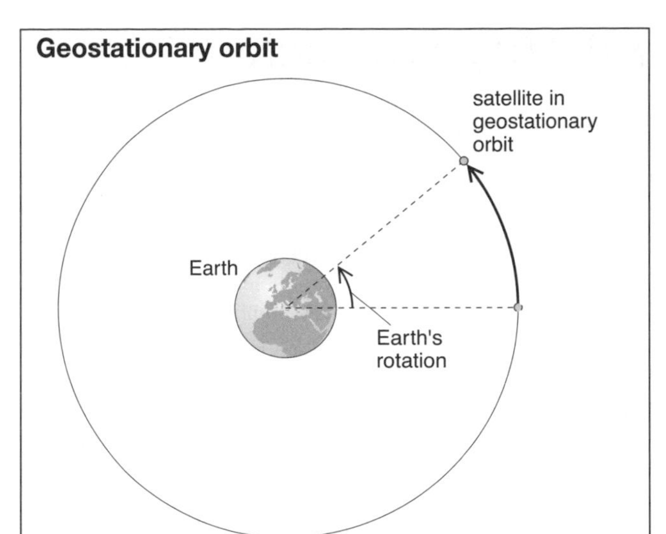

If a satellite is in a ***geostationary orbit***, then viewed from Earth, it appears to be in a fixed position in the sky. This is because the period of its orbit exactly matches the period of the Earth's rotation (24 hours). Communications satellites are normally in geostationary orbits.

Equation (3) can be used to calculate the value of r needed for a geostationary orbit. ω is found using $\omega = 2\pi/T$, with $T = 24 \times 3600$ s. GM is found using equation (2). r works out at 4.23×10^7 m.

Sun, planets, and moons

	Distance from Sun/ $\times 10^7$ km	Period of orbit/ days
Mars	22.8	687.0
Earth	15.0	365.3
Venus	10.8	224.7
Mercury	5.8	88.0
Sun		

Planets are natural satellites of the Sun. Most are in approximately circular orbits, so equation (4) also applies to them (except that M in the equation now becomes the mass of the Sun).

So (period of planet)$^2 \propto$ (distance from Sun)^3

This is called ***Kepler's third law***.

T^2/r^3 has the same value for all the planets around the Sun. If this value is known from astronomical data, equation (4) can be used to calculate the mass of the Sun (M).

Moons are natural satellites of planets. So the above method can also be used to find the mass of a planet.

H6 Magnetic fields

Magnets and fields

Strongly magnetic materials, such as iron and steel, are called ***ferromagnetic*** materials. They feel forces from other magnets and can be made into magnets.

The forces from a magnet seem to come from ***magnetic poles*** near its ends. In reality, every particle in a magnet acts like a tiny magnet. The two poles are the combined effect of all these tiny magnets lined up.

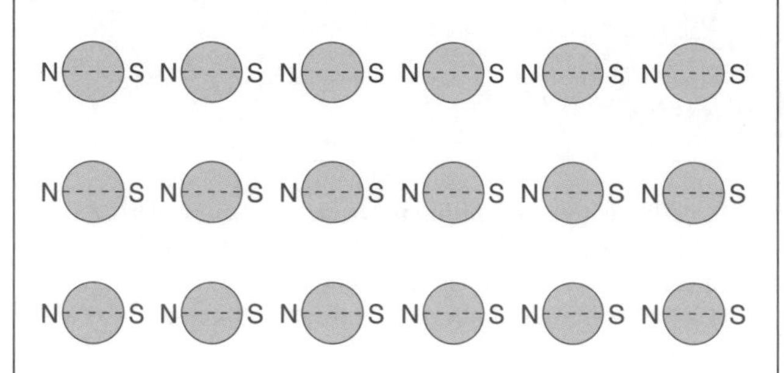

There are two types of pole: ***north*** (**N**) and ***south*** (**S**):
Unlike poles attract. Like poles repel.

The Earth is a weak magnet. As a result, a freely suspended bar magnet turns until it ends are pointing roughly north–south – which is how the two types of pole got their names. However, the north end of the Earth is, magnetically, a south pole because it attracts the north pole of a magnet.

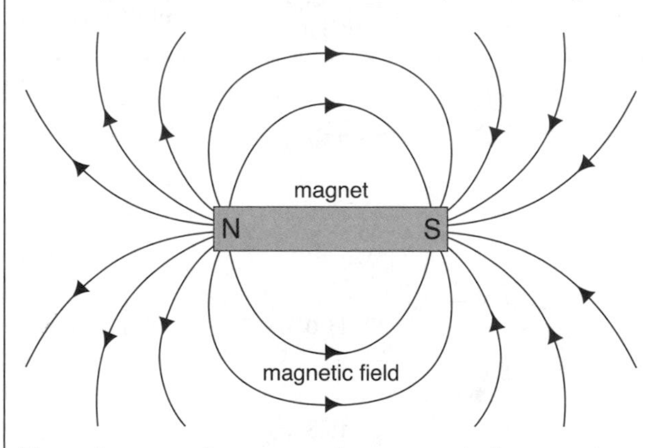

The region around a magnet where magnetic forces act is sometimes called a ***magnetic field***. However, see E6 for a more precise definition of a magnetic field. In diagrams, a magnetic field is represented by ***field lines***. The stronger the field, the closer the lines. The direction of the field is the direction in which a 'free' N pole would move (though in reality, magnetic poles always exist in pairs).
When a strongly magnetic material, such as iron or steel, is

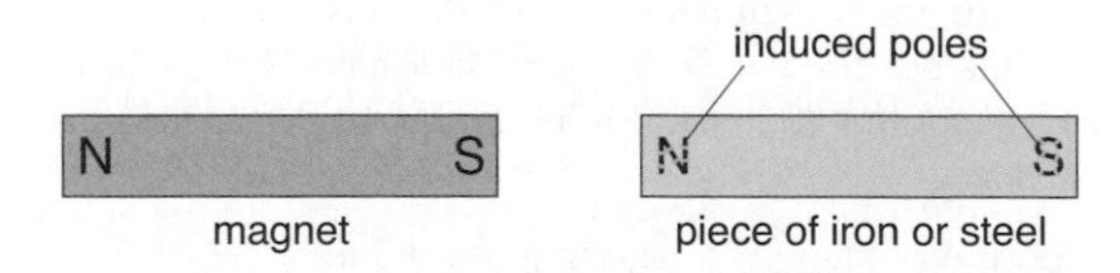

placed in a magnetic field, it becomes magnetized, as shown above. This is called ***induced magnetism***. A magnet attracts iron or steel because the direction of the induced magnetism means that two unlike poles are close together.

When a magnetic field is removed, iron quickly loses its magnetism, but steel becomes permanently magnetized.

Magnetic fields from currents

A current has a magnetic field around it. The greater the current, the stronger the field. The field round a current-carrying wire is shown on the right. The field direction is given by ***Maxwell's screw rule***. Imagine a screw moving in the conventional direction. The field turns the same way as the screw.

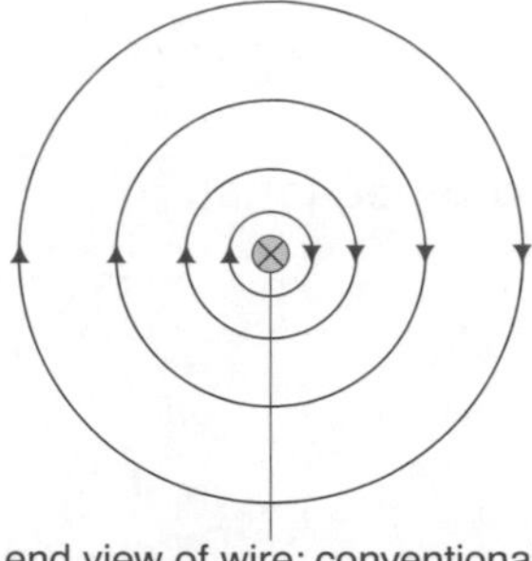

The field round a current-carrying coil (below) is similar to that round a bar magnet. The field direction can be worked out using either the screw rule or the ***right-hand grip rule***. Imagine gripping the coil with your right hand so that your fingers curl the same way as the conventional direction. Your thumb is then pointing to the N pole.

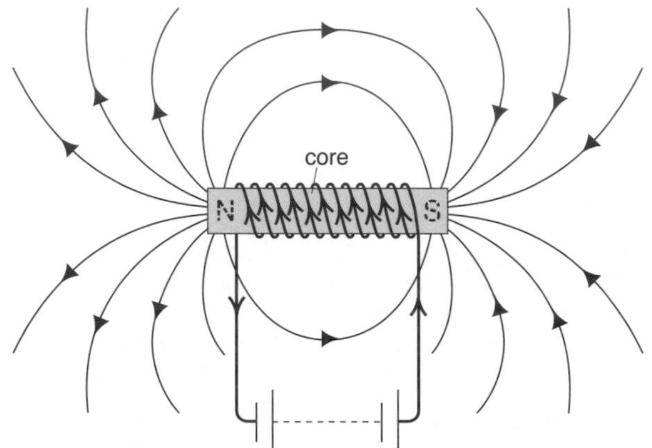

The field is very much stronger if the coil has an iron ***core***. Together, the coil and core form an ***electromagnet***.
With an iron core, the field vanishes when the current is turned off. However, a steel core keeps its magnetism. Permanent magnets are made using this principle.

Orbiting electrons in atoms are tiny currents. They are the source of the fields from magnets.

Magnetic field patterns

Any electric current has a magnetic field around it. Examples are shown in the cross-sectional diagrams below.

Single flat coil
Near any section of wire, the field direction is given by Maxwell's screw rule.

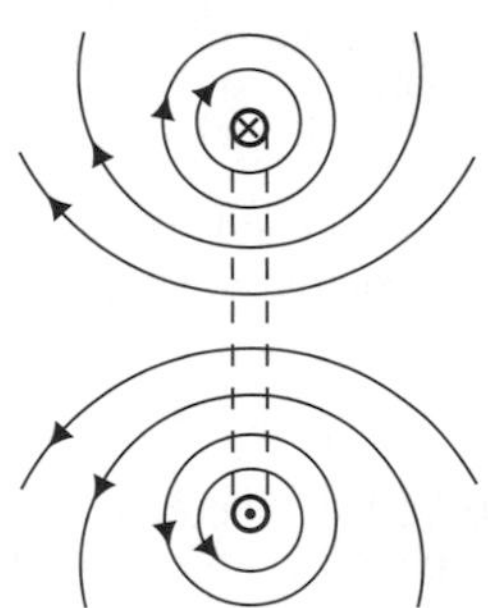

Opposing flat coils
A ***neutral point*** (N on the right) is the point where one magnetic field cancels another.

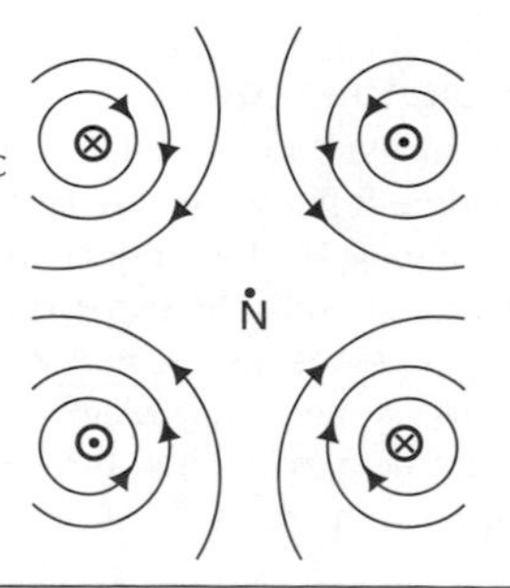

H7 Magnets and currents

A simple electric motor

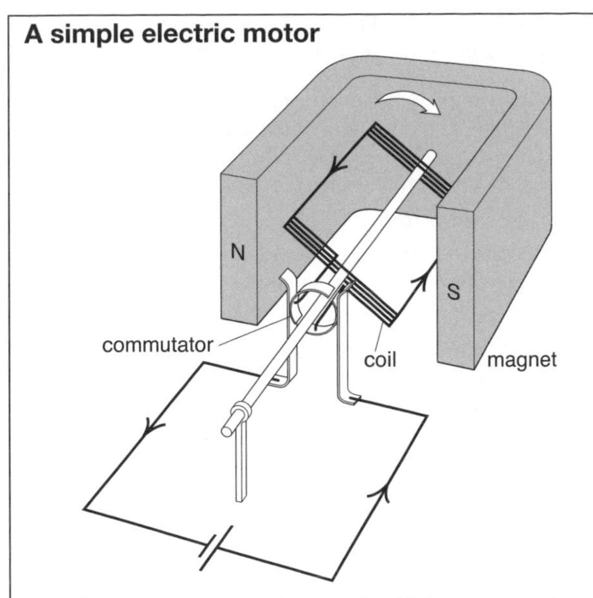

The electric motor above has a coil which can rotate in a magnetic field. When a current passes through the coil, the left side is pushed up, and the right side is pushed down, so there is a turning effect. A switching mechanism called a ***commutator*** keeps the left- and right-side current directions the same, whatever the position of the coil. So the turning effect is always the same way.

Magnetic force and flux density

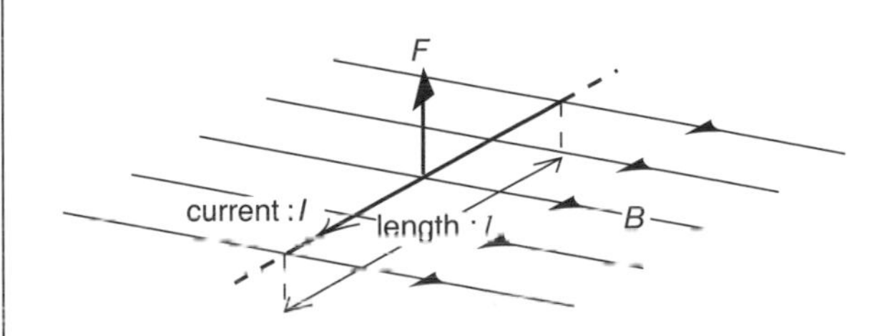

Above, a current-carrying wire is at right-angles to a uniform magnetic field. The field exerts a force on the wire. The direction of the force is given by Fleming's left-hand rule (see E5). The size of the force depends on the current I, the length l in the field, and the strength of the field. This effect can be used to define the magnetic field strength, known as the ***magnetic flux density***, B:

$$F = BIl \quad (1)$$

B is a vector. The SI unit of B is the ***tesla*** (T). For example, if the magnetic flux density is 2 T, then the force on 2 m of wire carrying a current of 3 A is $2 \times 2 \times 3 = 12$ N.

If a wire is not at right angles to the field, then the above equation becomes

$$F = BIl \sin \theta$$

where θ is the angle between the field and the wire. As θ becomes less, the force becomes less. When the wire is *parallel* to the field, $\sin \theta = 0$, so the force is zero.

Magnetic force on a current

There is a force on a current in a magnetic field. Its direction is given by ***Fleming's left-hand rule***:

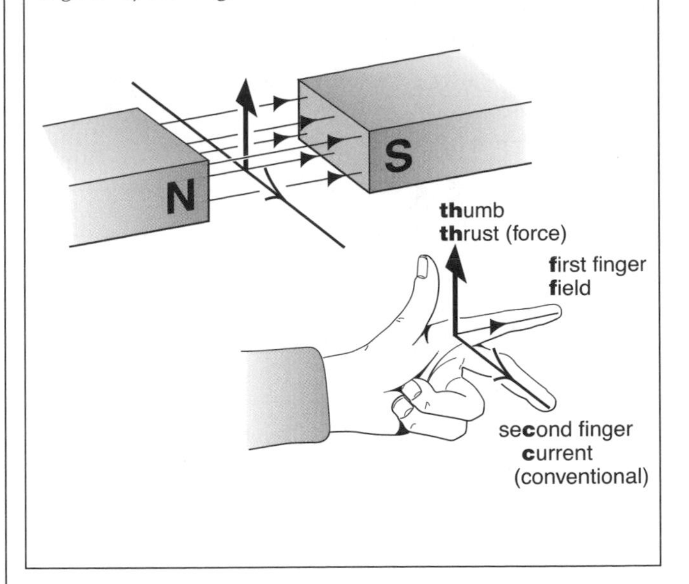

Torque on a current-carrying coil

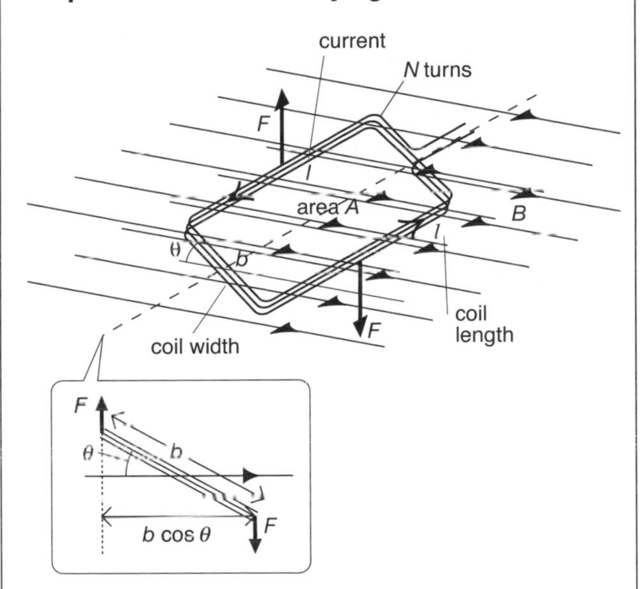

Above, applying equation (1), there is a total upward force F of $BIl \times N$ on one edge of the coil and the same downward force on the other. These form a couple of torque T:

$T = Fb \cos \theta$ (see B4) So $T = BIlNb \cos \theta$

But $lb = A$, the coil's area.

So $T = BIAN \cos \theta$

Note:

- The equation also applies to a non-rectangular coil.
- When the plane of the coil is parallel to the field, θ is zero, so $\cos \theta = 1$, and the torque is a maximum, $BIAN$.
- When the plane of the coil is at right-angles to the field, the torque is zero.
- T for torque should not be confused with T for tesla.
- Forces on the near and far ends of the coil do not have moments about the turning axis.

H8 Electromagnetic induction

Magnetic flux

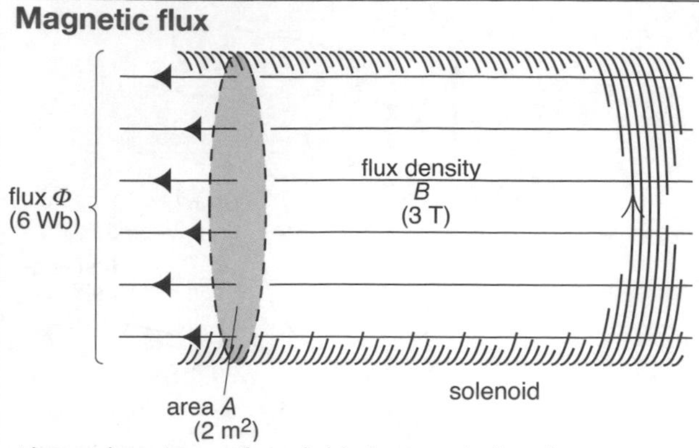

Above, there is a uniform field of magnetic flux density B inside a solenoid of cross-sectional area A. The ***magnetic flux*** Φ is defined by this equation:

flux = flux density × area $\quad \Phi = BA$

The SI unit of magnetic flux is the ***weber*** (**Wb**). For example, if B is 3 T and A is 2 m^2, then Φ is 6 Wb.

Note:

- Field lines do not really exist, but they can help you visualize what magnetic flux means. In the diagram above, each field line represents a flux of 1 Wb. There are 6 lines altogether, so the flux is 6 Wb. But there are 3 lines per m^2, so the flux density is 3 T.
- With flux, 'density' implies 'per m^2' and not 'per m^3'.
- 1 tesla = 1 weber per metre^2 i.e. 1 T = 1 Wb m^{-2}.

Faraday's law

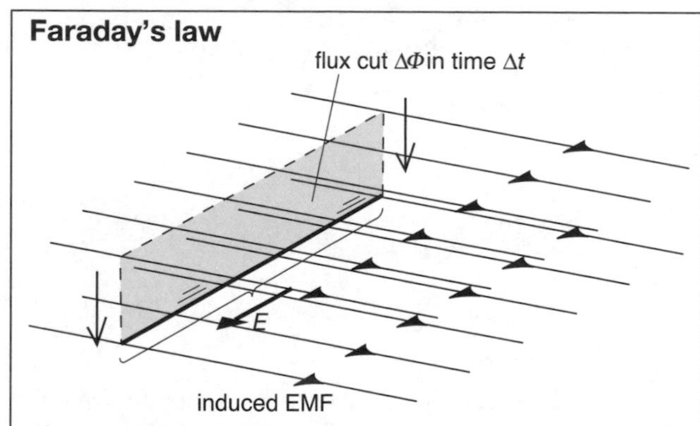

Above, a conductor is moving at a steady speed through a magnetic field. It cuts through flux $\Delta\Phi$ in time Δt. As a result, an EMF E is induced in the conductor. According to ***Faraday's law of electromagnetic induction***, the induced EMF is proportional to the rate of cutting flux:

$$\text{induced EMF} \propto \frac{\text{flux cut}}{\text{time taken}} \qquad E \propto \frac{\Delta\Phi}{\Delta t}$$

With a constant, this can be turned into an equation. The constant is 1 because of the way the units are defined. So

$$\text{induced EMF} = -\frac{\text{flux cut}}{\text{time taken}} \qquad E = -\frac{\Delta\Phi}{\Delta t}$$

For example, if 6 Wb of flux are cut in 2 s, E is 3 V.

There is a calculus version of the above equation which also applies if flux is not cut at a steady rate:

$$E = -\frac{d\Phi}{dt} \qquad (1)$$

The significance of the minus sign is explained on the right.

Lenz's law

According to this law,

if an induced current flows, its direction is always such that it will oppose the change in flux which produced it.

For example, in the diagram above, the induced current causes a N pole at the left end of the coil so that the approaching magnet is repelled. If the magnet is moved the other way, the induced current direction reverses so that there is a pull on the magnet to oppose its motion.

The minus sign in equation (1) comes about because the induced EMF opposes the flux change.

Lenz's law follows from the law of conservation of energy. Energy must be transferred to produce an induced current. So work must be done to make the change which causes it.

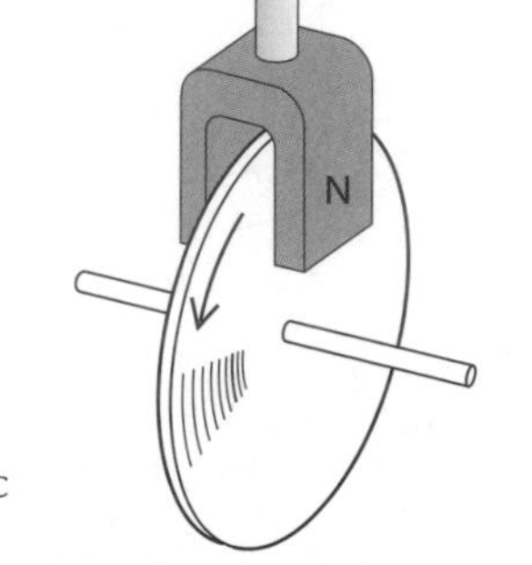

Eddy currents When the aluminium disc on the right is spun between the magnetic poles, ***eddy currents*** are induced in it. These set up a magnetic field which pulls on the poles and opposes the motion. So the disc quickly comes to a halt. Electromagnetic braking systems use this effect.

Currents from magnetic fields

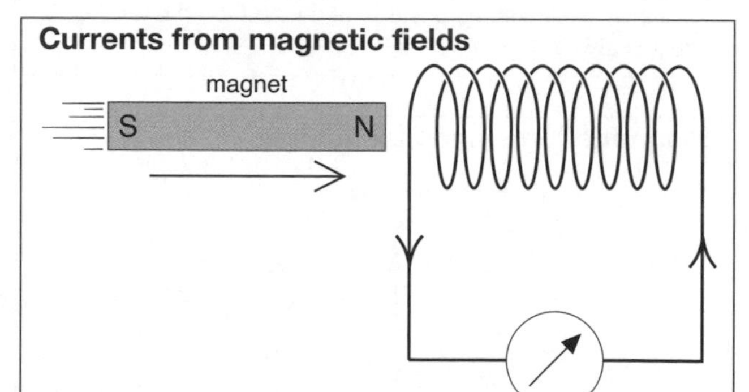

If one end of a bar magnet is moved into a coil, as above, an EMF (voltage) is generated in the coil. This effect is called ***electromagnetic induction***. The EMF makes a current flow in the circuit.

The induced EMF (and the current) is increased if:

- the magnet is moved faster,
- there are more turns on the coil,
- a magnet giving a stronger field is used.

If the magnet is moved out of the coil, the EMF is reversed. If the magnet is stationary, the EMF is zero.

Varying a magnetic field can have the same effect as moving a magnet. Below, an EMF is induced in the right-hand coil whenever the electromagnet is switched on or off. The current flows one way at switch-on and the opposite way at switch-off.

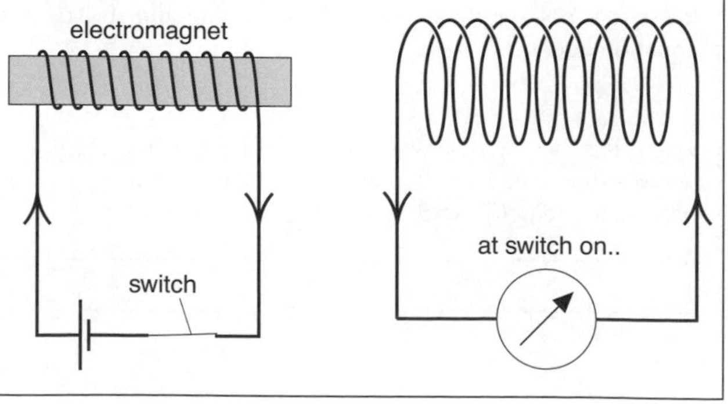

H9 More electromagnetic induction

Flux change and flux linkage

Changing flux has exactly the same effect as cutting flux, and the induced EMF is calculated in the same way.

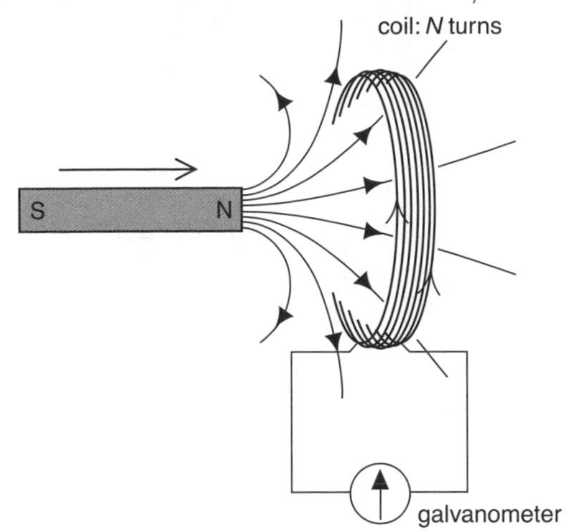

Above, a magnet is moved into a coil. If the flux through the coil changes (at a steady rate) by $\Delta\Phi$ in time Δt, then an EMF of $\Delta\Phi/\Delta t$ is induced *in each turn*. But there are N turns in series. So, the *total* induced EMF E is as follows:

$$E = -N\frac{\Delta\Phi}{\Delta t}$$

For example, if the flux changes by 6 Wb in 2 s, and the coil has 100 turns, then the total induced EMF is 300 V.

If there is a flux Φ through a coil of N turns, then $N\Phi$ is called the ***flux linkage***. $N\Delta\Phi$ is the *change* in flux linkage (600 Wb turns, in the previous example). So, the previous equation can be written as follows:

$$\text{induced EMF} = -\frac{\text{change in flux linkage}}{\text{time taken}}$$

Calculating an induced EMF

The conductor below is cutting magnetic flux. It moves a distance of 4 m in 6 s, at a steady speed. From the data supplied, the PD across the ends of the conductor can be calculated. (The PD is equal to the induced EMF E.)

area of flux cut = 3 × 4 = 12 m^2

∴ flux cut = BA = 2 × 12 = 24 Wb

∴ PD = $E = -\Delta\Phi/\Delta t = -24/6 = -4$ V

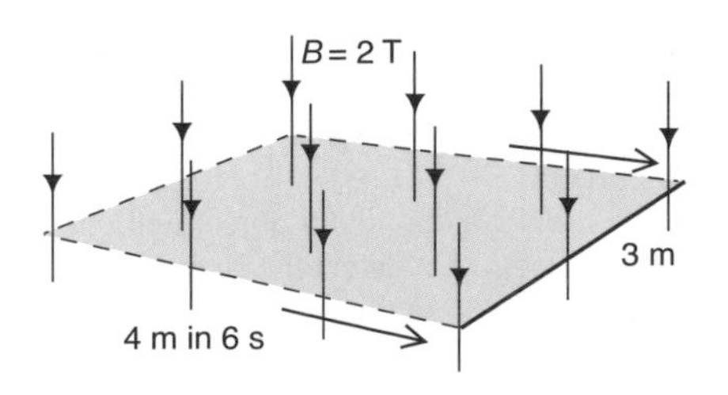

Induced EMF in a rotating coil

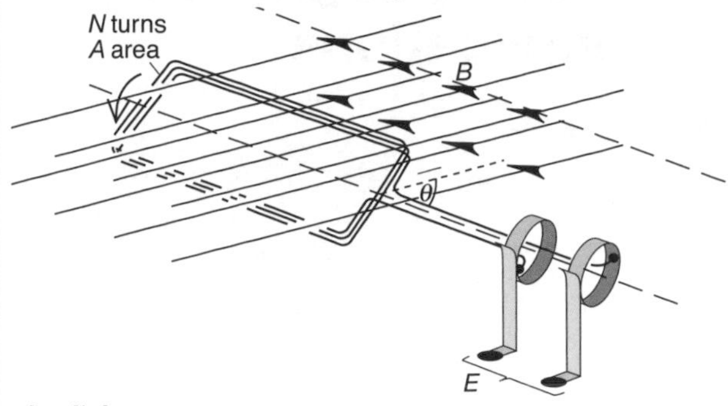

Flux linkage

When a coil, of N turns and area A, is stationary in a vertical position in a magnetic field of flux density B, the flux linkage Φ = BAN.

When the normal to the coil makes an angle θ with the magnetic field, the flux linkage is:

$\Phi = BAN\cos\theta$

If a coil is rotating with a steady angular velocity ω, so that in time t the coil has turned through an angle $\theta = \omega t$, the flux linkage at time t is:

$\Phi = BAN\cos\omega t$

The induced EMF E can be found by working out the rate of change of flux linkage, using calculus.

Using Lenz's law $E = -\frac{d\Phi}{dt}$

$$E = BAN\omega \sin \omega t$$

The graph shows how E varies with t. The ***peak*** (maximum) EMF, E_0, occurs when the coil is horizontal and $\sin\omega t = 1$.

So $E_0 = BAN\omega$

Increases in B, A, N, and ω all give an increased EMF.

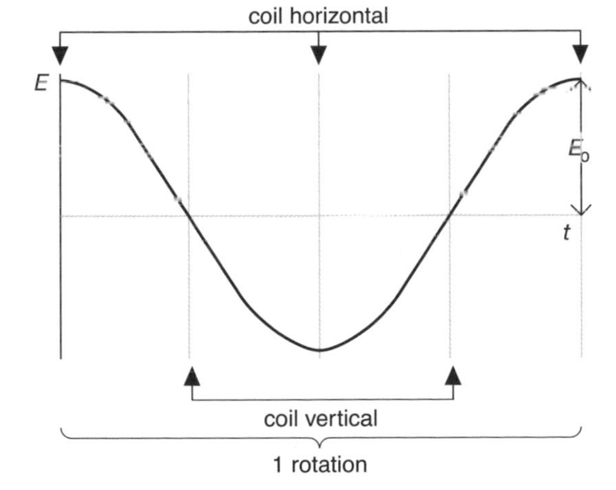

The output from an alternator (see H10) is as shown in the graph above. The current is ***alternating current*** (**AC**).

Induced EMF in an electric motor (See H7 for motor diagram.) When the coil of a motor is turning, it is cutting magnetic flux. As a result, an EMF is induced which, by Lenz's law, opposes the current. This is known as a ***back EMF***. At switch-on, when the coil is stationary, there is no back EMF, so the current is high. As the motor speeds up, the back EMF rises and the current drops.

For higher B, the coil of a motor is normally wound on an iron-based *armature*. This has a laminated (layered) structure to reduce eddy currents, which waste energy.

H10 Electromagnetic applications

Generating AC

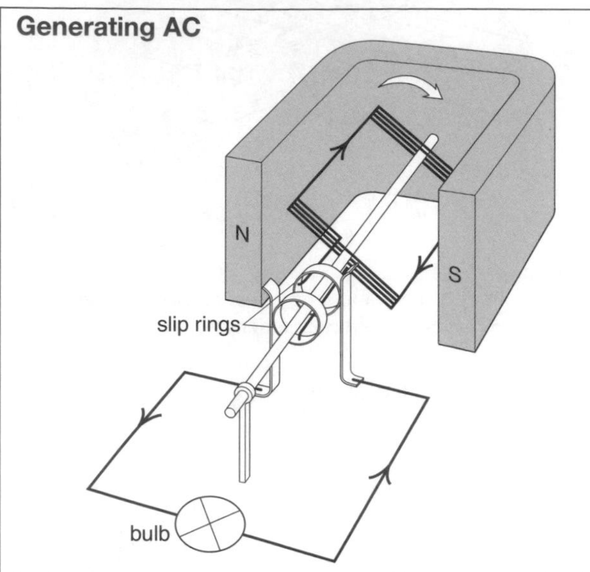

In the generator above, a coil is rotated in a magnetic field. This induces an EMF in the coil, so a current flows. The current keeps changing direction as the coil faces first one way and then the other: it is ***alternating current*** (**AC**). A generator which produces AC is called an ***alternator***.

'One way' current from a battery is ***direct current*** (**DC**).

How Science Works

Magnetic surveys

When a piece of iron is placed in a magnetic field, the field is noticeably changed. Different materials all have different magnetic properties. They also alter the magnetic field, but the effect is much smaller. A change in the material in an area can be detected as a change the Earth's magnetic field. Using sensitive magnetometers, local disturbances in the Earth's magnetic field can be used to detect archaeological features hidden below the ground. Examples are features of iron, burned soil, or brick.

The proton magnetometer can measure very small variations in the Earth's magnetic field. It uses the fact that the proton (see J1) behaves as a very small magnet.

A direct current in a coil creates a strong magnetic field. A fluid containing hydrogen atoms is placed in the field, and the hydrogen nuclei (protons) line up with the field. When the current is switched off, the protons realign with the Earth's field. As they move they produce a weak alternating magnetic field, which can be detected.

Transformers

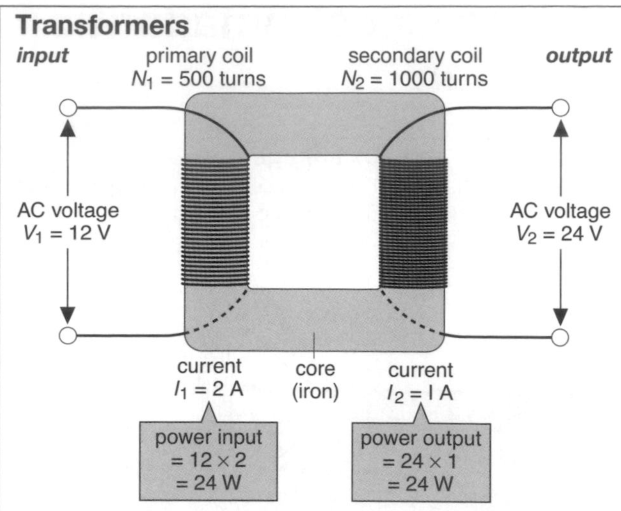

In a ***transformer***, an alternating current in the ***primary*** (input) coil creates a changing magnetic field in the core, which induces an alternating voltage in the ***secondary*** (output) coil. A ***step-up*** transformer, as above, increases the voltage. A ***step-down*** transfomer reduces it.

There is a link between the output and input voltages (V_2 and V_1) and the numbers of turns (N_2 and N_1) on the coils:

$$\frac{V_2}{V_1} = \frac{N_2}{N_1}$$

For the transformer in the diagram,

$$\frac{N_2}{N_1} = \frac{1000}{500} = 2 = \frac{24}{12} = \frac{V_2}{V_1}$$

If no power is wasted in the coils or core,

power output = power input

$$V_2 I_2 = V_1 I_1$$

This means that a transformer which *increases* voltage will *decrease* current, and vice versa.

On a circuit diagram, a transformer symbol is:

In practice, transformers waste energy as heat, due to:

- resistance of the coils
- eddy currents produced by the changing flux (see H8).

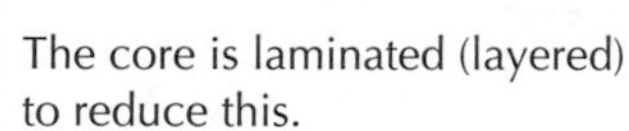

The core is laminated (layered) to reduce this.

For a practical transformer $V_2 I_2 = eV_1 I_1$

where e is the efficiency (typically over 0.95).

How Science Works

Metal detectors

One popular design of metal dectector uses two rings: a transmitter and a receiver.

The transmitting ring and the receiving ring both contain coils. These are the steps in the process:

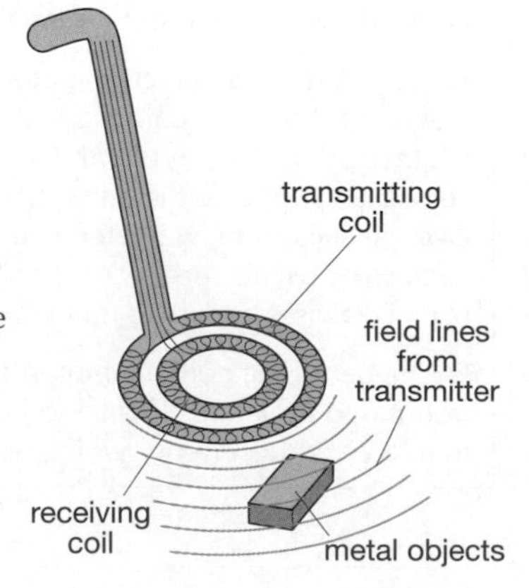

- AC in the transmitting coil produces a changing magnetic flux.
- Any metal object nearby will be cut by this changing magnetic flux, so there will be an induced EMF in the metal object.
- This changing EMF produces a changing current in the metal object.
- The changing current will cause a changing magnetic flux in the metal object.
- This changing flux from the metal object will cut the receiving coil of the metal detector.
- This will induce an EMF in the receiving coil so that current flows in the coil.

The receiver is shielded from the transmitter, so that when there is a current in the receiving coil it is a signal that there is a metal object nearby.

H11 Charged particles in motion

Electron beams

Read E2 first

Beams of electrons and other charged particles are deflected by magnetic and electric fields. The unit of energy used is the electronvolt (eV), and energies are often given in MeV and GeV.

Note: in all the diagrams in this section, the equipment is enclosed in a vacuum tube.

Deflection of electrons by an electric field

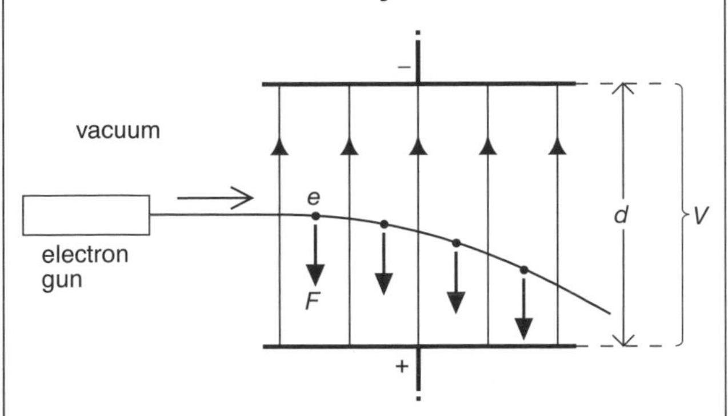

Above, electrons pass between two horizontal plates. The electric field strength between the plates is V/d (see I11).

force F on electron = electric field strength × charge

So $$F = \frac{Ve}{d}$$

Note:

- The force on the electron does not depend on its speed.
- The force is always in line with the electric field.
- The path of the electrons is a *parabola* (just as it is for the thrown ball in B5).

Deflection of electrons by a magnetic field

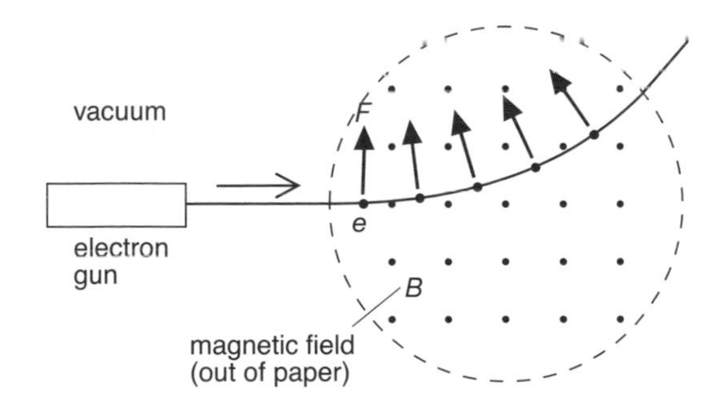

Above, electrons travel at right angles to a uniform magnetic field. The force F on an electron is found by putting Q equal to e in equation (1) on the left. So

$$F = Bev$$

Note:

- The force on the electron increases with speed.
- The force is always at right-angles to the direction of motion, as predicted by Fleming's left-hand rule. But in applying this rule, remember that the electron has negative charge, so electron motion to the *right* represents a conventional current to the *left*.
- The path of the beam is *circular* (see next page).

Magnetic force on a moving charge

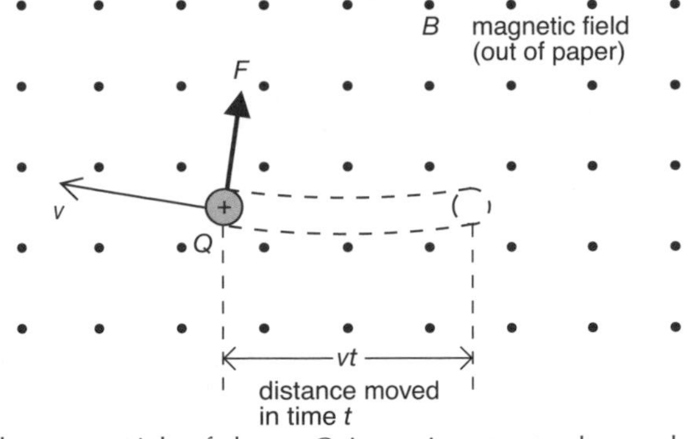

Above, a particle of charge Q, is moving at a steady speed through a uniform magnetic field. In time t, it travels a distance vt. So it is equivalent to a current Q/t in a wire of length vt. According to equation (1) in H7, the force on a current-carrying wire is BIl. Applying this to the above:

force $$F = B \times \frac{Q}{t} \times vt$$

So $$F = BQv \qquad (1)$$

Note:

- As the speed increases, the force increases.
- The direction of the force is given by Fleming's left-hand rule (see H6).
- If the particle is travelling at an angle θ to the field, then the above equation becomes

$$F = BQv \sin \theta. \qquad (2)$$

Mean what you say – say what you mean

For charged particles in magnetic fields, if the direction of the particle is towards or away from a magnetic pole (so that the field is in line with the motion of the particle), the force on the particle is zero. There is no deflection. This is shown by equation (2), when $\theta = 0$ or $180°$, $\sin \theta = 0$. It is incorrect to say that the charged particle is 'attracted' or 'repelled' by the magnetic field.

The maximum deflection occurs when the field and the particle motion are perpendicular to each other. Describing the directions clearly can be difficult. If you write the force is 'up' or 'down' do you mean towards the top or bottom of the paper, or do you mean into or out of the paper?

Make sure your answers are not ambiguous. Use diagrams if these help.

A beam of both positive and negative particles can be separated by a magnetic field. The force on the positive particle is in the opposite direction to the force on the negative particle, so the ***deflection*** will be in opposite directions initially, as shown below. It is incorrect to say that the particles will ***move*** in opposite directions.

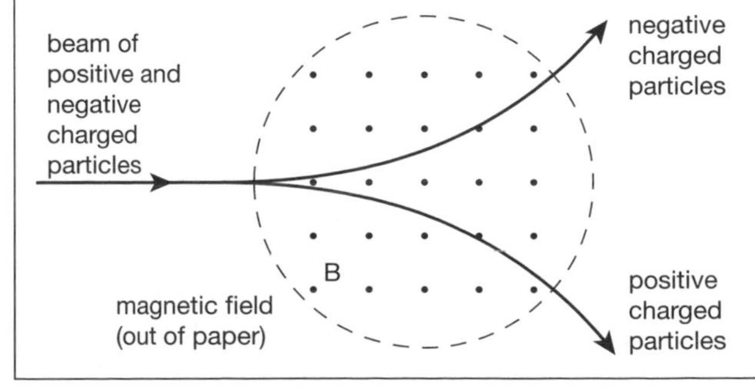

Circular motion in a magnetic field

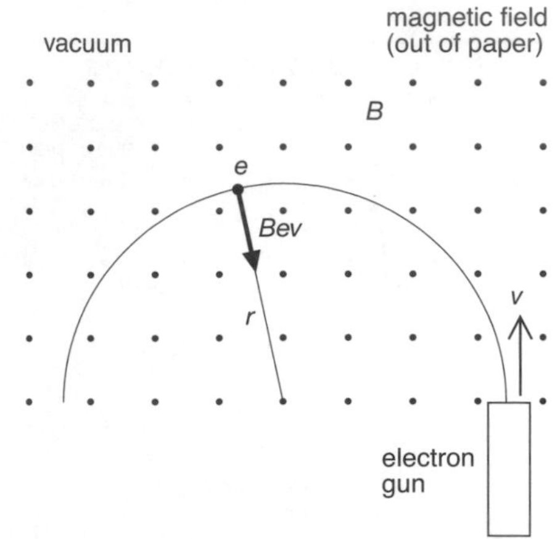

Above, electrons leave the electron gun at speed v. They move in a circle because the magnetic force Bev supplies the necessary centripetal force (see B14). So

$$Bev = \frac{m_e v^2}{r} \quad \text{So} \quad r = \frac{m_e v}{eB} \tag{3}$$

Note:

- The radius of the circle is proportional to the speed.
- Increasing B decreases the radius.

Specific charge of an electron This an electron's charge per unit mass, e/m_e. Its value is -1.8×10^{11} C kg^{-1}. Methods of measuring it make use of the above equation.

Ion beams If atoms lose electrons, they become positive ions. These particles also have circular paths in a magnetic field. If Q is the charge on a particle, and m the mass, then equation (3) can be written in this more general form:

$$r = \frac{mv}{QB} \tag{4}$$

Note:

- The radius depends on the specific charge (Q/m).

How Science Works

Mass spectrometer

From equation (4), the radius depends on the specific charge Q/m. This idea is used in the mass spectrometer to separate nuclei with different specific charges:

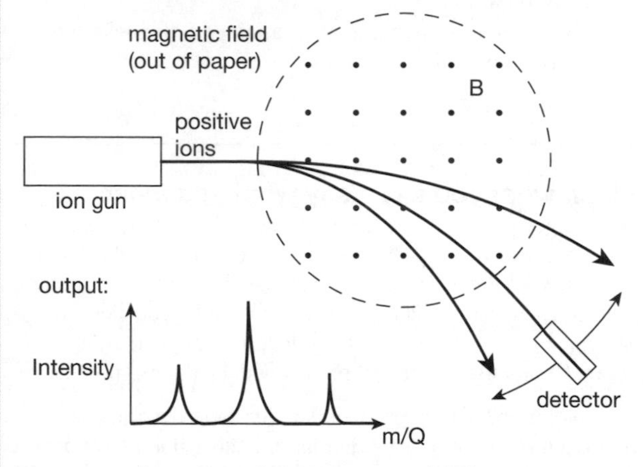

The output shows the relative intensity of different isotopes present in a sample. (Usually all the ions have charge +1. If, for example, some are +2, this must be taken into account.)

The ratio of different isotopes is used in some dating techniques, e.g. carbon dating.

Speed selection

An ion beam may contain ions with a range of speeds. Ions of only one speed can be selected by the method shown below. This uses the principle that the magnetic force on an ion depends on its speed, while the electric force does not:

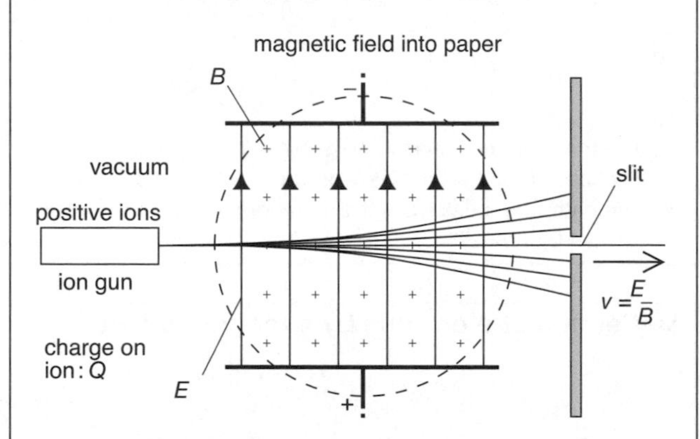

The magnetic field produces a force towards the top of the page, BQv. The electric field produces a force towards the bottom of the page, EQ. The only ions to pass through the slit are those for which these forces are equal. So, if $EQ = BQv$, the selected speed $v = E/B$. Faster ions are deflected upwards (because BQv is more); slower ions are deflected downwards.

H12 Particle accelerators

Probing the nucleus

To investigate the nucleus of the atom, scientists have broken it into bits using beams of high-energy particles (such as protons), from ***particle accelerators***.

Accelerators can supply charged particles with the energy needed to create new matter in collisions.

Protons must be accelerated to very high speeds to penetrate the nucleus. They have to overcome the electric repulsion of the protons there. This is called a ***Coulomb repulsion*** (from Coulomb's inverse square law: see H1).

Being uncharged, neutrons can penetrate the nucleus more easily. But they cannot be directed and controlled by electric and magnetic fields.

Ordinary matter is made up of protons, neutrons, and electrons. But collision experiments with accelerators have produced hundreds of other 'elementary' particles as well. These are described in K2.

Relativistic effects

The mass of a particle when at rest to an observer is its ***rest mass***, m_0. As the particle gains speed (and therefore energy), its mass increases (see J6). Its total observed mass is called its ***relativistic mass***, m. For a particle at a speed v relative to the observer (c is the speed of light),

$$m = \frac{m_0}{\sqrt{1 - v^2/c^2}} \qquad (1)$$

At the speeds achieved in accelerators, the mass increase can be significant. For example, for an electron travelling at 90% of c, m is more than twice m_0.

Energy and mass According to Einstein (see J6), a mass m has an ***energy equivalent*** E as given by this equation:

$$E = mc^2 \qquad (2)$$

For a particle at rest, $E_0 = m_0c^2$. This is its ***rest energy***.

Energy and momentum If a particle's momentum is p, then $p = mv$. So, from equation (2), it follows that

$$p = Ev/c^2 \qquad (3)$$

Using equations (1), (2), and (3), it can be shown that

$$E^2 = m_0^2c^4 + p^2c^2$$

Note:

- If E is much greater than E_0, then $E \approx pc$.

Units The energies of particles from accelerators are often measured in GeV: 1 GeV = 1000 MeV = 10^9 eV.

As energy, mass, and momentum are linked, the masses and momenta of particles can be expressed in energy-related units. For example,

mass can be measured in GeV/c^2 (from equation 2)

momentum can be measured in GeV/c (from equation 3).

Cyclotron

In a cyclotron ions (e.g. alpha particles) are injected at a point near the centre. A potential difference between the 'dee'-shaped electrodes accelerates the particles. A magnetic field vertically into the dees causes the particles to move in a circular path.

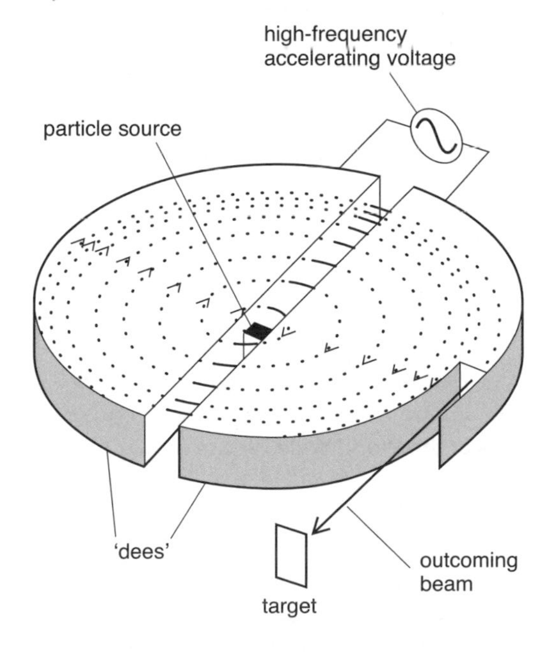

The frequency of the accelerating voltage is adjusted so that the time taken for the particle to travel in a semicircle in a dee is equal to half the period. The time for one revolution of the particle is then equal to the period of the accelerating PD. This condition results in the particle being accelerated every time it crosses from one dee to the other. After each acceleration the particle moves in an orbit of slightly larger radius. When it reaches the outer edge of the cyclotron the particle beam is extracted and used in experiments. The frequency of the orbit remains constant as the particle accelerates (see below) so that there is no need to adjust the accelerating frequency or the magnetic field.

Cyclotron frequency

Whilst travelling in the dees,

$$Bqv = mv^2/r$$

where B is the magnetic flux density, q is the charge on an ion, m is the mass of an ion, v is the velocity of an ion, and r is the radius of the path.

It follows that $v = Bqr/m$

The velocity is also given by $v = 2\pi r/T = 2\pi rf$

so that $$f = \frac{Bq}{2\pi m}$$

This is the cyclotron frequency. Notice that the frequency does not depend on r or the velocity of the particle.

Colliding beams

One way of conducting nuclear experiments is to allow an accelerated beam to collide with a stationary target.

To obtain the higher-energy collisions that are needed to explore the structure of matter in greater detail two beams of particles are made to collide head on. This doubles the energy involved in the collision.

The accelerator first accelerates a group of one type of particle (e.g. electrons) to a high energy. This is then stored in a storage ring while a group of particles of another type (e.g. positrons) is accelerated. The beams are then steered so that they interact.

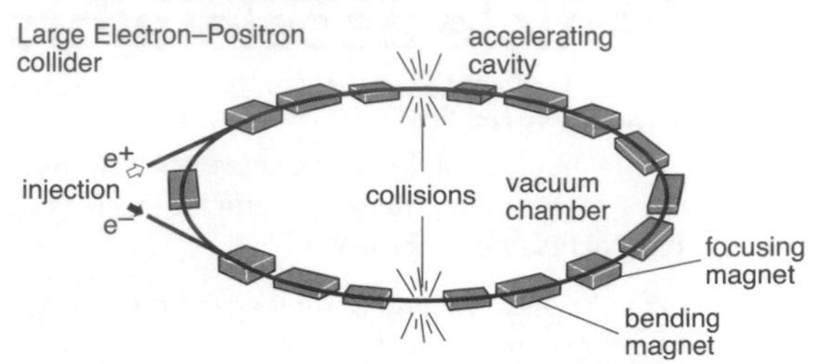

At CERN, between 1989 and 2000 in the LEP collider, the energy of each beam was about 90 GeV, giving total collision energy of 180 GeV. This enabled the production of new particles as the energy appeared as mass (from $E = mc^2$).

Linear accelerator (up to 20 GeV)

Charged particles (e.g. electrons or protons) in a vacuum pipe are accelerated through a series of electrodes by an alternating voltage. The frequency is carefully chosen so that, as each electrode goes alternately + and –, particles leaving one electrode are always pulled towards the next. The beam of particles is directed at a target or into a ***synchrotron***.

Synchrotron (1000 GeV or more)

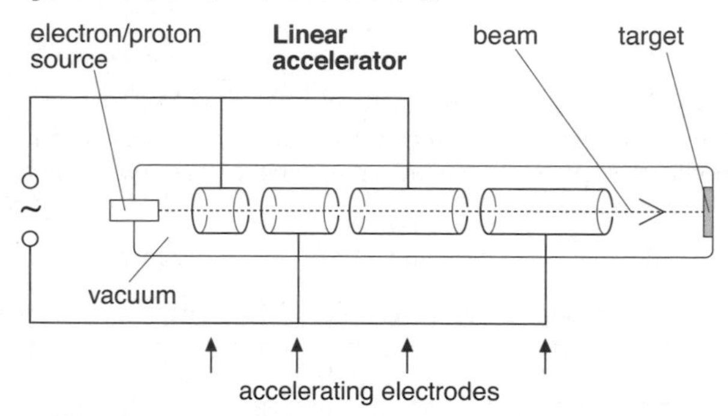

In effect, this is a linear accelerator, bent into a ring so that the charged particles can be given more energy each time they go round. Electromagnets keep the particles in a curved path. As the speed increases, the magnetic field strength is increased As the speed increases and relativistic effects cause the mass of the particles to increase, a larger force is needed to accelerate them (to speed them up and keep them in a circular path), so the magnetic field strength is increased. As the speed increases and relativistic effects cause the mass of the particles to increase, a larger force is needed to accelerate them (to speed them up and keep them in a circular path), so the magnetic field strength is increased.

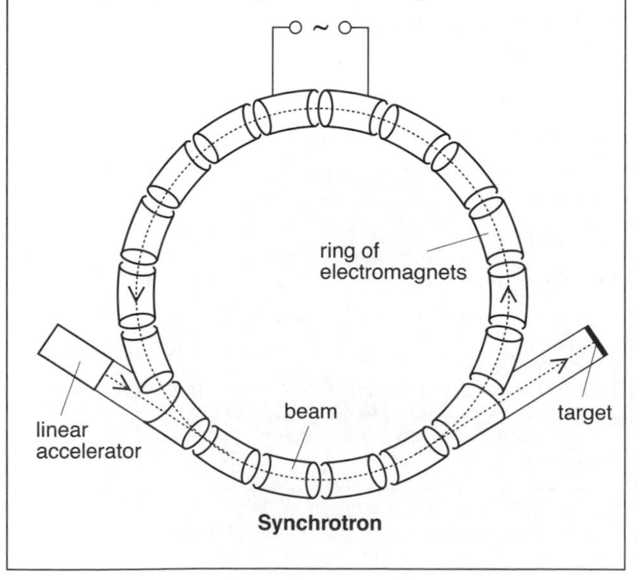

Synchrotron

How Science Works

CERN

CERN is the European particle physics laboratory. It is near Geneva in Switzerland and was established in 1954. Twenty European countries collaborate in funding and running CERN. About 3000 people work there, but the 6500 visiting scientists represent 80 nationalities. They have a number of accelerators. The Large Electron Positron (LEP) collider was in a circular underground tunnel with a 27 km circumference. It was dismantled in 2000 so that the Large Hadron Collider (LHC) could be built in the tunnel.

The Large Hadron Collider

The LHC uses superconducting magnets to steer a beam of either protons or of lead ions in a circle. There are two beams travelling in opposite directions. They circulate for 20 minutes to reach maximum speed. The beams collide at high speed, and detectors record the particles produced. The first beam was switched on in 2008.

The World Wide Web

Scientists working at CERN have a large amount of information to send to each other. In 1989 Time Berners-Lee wrote a proposal for an information system, and by the end of 1990 the World Wide Web was up and running. Developed by CERN, and given free to the world, this is a good example of an unexpected spin-off resulting from scientific research.

Benefits of CERN

- sharing money for research between countries is economical
- sharing ideas between scientists can speed progress
- working with people of other nationalities promotes understanding
- technology can be developed that has other benefits (spin-offs), e.g. detectors can be used in other applications such as medicine
- unexpected beneficial spin-offs can occur, such as the World Wide Web
- jobs for local people

Risks of CERN

- expensive – money spent on other research, or to directly benefit people, could have more useful results.
- the local economy is dependent on CERN and may be badly affected if funding is cut or if CERN closes

Benefits of the World Wide Web

- education
- access to information
- putting special interest groups and support groups in-touch with each other
- shopping and banking

Risks of the World Wide Web

- criminal activities, e.g. fraud, downloading copyright material without paying, paedophile activity
- health risks associated with too much time online

I1 Heat and temperature

Solids, liquids, and gases

According to the ***kinetic theory***, matter is made up of tiny, randomly moving particles. Each particle may be a single atom, a group of atoms called a ***molecule***, or an ion. The three normal ***phases*** of matter are solid, liquid, and gas.

Solid The particles are held close together by strong forces of attraction. They vibrate, but about fixed central positions, so a solid keeps a fixed shape and volume.

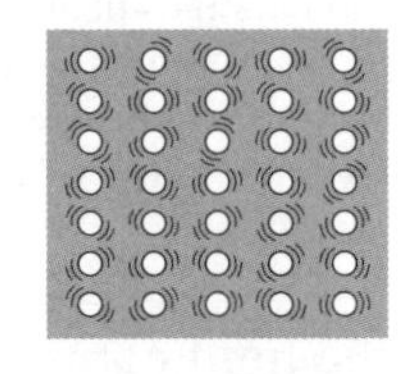

Liquid The particles are held close together. But the vibrations are strong enough to overcome the attractions, so the particles can change positions. A liquid has a fixed volume, but it can flow to fill any shape.

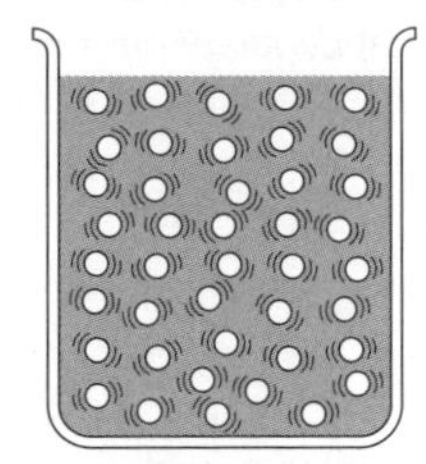

Gas The particles move at high speed, colliding with each other and with the walls of their container. They are too spread out and fast-moving to stick together, so a gas quickly fills any space available. Its pressure is due to the impact of its particles on the container walls.

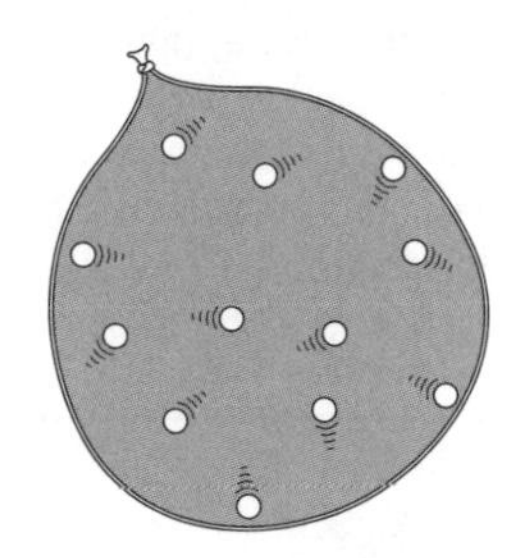

Temperature

The particles in, for example, a gas move at a range of speeds. However, the higher the temperature, the faster the particles move on average.

If two objects at the same temperature are in contact, there is no flow of heat between them. This is because the average kinetic energy of each particle due to its vibrating or speeding motion is the same in each object, so there is no overall transfer of energy from one object to the other.

Celsius scale On this scale, pure water freezes at 0 °C and boils at 100 °C (under standard atmospheric conditions).

Kelvin scale This has the same sized 'degree' as the Celsius scale, but its 'zero' is ***absolute zero*** (–273.15 °C), the temperature at which particles have the minimum possible kinetic energy. (The laws governing the behaviour of atoms do not permit zero energy.)

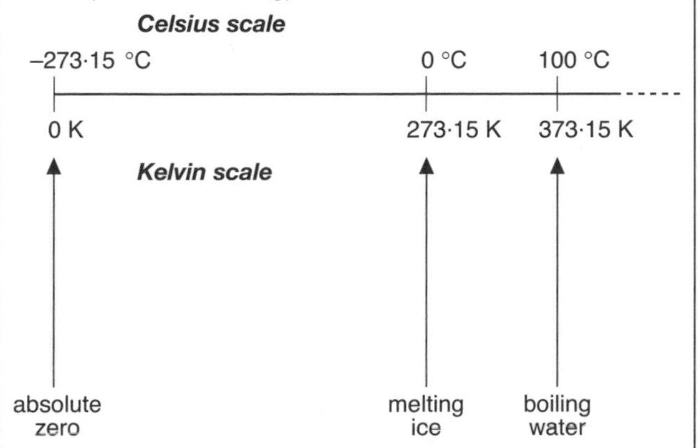

Note: using 273 instead of the value of 273·15 gives enough accuracy in most cases.

Heat transfer

Heat can be transferred by ***conduction***, ***convection***, and ***radiation***, as well as by evaporation.

Conduction In all materials, fast-moving particles in one region can gradually pass on energy to neighbouring particles, and hence on to all the particles.

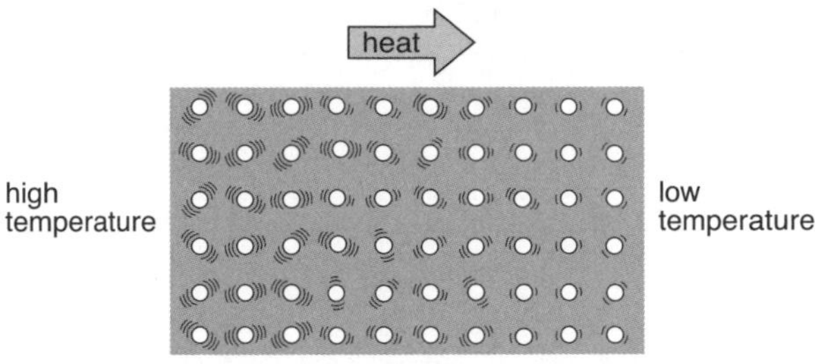

Metals are the best conductors of heat. This is because they have free electrons which can transfer energy rapidly from one part of the material to another. These same electrons also make metals good conductors of electricity.

Non-metal solids and liquids are normally poor conductors of heat because they do not have free electrons. Bad conductors are called ***insulators***. Gases are especially poor conductors: most insulating materials rely on tiny pockets of trapped air for their effect.

Convection Heat is carried by a circulating flow of particles in a liquid or gas.

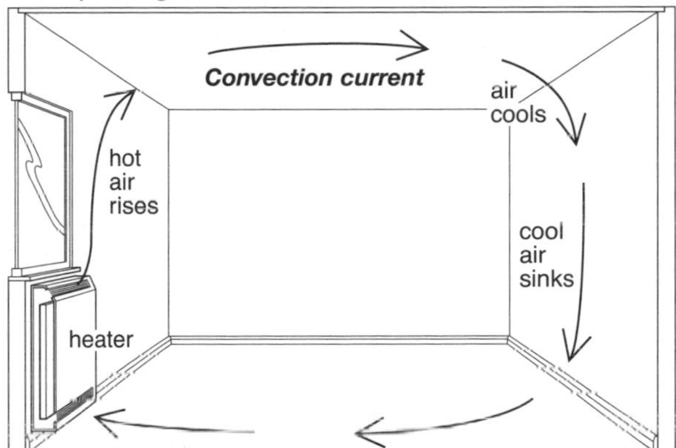

Most room heaters rely on convection. Hot air from the heater expands and floats upwards through the cooler air around it. Cooler air sinks to replace the hot air which has risen. In this way, a ***convection current*** is set up.

Radiation Hot objects radiate energy in the form of electromagnetic waves such as infrared (see D1). The higher the temperature, the more they emit. When this radiation is absorbed by other things, it produces a heating effect. So it is known as ***thermal radiation***.

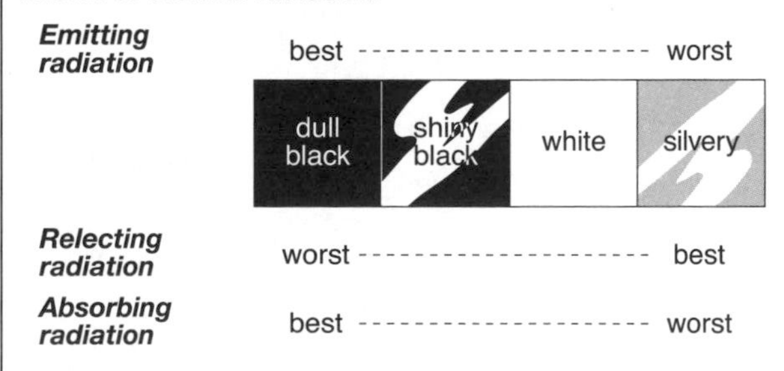

Black surfaces are the best emitters of thermal radiation and also the best absorbers. (They look black because they absorb light.)

Shiny surfaces are poor emitters and also poor absorbers. They reflect most of the radiation that strikes them.

Temperature and thermal equilibrium

Objects A and B are in contact. If heat flows from A to B, then A is at a higher ***temperature*** than B.

When the heat flow from A to B is zero, the two objects are in ***thermal equilibrium*** and at the same temperature.

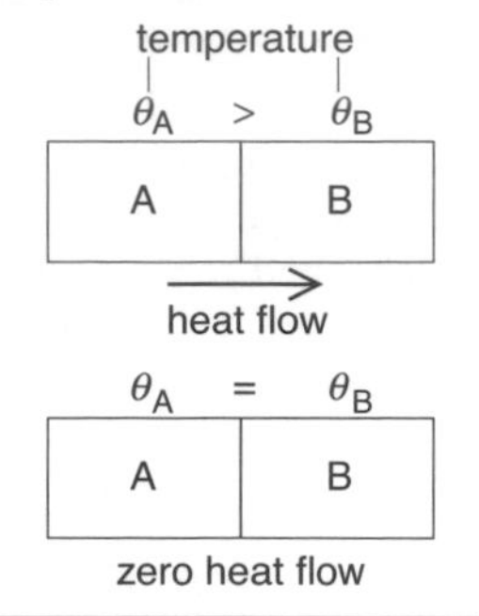

The zeroth law of thermodynamics states that if objects A and B are each in thermal equilibrium with an object C, then they are also in thermal equilibrium with each other.

Note:

- Any objects in thermal equilibrium are at the same temperature.
- Thermodynamics deals with the links between heat and other forms of energy. The zeroth law is so named because the first and second laws of thermodynamics (see I5) had already been stated when the need for a more basic law was realised.

Internal energy

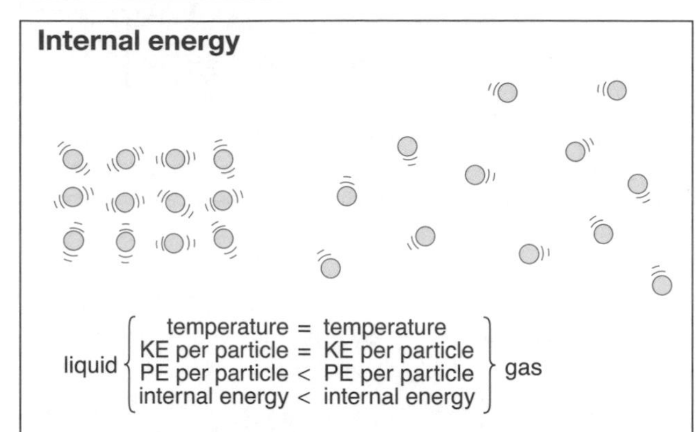

liquid { temperature = temperature; KE per particle = KE per particle; PE per particle < PE per particle; internal energy < internal energy } gas

In the liquid, above left, the particles (e.g. molecules) are in motion, so they have kinetic energy (KE). This motion has moved them apart, against the forces of attraction, so they also have potential energy (PE). The total of their KEs and PEs is the ***internal energy*** of the liquid.

Above right, the liquid has become a gas. The temperature is the same as before. So the average KE of each particle due to its linear motion is the same. However, the average PE is more because of the increased separation of the particles. The gas has more internal energy than the liquid.

Thermodynamic temperatures

The ***Kelvin scale*** is a thermodynamic scale, related to the average kinetic energy per particle (e.g. molecule).

On the Kelvin scale ***absolute zero*** is 0 kelvin (0 K). This is the temperature at which all substances have minimum internal energy. One advantage of using this scale is that the temperature in Kelvin is proportional to the average kinetic energy per particle.

Heat capacity

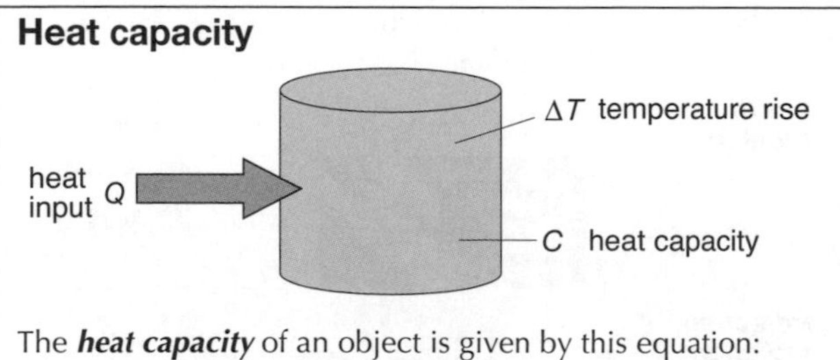

The ***heat capacity*** of an object is given by this equation:

$$\text{heat capacity} = \frac{\text{heat input}}{\text{temperature rise}} \qquad C = \frac{Q}{\Delta T}$$

For example, if a heat input of 4000 J causes a temperature rise of 2 K, then the heat capacity is 2000 J K^{-1}.

Specific heat capacity

The heat capacity per unit mass is called the ***specific heat capacity***. If a substance's specific heat capacity is c, then, for a mass m:

$$Q = mc\Delta T \qquad (1)$$

Water has a high specific heat capacity (4200 J kg^{-1} K^{-1}). This makes it a good 'heat storer'. A relatively large heat input is needed for any given temperature rise, and there is a relatively large heat output when the temperature falls.

Measuring *c* for a liquid (e.g. water) This can be done using the equipment below. The principle is to supply a measured mass of liquid with a known amount of heat from an electric heating coil, measure the temperature rise, and calculate c using equation (1).

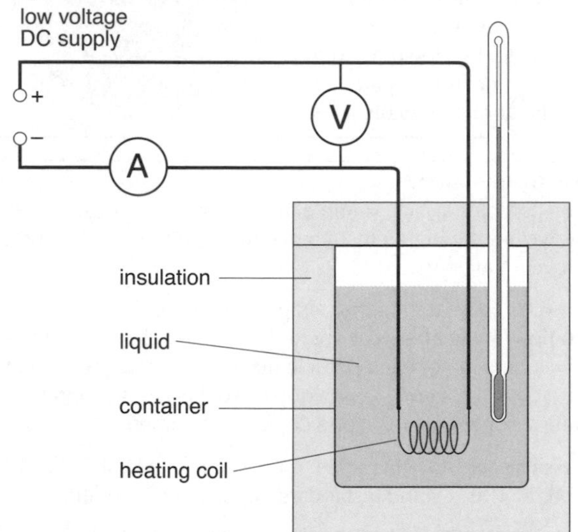

If the PD across the coil is V, and a current I passes for time t, then the electrical energy supplied = VIt (see D2).
If all this energy is supplied as heat, and none is lost:

$$VIt = mc\Delta T$$

Knowing the mass m and temperature rise ΔT of the liquid, its specific heat capacity c can be calculated.

Note:

- When the water heats up, its container does as well. For greater accuracy, this must be allowed for.
- Some heat is lost, despite the insulation. However, there are experiments in which, using different sets of results, heat losses can be eliminated from the calculation.

Measuring *c* for a solid (e.g. a metal) The method is essentially the same as that shown above, except that a solid block is used instead of the liquid. The block has holes drilled in it for an electric heater and a thermometer.

Copper has a ***specific heat capacity*** of 390 J kg^{-1} K^{-1}. This means that 390 J of energy are required to raise the temperature of 1 kg of copper by 1 K. To raise the temperature of 2 kg of copper by 10 K, the heat input required = 2 × 390 × 10 = 7800 J.

I2 Change of phase

Changing phase

The graph shows what happens when a very cold solid (ice) takes in heat at a steady rate. Melting and boiling are both examples of a change of ***phase***.

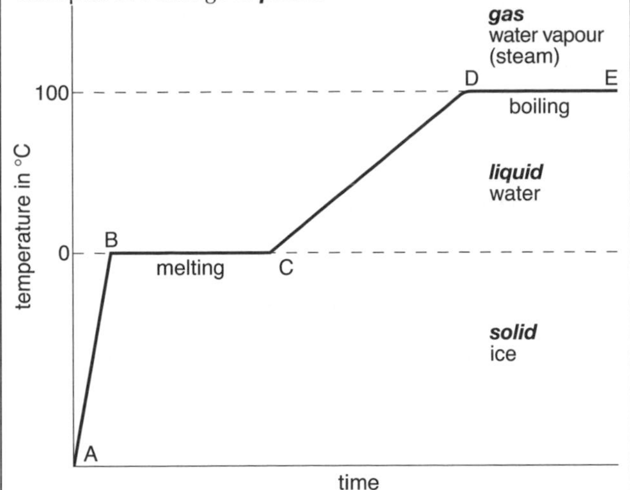

A to B The temperature rises until the ice starts to melt.

B to C Heat is absorbed, but with no rise in temperature. The energy input is being used to overcome the attractions between the particles as the solid changes into a liquid.

C to D The temperature rises until the water starts to boil.

D to E Heat is absorbed, but with no rise in temperature. The energy input is being used to separate the particles as the liquid changes into a gas (water vapour).

A liquid, such as water, starts to turn to gas well below its boiling point. This process is called ***evaporation***. It happens as faster particles escape from the surface.

Boiling is a rapid type of evaporation in which vapour bubbles, forming in the liquid, expand rapidly because their pressure is high enough to overcome atmospheric pressure.

The heat required to change a liquid into a gas (or a solid into a liquid) is called ***latent heat***. When water evaporates on the back of your hand, it takes the latent heat it needs from your hand. That is why there is a cooling effect.

Latent heat is released when a gas changes back into a liquid (or a liquid changes back into a solid).

Evaporation

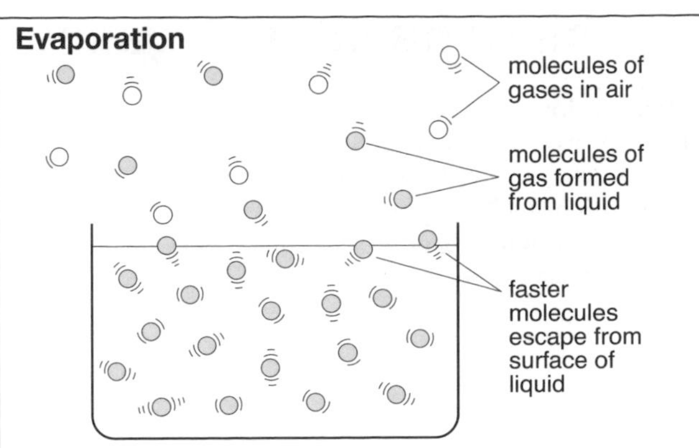

When a liquid evaporates, molecules escape from its surface and move about freely as a gas.

In a liquid, the vibrating molecules keep colliding with each other, some gaining kinetic energy and others losing it. At the surface, some of the faster, upward-moving molecules have enough kinetic energy to overcome the attractions from other molecules and escape from the liquid. With these faster molecules gone, the average KE of those left behind is reduced i.e. the temperature of the liquid falls. That is why evaporation has a cooling effect.

The rate of evaporation (and therefore the rate at which heat is lost from a liquid) is increased if:

- the surface area is increased (more of the faster molecules are near the surface),
- the temperature is increased (more of the molecules have enough kinetic energy to escape),
- the pressure is reduced (escaping molecules are less likely to rebound from other molecules back into the liquid),
- there is a draught across the surface (escaping molecules are removed before they can rebound),
- gas is bubbled through the liquid.

Specific latent heat of fusion

The ***specific latent heat of fusion*** of a substance is the heat which must be supplied per unit mass to change a solid into a liquid, without change in temperature.

If Q is the heat supplied, m is the mass, and l_f is the specific latent heat of fusion, then

$$Q = ml_f$$

The specific latent heat of fusion of water is 3.3×10^5 J kg^{-1} (about $\frac{1}{7}$ of its specific latent heat of vaporization).

Specific latent heat of vaporization

Energy is needed to change a liquid into a gas, even though there is no change in temperature (see I1).

- Most of the energy is needed as extra internal energy (separating the particles means giving them more PE).
- Some energy is needed to do work in pushing back the atmosphere (because the gas takes up more space).

The ***specific latent heat of vaporization*** of a substance is the heat which must be supplied per unit mass to change a liquid into a gas, without change in temperature.

If Q is the heat supplied, m is the mass, and l_v is the specific latent heat of vaporization, then

$$Q = ml_v \qquad (2)$$

The specific latent heat of vaporization of water is 2.3×10^6 J kg^{-1}. So, to turn 2 kg of water into water vapour (at the same temperature) would require 4.6×10^6 J of heat.

Measuring l_v This can be done using the equipment below. The principle is to supply boiling liquid (e.g. water) with a known amount of heat from an electrical heater, find the mass of vapour formed as a result, and calculate l_v using equation (2).

The vapour is cooled, condensed, and collected as a liquid so that its mass can be measured.

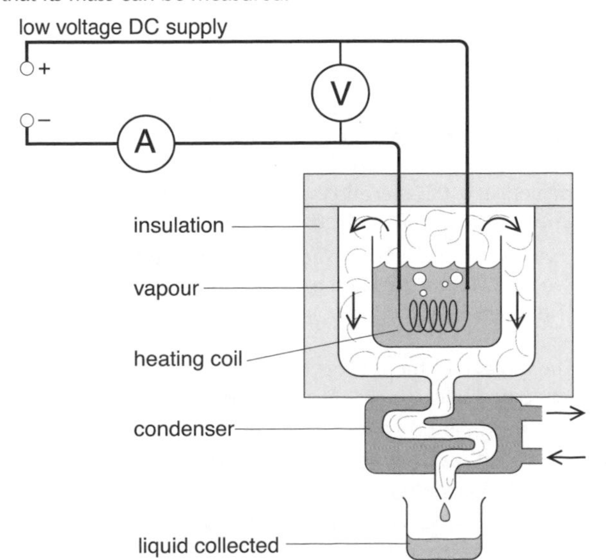

I3 The behaviour of gases

Boyle's law

Experiments show that, for a fixed mass of gas at constant temperature, the pressure p decreases when the volume V is increased. A graph of p against V for air is shown on the right.

According to ***Boyle's law***, for a fixed mass of gas at constant temperature,

pV = constant

If pV = constant, then $p \propto 1/V$. So a graph of p against $1/V$ is a straight line through the origin.

Note:

- The value of the constant depends on the mass of the gas and on its temperature. The dashed lines show the effects of raising the temperature.
- Under some conditions, the behaviour of real gases departs from that predicted by Boyle's law (see below).

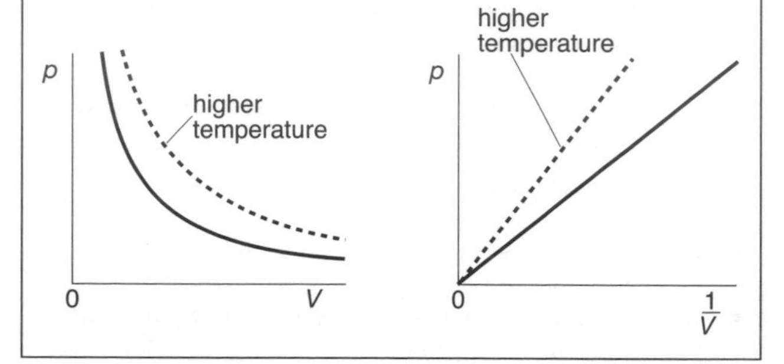

The pressure law

Before reading this panel, see I1 on temperature.

According to the ***pressure law***, for a fixed mass of gas at constant volume,

$\frac{p}{T}$ = constant

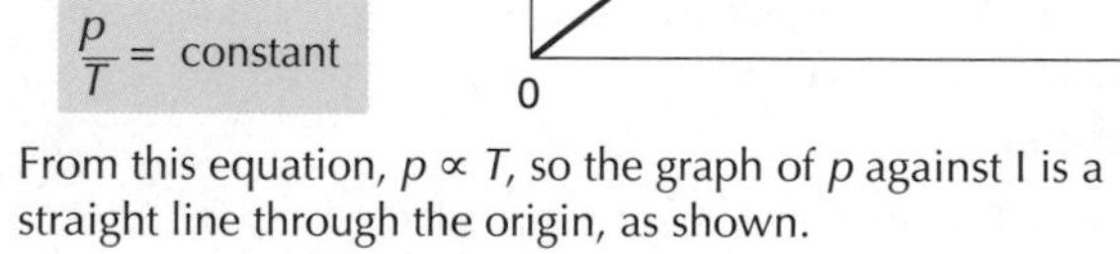

From this equation, $p \propto T$, so the graph of p against I is a straight line through the origin, as shown.

Note:

The pressure law predicts that the pressure of any ideal gas should be zero at absolute zero. This concept is used to define the zero point (0 K) on the Kelvin scale and to find its Celsius equivalent (–273.15 °C).

Real and ideal gases

Most common gases are made up of molecules. For convenience, in this unit and the two following, the particles of all gases will be called 'molecules', even if they are single atoms.

An ***ideal gas*** is one which exactly obeys Boyle's law. It can be shown (see I4) that, for such a gas:

- The forces of attraction between the molecules are negligible.
- The molecules themselves have a negligible volume compared with the volume occupied by the gas.

Ideal gases do not exist. However, real gases approximate to ideal gas behaviour at low densities and at temperatures well above their liquefying points.

Ideal gas behaviour is assumed in the rest of this unit.

Charles's law

For a fixed mass of gas at constant pressure, the volume V increases with the Kelvin temperature T, as on the right.

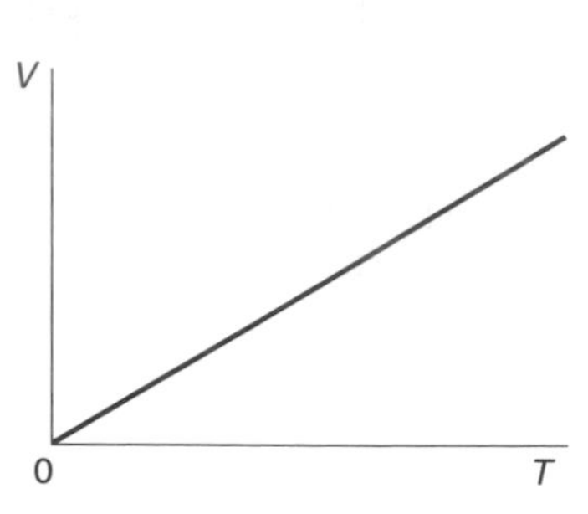

According to ***Charles's law***, for a fixed mass of gas at constant pressure,

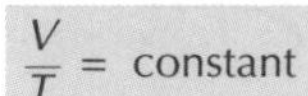

$\frac{V}{T}$ = constant

From this equation, $V \propto T$, which is why the graph is a straight line through the origin.

Note:

- Charles's law predicts zero volume at absolute zero. However no real gas behaves like an ideal gas at near-zero volume and temperature.

Equation of state

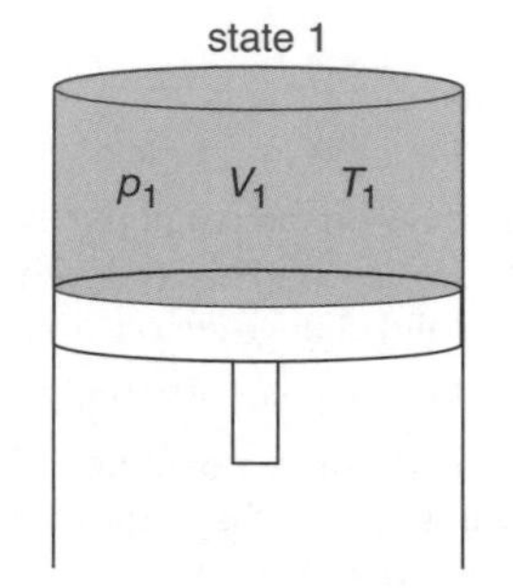

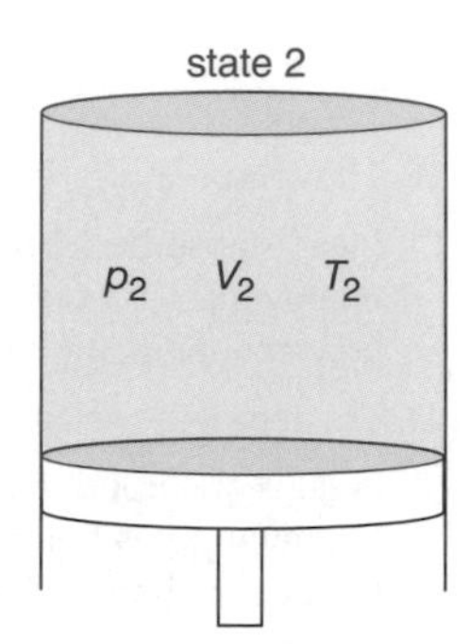

The three gas laws can be combined in a single equation.

If a fixed mass of gas changes from state 1 to state 2, at a different pressure, volume, and Kelvin temperature, as above, then

$$\frac{p_1 V_1}{T_1} = \frac{p_2 V_2}{T_2}$$

This is called the ***equation of state*** for an ideal gas.

Note, in the above equation:

- If $V_1 = V_2$, then $p_1 V_1 = p_2 V_2$. This is Boyle's law.
- If $V_1 = V_2$, then $p_1/T_1 = p_2/T_2$. This is the pressure law.
- If $p_1 = p_2$, then $V_1/T_1 = V_2/T_2$. This is Charles's law.

The ideal gas equation

From the equation of state, pV/T = constant. The constant can have different values depending on the type and mass of gas. However, if the amount of gas being considered is one mole (6.02×10^{23} molecules) (see A1), then the constant is the same for all gases:

for one mole of any gas $pV = RT$

R is called the ***universal molar gas constant***. Its value is 8.31 J mol^{-1} K^{-1}.

for n moles of any gas $pV = nRT$

Note:

- The number of moles $n = m/M$, where m is the mass of the gas and M is its ***molar mass*** (the mass per mole).

Molar masses of some common gases, in kg mol^{-1}	
hydrogen gas	2×10^{-3}
nitrogen gas	28×10^{-3}
oxygen gas	32×10^{-3}

I4 Kinetic theory

Molecules in motion

According to the ***kinetic theory***, matter is made up of randomly-moving particles (e.g. molecules). The following effects provide evidence to support this theory for gases:

Brownian motion Smoke from a burning straw is mainly oil droplets which drift through the air. When illuminated, these oil droplets are just big enough to be seen as points of light, but small enough to be affected by collisions with molecules in the air. Observed through a microscope, the droplets wander in random, zig-zag paths as they are bombarded by the molecules of the air around them. These random wanderings are called ***Brownian motion***.

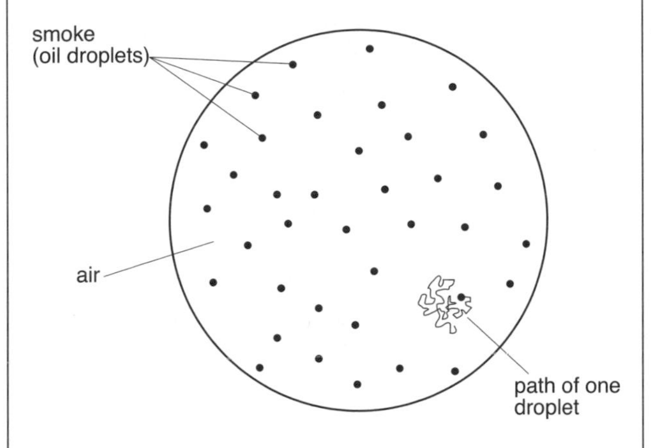

Diffusion If a phial of liquid bromine is broken in a tube of air (sealed for safety), brown bromine gas slowly spreads through the air. This spreading effect is called ***diffusion***. It happens because the bromine molecules keep colliding with the molecules of the air around them.

If there is no air in the tube (i.e. there is a vacuum), the bromine gas almost instantly fills the container when the phial is broken. This suggests that some bromine molecules travel at very high speeds.

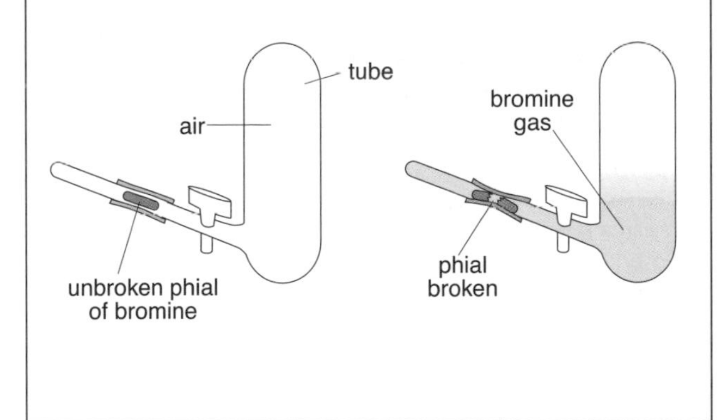

Size of a molecule

An estimate of molecular size can be obtained by putting a tiny drop of olive oil, of measured diameter, onto clean water, lightly covered with lycopodium powder. The oil spreads to form a very thin, circular film whose edge is made visible by the powder. Knowing the volume of oil, and the diameter of the film, the film's thickness can be calculated. Assuming that it is just one molecule thick, one molecule works out to be about 10^{-9} m. (Note: molecules vary considerably in size. Those in olive oil are relatively large.)

Kinetic theory for an ideal gas

The laws governing the behaviour of ideal gases can be deduced mathematically from the kinetic theory, as shown on the next page. In using the theory, the following assumptions are made:

- The motion of the molecules is completely random.
- The forces of attraction between the molecules are negligible.
- The molecules themselves have a negligible volume compared with the volume occupied by the gas.
- The molecules make perfectly elastic collisons (see I2) with each other and with the walls of their container.
- The number of molecules is so large that there are billions of collisions per second.
- Each collision takes a negligible time.
- Between collisions, each molecule has a steady speed.

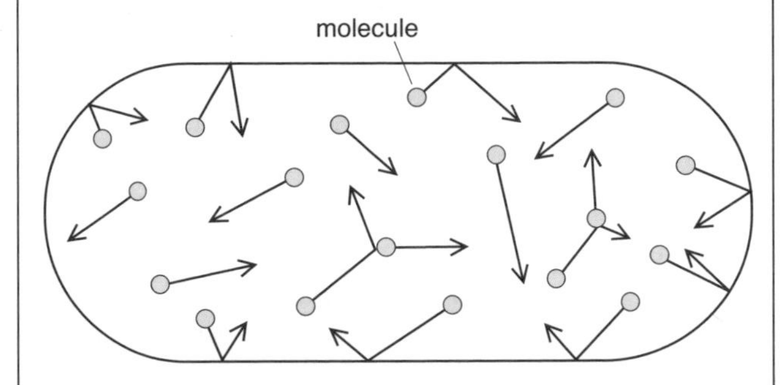

The diagram above shows a simple model of the moving molecules in a gas. The gas exerts a pressure on the walls of its container because its molecules are continually bombarding the surface and rebounding from it. The panel on the right shows how this pressure can be calculated.

Molecular speeds in a gas

In any gas, the molecules randomly collide with each other. In these collisions, some molecules gain energy (and therefore speed) while others lose it. As a result, at any instant, the molecules have a range of speeds, as shown in the distribution graph below.

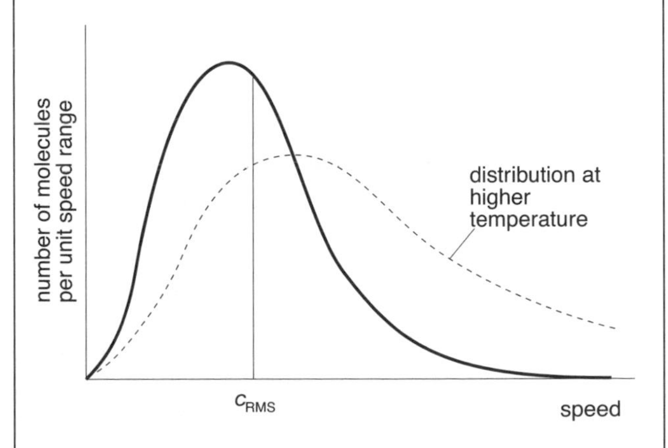

The temperature and the pressure of a gas depend, not on the average speed of the molecules, but on the average of (speed)^2, as explained on the next page. For this reason, it is useful to define the ***mean square speed*** of the molecules. This is the average of (speed)^2 for all the molecules. Its square root is called the ***root mean square speed***, or ***RMS speed***.

RMS speed is represented by the symbol c_{RMS}, or alternatively by $\sqrt{\overline{c^2}}$ or $\sqrt{\langle c^2 \rangle}$.

Pressure due to an ideal gas

The pressure of an ideal gas can calculated by considering the motion of its molecules. This has been done below for a gas in a spherical container, though the result applies for any shape. To begin with, it is assumed that all the molecules have the same speed, v.

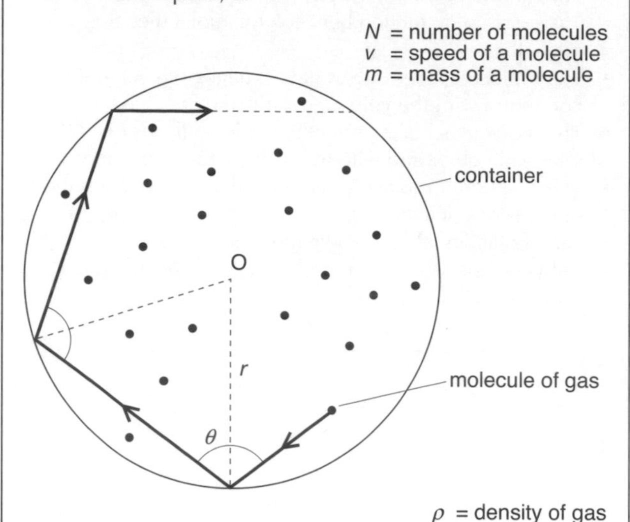

ρ = density of gas
V = volume of gas = $4\pi r^3/3$ (1)
surface area of container = $4\pi r^2$ (2)

Note: m has a different meaning from that used in F4.

The stages in the calculation are as follows:

- The labelled molecule above makes a series of collisions with the container, but always at the same angle θ. Just before each collison it has a component of momentum of $mv\cos\theta$ away from O. This is reversed by the collision. So, for each collision,

  change in momentum (away from O) = $2mv\cos\theta$ (3)

- For the labelled molecule, the distance between collisions is always $2r\cos\theta$. As time = distance/speed,

  time between collisions = $2r\cos\theta/v$ (4)

- The average force exerted by the molecule on the container can found by dividing the change in momentum (3) by the time (4) (see B10).

  This gives force due to one molecule = mv^2/r
  So total force due to all molecules = Nmv^2/r (5)

  The force does not depend on θ. A smaller angle gives a higher momentum change, but less frequent collisions.

- The pressure is found by dividing the force (5) by the area (2).

  This gives $p = Nmv^2/4\pi r^3$.
  Combining this with (1) gives $pV = Nmv^2/3$. (6)

- The pressure depends on v^2. But in reality molecules travel at a range of speeds, so v^2 should be replaced by the mean value of v^2 for all the molecules. This is $c_{RMS}{}^2$.
  So, equation (6) should be rewritten as follows:

  $$pV = \frac{1}{3}Nmc_{RMS}{}^2 \quad (7)$$

  For one mole of gas $pV = \frac{1}{3}N_A mc_{RMS}{}^2$ (8)

  where N_A is the Avogadro constant (6.02×10^{23}) (see A1).

- Nm is the total mass of gas. So Nm/V is its density. Equation (7) can therefore be rewritten

  $$p = \frac{1}{3}\rho\, c_{RMS}{}^2$$

Linking kinetic energy and temperature

In a gas, each molecule has kinetic energy because of its linear motion. This is called ***translational*** kinetic energy. Its average value depends on c_{RMS}. For simplicity, average translational kinetic energy will just be called KE.

So KE per molecule = $\frac{1}{2}mc_{RMS}{}^2$ (9)

This can be linked with the Kelvin temperature as follows.

- According to the ideal gas equation, for one mole of an ideal gas $pV = RT$ (see I3). This equation is used to define the ideal gas scale of temperature.
- Combining $pV = RT$ with (8) gives

  $$\frac{1}{3}N_A mc_{RMS}{}^2 = RT \quad (10)$$

- Combining the above equation with (9) gives

  $$\text{KE per molecule} = \frac{3}{2}\frac{R}{N_A}T$$

- R/N_A is the gas constant per molecule. Known as the ***Boltzmann constant***, k, its value is 1.38×10^{-23} J K^{-1}. Using k, the above equation can be rewritten:

  $$\text{KE per molecule} = \frac{3}{2}kT$$

$pV = nRT$, where n is the number of moles, can be written

$pV = NkT$ (the number of molecules $N = nN_A$)

Deducing Boyle's law

According to equation (7) $pV = \frac{1}{3}Nmc_{RMS}{}^2$

On the right-hand side of this equation:
- If the mass of gas is fixed, N is constant.
- If the temperature is steady, the KE per molecule (9) is constant, so $mc_{RMS}{}^2$ is constant.

Therefore, it follows that, for a fixed mass of gas at steady temperature, pV is constant. This is Boyle's law (see I3).

Deducing Avogadro's law

According to ***Avogadro's law***, equal volumes of ideal gases under the same conditions of temperature and pressure contain equal numbers of molecules.

This law can be deduced from equation (7). If two gases are at the same temperature, then $mc_{RMS}{}^2$ is the same for each. If they are at the same pressure and volume, then pV is the same for each. So, from equation (7), N must also be the same for each.

Calculating c_{RMS}

$N_A m$ is the molar mass M of a gas (see I3). So equation (10) can be rewritten

$$\frac{1}{3}Mc_{RMS}{}^2 = RT$$

Rearranged, this gives $c_{RMS} = \sqrt{3RT/M}$.

With R, T, and M known, c_{RMS} can be calculated. For example, for nitrogen (the main gas in air) at a room temperature of 300 K, c_{RMS} works out at 517 m s^{-1}.

I5 Internal energy, heat and work

The first law of thermodynamics

An object can be given more internal energy:

- by supplying it with heat,
- by doing work on it (i.e. by compressing it) (see B6).

If ΔU is the increase in internal energy when heat Q is supplied to an object *and* work W is done on it, then according to the ***first law of thermodynamics***:

$$\text{increase in internal energy} = \text{heat supplied to object} + \text{work done on object}$$

In symbols $\Delta U = Q + W$

Note:

- The joule is the unit of internal energy, heat, and work.

Converting work into heat

Work can be completely converted into heat. For example, if a gas is compressed and then left to cool to its original temperature, its internal energy is unchanged, so $\Delta U = 0$. If W is the work done *on* the gas, then from the first law of thermodynamics, $0 = Q + W$, so $-Q = W$.

Q is the heat supplied *to* the gas, so $-Q$ is the heat given out *by* the gas. It is equal to the work done *on* the gas.

High and low grade energy

Some forms of energy are more useful for doing work than others. For example, in a large electric motor, electrical energy can be converted into work with very high efficiency. Electrical energy is ***high grade energy***.

By comparison, internal energy is ***low grade energy***. The materials around us contain huge amounts of it, but it is unavailable for useful work unless the material is at a higher temperature than its surroundings. Equation (3) shows that the lower the temperature difference, the more unavailable the energy becomes – the more ***degraded*** it is. The waste heat from an engine is very low grade energy.

Heat pumps

In a refrigerator, heat is absorbed when a coolant evaporates, and given out at the back when the vapour is condensed by compression. Work is done by the compressor.

A refrigerator is a ***heat pump*** – a heat engine (see right) in reverse. Work is done *on* it in order to transfer heat from a *low* temperature source to a *higher* temperature sink.

Some heating systems use heat pumps. The building is heated by cooling the air, ground, or nearby stream outside. Much less energy is needed than is given out as heat. But the disadvantages are that (1) the local environment is affected and (2) the system works *less* well on a *cold* day.

The coefficient of performance for a refrigerator is:

$$COP_{ref} = \frac{Q_{out}}{W} = \frac{Q_{out}}{Q_{in} - Q_{out}}$$

The coefficient of performance for a heat pump is:

$$COP_{HP} = \frac{Q_{in}}{W} = \frac{Q_{in}}{Q_{in} - Q_{out}}$$

Converting heat into work

Petrol, diesel, jet engines, and the boilers-plus-turbines in powers stations are all ***heat engines***. They convert heat into work. But the process can never be 100% efficient. Some heat must always be wasted. This idea is expressed by the ***second law of thermodynamics***. This can be stated in several forms, one of which is as follows:

No continually working heat engine can take in heat and completely convert it into work.

Heat naturally flows from a higher to a lower temperature. So, without a temperature difference, there is no flow of heat. All heat engines take heat from one material at high temperature (e.g. a burning petrol-air mixture) and pass on less heat to a material at a lower temperature (e.g. the atmosphere). The difference is converted into work.

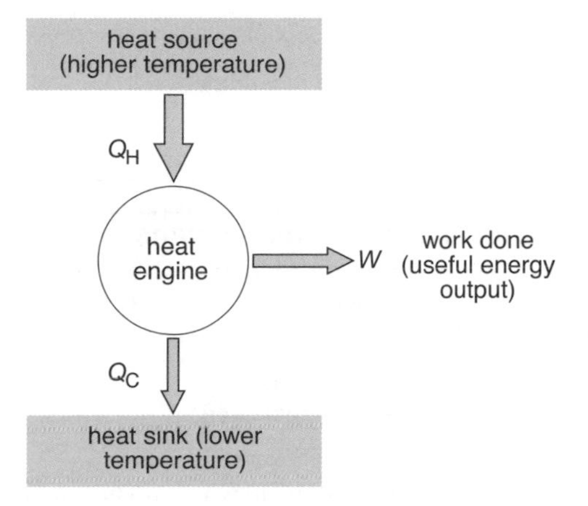

If an engine takes in heat Q_H from a ***heat source*** and puts heat Q_C into a ***heat sink***, then the work done $W = Q_H - Q_C$.

The engine's efficiency is calculated like this (see A2):

$$\text{efficiency} = \frac{\text{useful energy output}}{\text{energy input}} = \frac{W}{Q_H} = \frac{Q_H - Q_C}{Q_H}$$

An ideal heat engine is one which converts the maximum possible amount of heat into work, so there are no energy losses because of friction. For an engine like this, operating between a source temperature T_H (in K) and a sink temperature T_C, it can be shown that

$$\text{efficiency} = \frac{T_H - T_C}{T_H} \quad (3)$$

For an ideal heat engine, operating between, say, 1000 K (burning fuel) and 300 K (typical atmospheric temperature)

$$\text{efficiency} = \frac{1000 - 300}{1000} = 0.7, \text{ or } 70\%$$

Efficiencies of real engines are much less than this – for example, 30% for a petrol engine. So, in practice, engines waste more heat than they convert into work.

Work done by an expanding gas

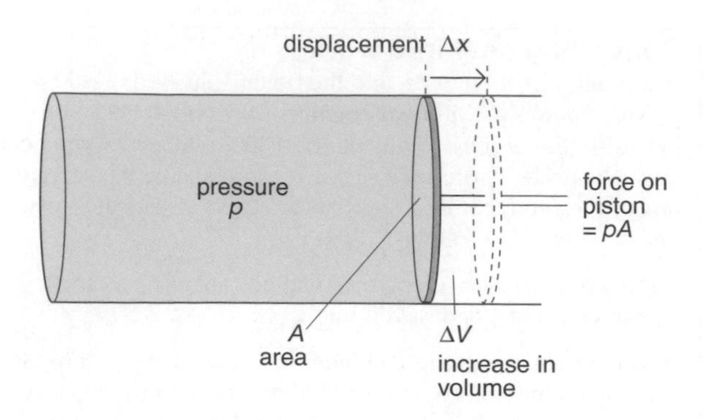

Above, a gas at pressure p exerts a force pA on the piston, and moves it a short distance Δx.
If the expansion of the gas is so small that the pressure does not change:

work done by gas = force × displacement = $pA\Delta x$

But $A\Delta x = \Delta V$, the increase in volume.

So work done by gas = $p\Delta V$

Note:

- For convenience, p has been called the pressure. Really, it is the pressure *difference* across the piston (i.e. gas pressure minus atmospheric pressure).
- If the volume of the gas is *decreased* by ΔV, then $p\Delta V$ is the work done *on* the gas.

The graph below left shows the expansion of a gas at constant pressure. The area under the graph gives the work done by the gas ($p\Delta V$). The same principle applies when the pressure is not constant, as shown below right.

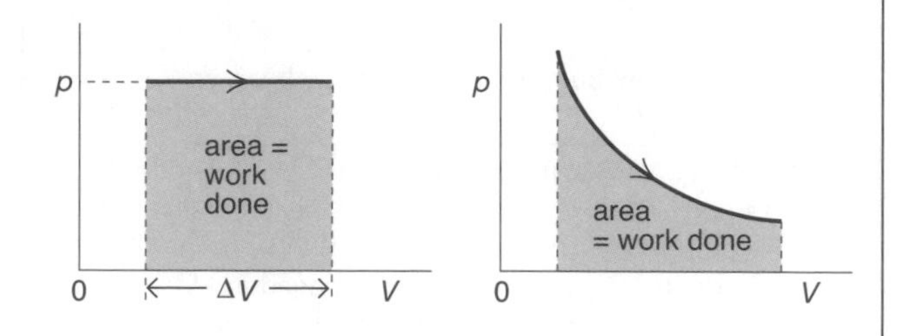

Adiabatic and isothermal expansion

According to the first law of thermodynamics, $\Delta U = Q + W$ (see left). However, when dealing with an expanding gas, it is useful to know that

Q = heat *taken in* by gas
$-\Delta U$ = *decrease* in internal energy
$-W$ = work done *by* gas

An ***adiabatic*** expansion is one in which no heat is taken in or given out, so $Q = 0$. If during an adiabatic expansion, a gas does work $-W$ it follows that $-W = -\Delta U$ (because $Q = 0$). So there is a decrease in internal energy equal to the work done. As a result, the temperature of the gas falls.

Note:
- Rapid expansions are adiabatic because the gas has negligible time to take in heat from its surroundings.
- Adiabatic *compression* produces a temperature *rise.*

An ***isothermal*** expansion is one in which the temperature is constant. There is no change in internal energy, so $\Delta U = 0$. As before, the gas does work $-W$. However, as $\Delta U = 0$, it follows that $-W = -Q$. So the gas takes in heat from it surroundings equal to the work done.

Note:
- For an isothermal expansion, a gas must stay in thermal equilibrium with its surroundings. In practice, this means a very slow expansion.
- During isothermal *compression,* a gas *gives out* heat.

Indicator diagrams

Pressure–volume graphs are called ***indicator diagrams.*** They can be used to show the cycle of changes taking place in an engine. The diagram below shows, in simplified form, what happens in a cylinder of a petrol engine, where there is compression and expansion of a gas as a piston goes up and down.

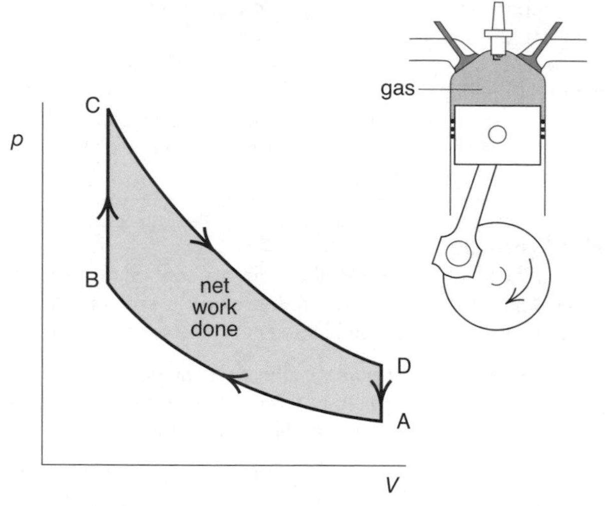

A to B Gas (air–petrol mixture) is compressed adiabatically by the rising piston. This causes a rise in temperature.
B to C Ignited by a spark, the mixture explodes. The further rise in temperature causes a further rise in pressure.
C to D The hot, high-pressure gas pushes down the piston as it expands adiabatically, and the temperature falls.
D to E The warm, waste gas is removed and replaced by cooler, fresh gas mixture, ready for the next cycle.

Note:
- From A to B, work is done *on* the gas. From C to D, work is done *by* the gas. The shaded area therefore represents the net work done during the cycle.

16 Thermal conductivity

Thermal conductivity

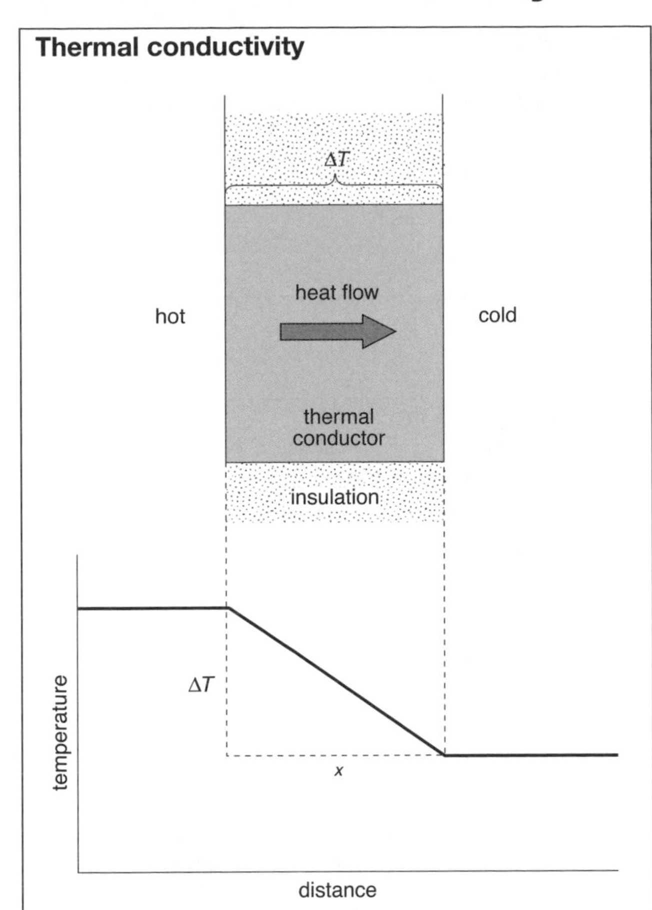

Above, there is a temperature difference ΔT across a block of material of thickness x. As a result, heat ΔQ flows through the material in time Δt. The flow is in the direction of *decreasing* temperature.

The graph shows how the temperature falls, through the block. $\Delta T/x$ is the temperature gradient. It is negative.

$\Delta Q/\Delta t$ is the rate of flow of heat. It is proportional to:

- the temperature gradient (a larger temperature difference or a thinner block give a greater heat flow),
- the cross-sectional area A (a larger area gives a greater heat flow).

The above principles can be expressed as an equation:

rate of flow of heat = $-k$ × area × temperature gradient $\qquad \dfrac{\Delta Q}{\Delta t} = -kA\dfrac{\Delta T}{x}$ (1)

k is called the ***thermal conductivity*** of the material. Some typical values of k are given below.

Note:

- In the above equations, the minus sign indicates that the heat flow is in the direction of *decreasing* temperature.
- Rate of flow of heat is the same as power. Its unit is the watt (W).
- As no heat escapes from the sides of the block, the rate of flow of heat is the same throughout.
- k is *defined* by the above equation. Good conductors have high k values. Good insulators have low k values.

Thermal conductivities, in W m^{-1}K^{-1}	
copper	400
aluminium	238
air (at normal temperature and pressure)	0.03

Thermal conduction through layers

On the right, a layer of brick is covered with insulating foam to reduce the heat flow. Knowing T_1 and T_2, x_1 and x_2, and k_1 and k_2, the temperature T at the boundary between the two materials can be found, and also the rate of flow of heat.

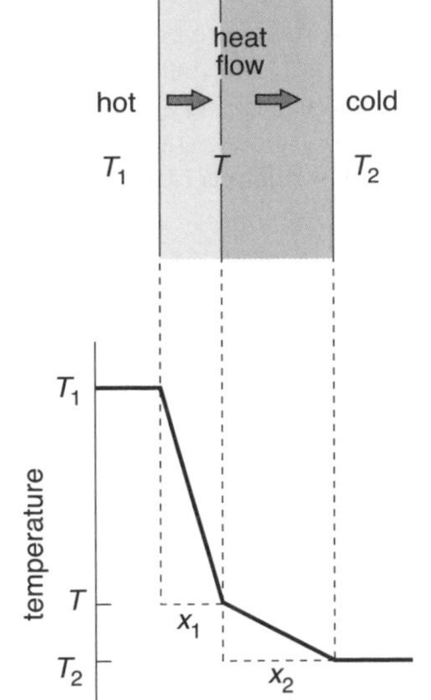

As no heat escapes from the sides, the rate of flow of heat, $\Delta Q/\Delta t$, must be the same through both layers. So, from (1),

$$\frac{\Delta Q}{\Delta t} = k_1 A\frac{(T_1 - T)}{x_1}$$

$$= k_2 A\frac{(T - T_2)}{x_2}$$

T can be found using the right-hand parts of the above equations, and rearranging. With T known, either of these parts gives the value of $\Delta Q/\Delta t$.

Note:

- In the above example $k_2 > k_1$. For the *same* rate of flow of heat, the material with the *lower* k must have the *higher* temperature gradient.

U-values

Heating engineers use ***U-values***, rather than k values, when calculating heat losses through walls, windows, and roofs. A U-value is defined by the following equation:

rate of flow of heat = U-value × area × temperature difference

Using the symbols in the panel on the left,

$$\frac{\Delta Q}{\Delta t} = \text{U-value} \times A\,\Delta T \qquad (2)$$

From (1) and (2), it follows that, for a material of thermal conductivity k, the U-value = k/x. So, unlike k, the U-value depends on thickness. For good insulation, a low U-value is needed. The requirements for this are a low k and a high thickness. Here are some typical U-values:

U-values in W m^{-2} K^{-1}	
single brick wall	3.6
double brick wall with air space	1.7
window, single glass layer	5.7
double-glazed window	2.7

How Science Works

Building design

By understanding how different materials lose heat, decisions can be made about how buildings (for example, houses) should be designed or modified to reduce wasted energy.

I7 Thermal radiation

Thermal radiation

Vibrating and spinning molecules in one object give off electromagnetic radiation whose energy can be absorbed by molecules in another object so that they speed up. This radiation is called ***thermal radiation***. From most warm or hot objects, it is mainly infrared (see D1).

Some surfaces are better absorbers of thermal radiation than others (see I1). A perfect absorber (i.e. one which reflects no radiation) is called a ***black body***. It is also the best possible emitter of thermal radiation. The Sun, odd though it may sound, is effectively a black body radiator.

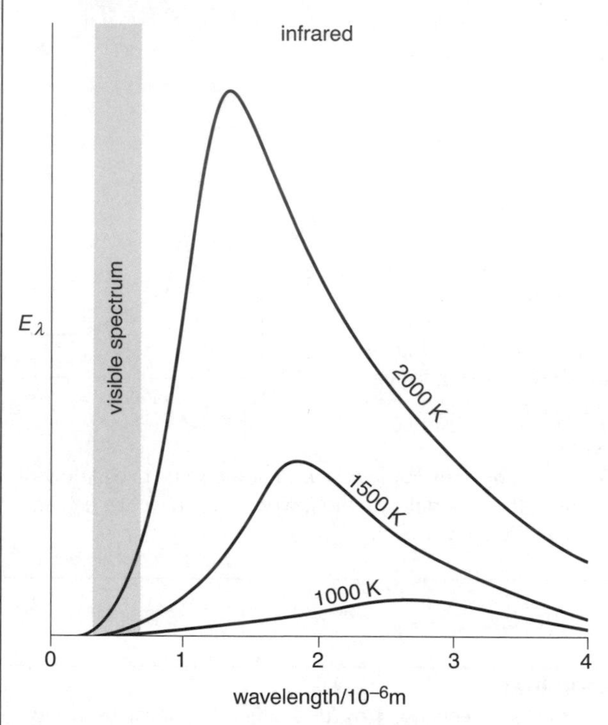

The graph above shows how radiated energy is distributed across different wavelengths for a black body radiator at various temperatures. The E_λ axis represents the relative energy output per second per unit range of wavelength.

Note:

- As the temperature increases, the total energy output per second increases (in proportion to T^4).
- As the temperature increases, the peak wavelength becomes less. By 1000 K some of the radiation has reached the red end of visible spectrum, so the object is glowing red hot. When the peak is within the visible spectrum, the object is glowing white hot.

For a radiator which is not a black body, the lines of the graph are of a similar form, but the peaks are lower.

Note: the shape of this curve cannot be predicted from classical physics. This is called the ***ultraviolet catastrophe***. Quantum theory was proposed by Planck to solve this problem (see H12).

Wien's law

In the diagram at left, the peak wavelength, λ_{max}, of the curve moves to a smaller wavelength as the temperature T increases.

According to Wien's law:

$\lambda_{max} T = 2.898 \times 10^{-3}$ m K

Note: T must be in Kelvin.

Luminosity and flux

The luminosity of a source of radiation is the rate at which it radiates energy. It is measured in W.

The energy spreads out in all directions, so at a distance d from the source, the energy will be spread over the surface area of a sphere of radius d.

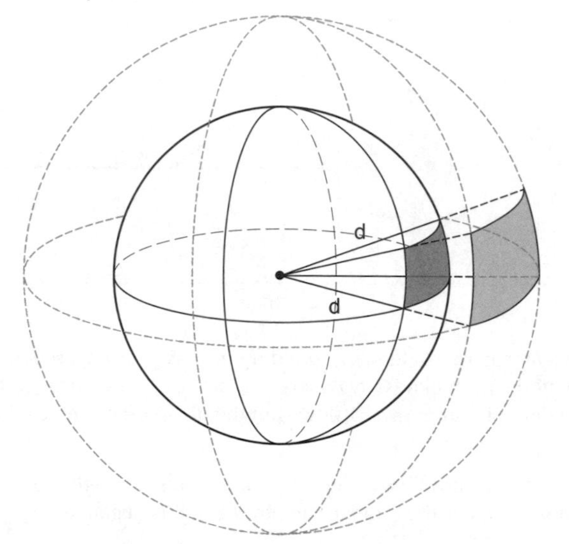

$$\text{The flux} = \frac{\text{total energy radiated by the source per second}}{\text{surface area of sphere}}$$

$F = \dfrac{L}{4\pi d^2}$ flux is measured in Wm^{-2}

The Stefan-Boltzmann law

For a black body, the relationship between the luminosity and the temperature is:

$L = \sigma T^4 \times$ surface area

For a sphere, radius r: $L = 4\pi r^2 \sigma T^4$

The Stefan-Boltzmann constant, $\sigma = 5.67 \times 10^{-8}$ W m^{-2} K^{-4}

J1 The structure of the atom

Atoms

All matter is made from ***atoms***. It would take more than a million million atoms to cover this full stop.

An atom has a tiny central ***nucleus*** made of ***protons*** and ***neutrons*** (apart from the simplest atom, hydrogen, whose nucleus is a single proton). Orbiting the nucleus are much lighter particles called ***electrons***.

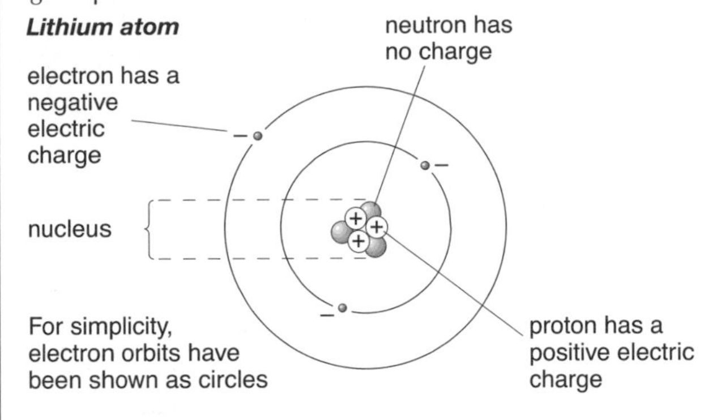

An atom has the same number of electrons as protons, so the amounts of negative and positive charge balance.

Unlike charges (– and +) attract each other. This ***electric force*** holds electrons in orbit around the nucleus.

Like charges (– and –, also + and +) repel each other. However, the particles in the nucleus are held together by a ***strong nuclear force***, which is strong enough to overcome the repulsion between the protons.

Atoms can stick together, in solids for example. The forces that bind them are attractions between opposite charges.

Moving electrons In metals, some of the electrons are only loosely held to their atoms. These ***free electrons*** can drift between the atoms. The electric current in a wire is a flow of free electrons.

If an atom gains or loses electrons, it is left with an overall – or + charge. Charged atoms are called ***ions***.

Early models of the atom

In 1810 the model of the atom was John Dalton's '***Billiard Ball model***'. According to this model, the atom was the smallest piece of an element that was possible. It was imagined to be a small solid sphere.

In this model, atoms
- of an element have identical masses
- of different elements have different masses
- combine only in small, whole number ratios (for example, 1:1, 1:2, 2:3)
- cannot be created or destroyed.

In 1897 Joseph John Thomson used electric and magnetic fields to deflect electron beams (see H11) and showed that the electron was a negative particle.

He measured a value for e/m. This led to the '***Plum Pudding model***' of the atom. According to this model, the atom was a positive material, or 'pudding', in which the electrons were negative particles, or 'plums'. Thomson imagined them circulating inside the positive material. The bullet points above were still part of the new model.

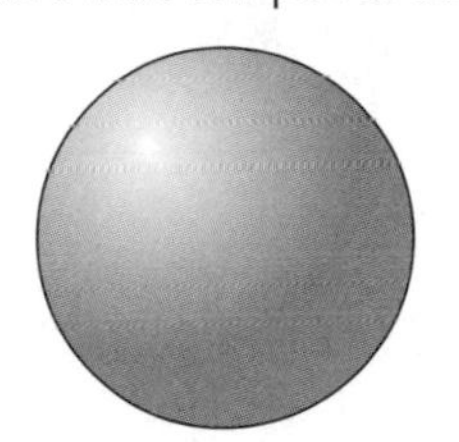

The Billiard Ball model

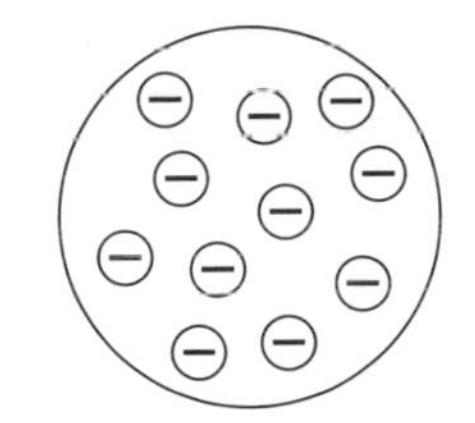

The Plum Pudding model

Evidence for a nucleus

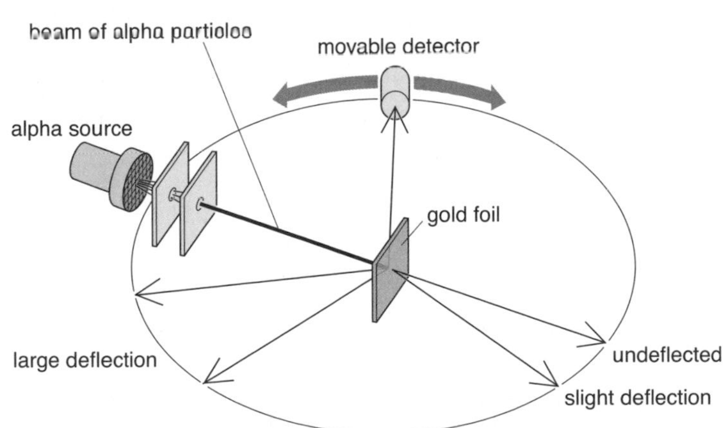

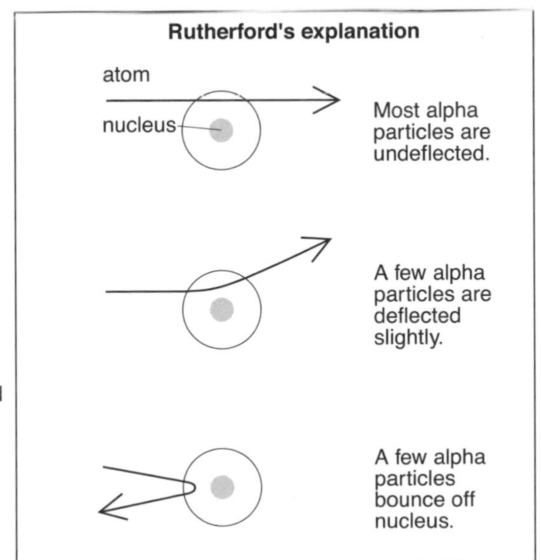

This experiment was carried out by Hans Geiger and Ernest Marsden under the supervision of Ernest Rutherford.

A thin piece of gold foil was bombarded with alpha particles, which are positively charged (see J3). Most of the alpha particles passed straight through the gold atoms. But a few were repelled so strongly that they bounced back or were deflected through large angles. These results came as a big surprise, and in 1911 Rutherford proposed a new model, the '***Nuclear model***', to take account of them. According to this model, virtually all the mass of the atom is concentrated at the centre in a positively charged nucleus, with much lighter negatively charged electrons in orbit around it. Rutherford calculated the scattering this would cause and showed that the model was consistent with the experimental results.

How Science Works

Recent models of the atom

In 1913 the '***Rutherford–Bohr***' model took account of the work of Henry Moseley, who showed that the atomic number corresponded to the number of positive charges (protons) in the nucleus, and the work of Niels Bohr, who suggested that the electrons have certain allowed circular orbits, or shells.

Since then changes to the model have included

- the 'wave mechanical' model, in which the electrons set up stationary waves around the nucleus (see J2)
- the idea of neutrons in the nucleus, following the discovery of the neutron by James Chadwick in 1932
- the idea that neutrons and protons are made of quarks
- the existence of other fundamental particles (see K2).

Our understanding of the atom is an example of the tentative nature of scientific knowledge. The model we use is continually being refined as we discover more about the atom and its properties. We can expect that this process may continue in the future.

Atomic measurements

mass of proton	1.673×10^{-27} kg
mass of neutron	1.675×10^{-27} kg
mass of electron	9.110×10^{-31} kg
charge on proton	$+1.60 \times 10^{-19}$ C
charge on electron	-1.60×10^{-19} C
diameter of an atom	$\sim 10^{-10}$ m
diameter of a nucleus	$\sim 10^{-14}$ m

Note:

- The proton and neutron have approximately the same mass – about 1800 times that of the electron.
- ~ means 'of the order of' i.e. 'within a factor ten of'.
- The diameter of an atom is $\sim 10^4$ times that of its nucleus. (Atom size varies from element to element.)
- Confusingly, the symbol *e* may be used to represent the charge on an electron (–) or a proton (+). In this unit, the charge on an electron will be called –*e*.

Elements, nuclides, and isotopes

Everything is made from about 100 substances called ***elements***. For most elements, a sample contains a number of different versions, called ***isotopes***. These have the same number of protons (and electrons), but different numbers of neutrons. This table shows some examples (italic numbers are for rarer isotopes).

Element	Electrons	Protons	Neutrons
hydrogen	1	1	0 or *1* or *2*
helium	2	2	*1* or 2
lithium	3	3	*3* or 4
carbon	6	6	6 or *7* or *8*
uranium	92	92	*142* or *143* or 146

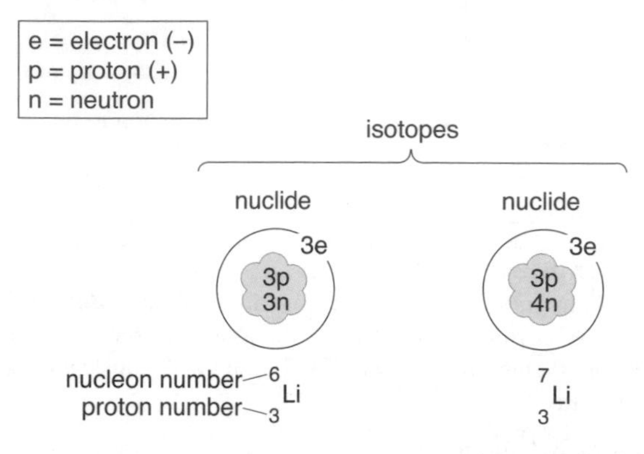

Nuclide This is any particular version of an atom. Above are simple models of the two naturally occurring nuclides of lithium, along with the symbols used to represent them.

Nucleon number A As protons and neutrons are called ***nucleons***, this is the total number of protons plus neutrons in the nucleus. It was once called the ***mass number***. It is used when naming different isotopes – for example, carbon-12, carbon-13, carbon-14.

Proton number Z This is the number of protons in the nucleus (and therefore the number of electrons in a neutral atom). It was once called the ***atomic number***.

Isotopes These are atoms with the same proton number but different nucleon numbers. They have the same electron arrangement and, therefore, the same chemical properties.

The following statements illustrate the meanings of the terms *element*, *nuclide*, and *isotope*.

- Lithium is an element.
- Lithium-6 is a nuclide; lithium-7 is a nuclide.
- Lithium-6 and lithium-7 are isotopes.

Note:

- A nuclide is commonly referred to as 'an isotope', though strictly speaking, this is incorrect.

Radioactive isotopes These have atoms with unstable nuclei. The nuclei break up, emitting ***nuclear radiation***. The three main types of nuclear radiation are ***alpha*** particles, ***beta*** particles, and ***gamma*** waves (see D1).

J2 Electrons in atoms

Electron stationary waves in atoms

Matter waves can set up stationary waves in a similar way to those set up in a stretched string. This formed the basis of Erwin Schrödinger's use of wave mechanics to predict the energy levels in a hydrogen atom.

This is a simple treatment that outlines the principle of the process in one dimension as follows. It treats the electron as if it were trapped in a box of size equal to the atomic radius.

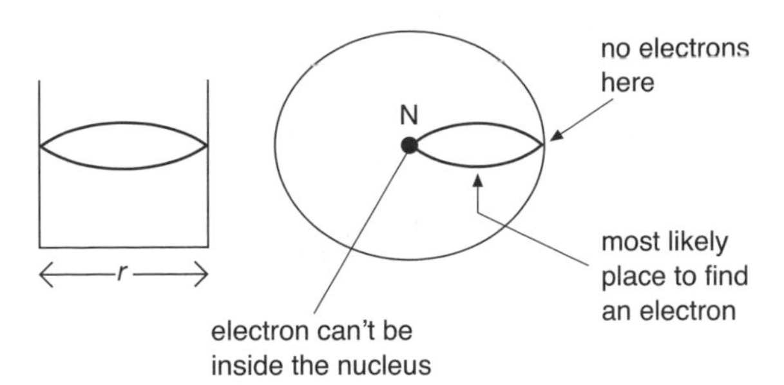

Consider the following 'facts':

- The electron in an atom cannot be in the nucleus or outside the edge so at the nucleus and the edge the amplitude of the wave is zero.
- The electron wave is confined between these points, reflecting backwards and forwards to produce the standing wave.
- The nucleus and the edge of the atom are the nodes of the stationary wave.
- The distance between the nucleus and the edge of the atom is about 1×10^{-10} m.

The possible stationary waves for the electron are as follows:

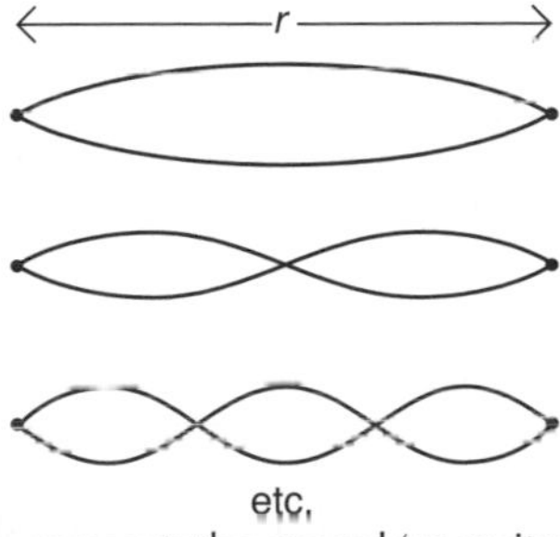

The single loop represents the ground (unexcited) state of the electron. The model suggests that the electron is most likely to be found half way between the nucleus and the edge of the atom.

Energy levels in atoms: excited states

This simple model cannot accurately predict energy levels as no account is taken of the changes in potential energy of the electron at different distances from the nucleus. However, the following shows how one can begin to see why there are discrete energies for the electron.

For one loop wavelength $= 2r = \frac{2r}{1}$

For two loops wavelength $= r = \frac{2r}{2}$

For three loops wavelength $= \frac{2r}{3}$

etc.

Generally wavelength $= \frac{2r}{n}$

where n is the number of loops in the standing wave.

The possible momenta of the electron $= \frac{nh}{2r}\left(\text{i.e. } \frac{h}{\lambda}\right)$

Electron collisions with atoms

Electrons can collide with atoms elastically or inelastically.

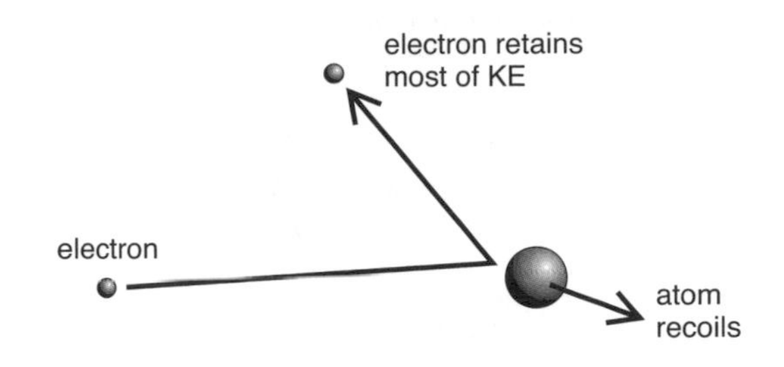

In the elastic collisions (as in all such collisions) kinetic energy is conserved. Since the electron is so much less massive than any atom the electron effectively bounces off the atom, giving up little of its energy to the recoiling atom.

Kinetic energy may be lost in inelastic collisions because of

- ionization, in which some of the energy is used to remove an electron from the atom,

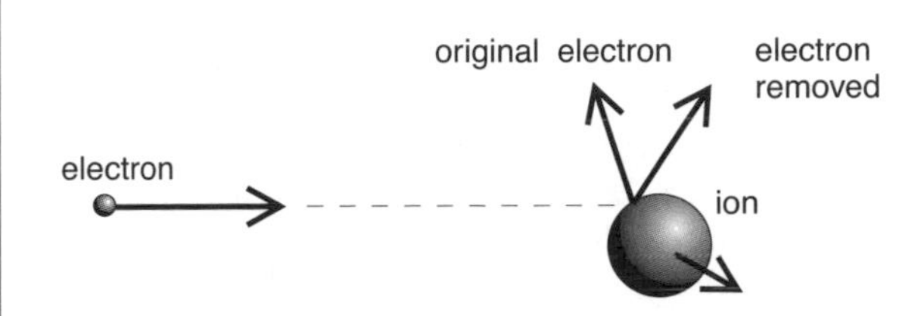

- excitation, in which some energy is used to raise an atomic electron from a ground state to an excited state.

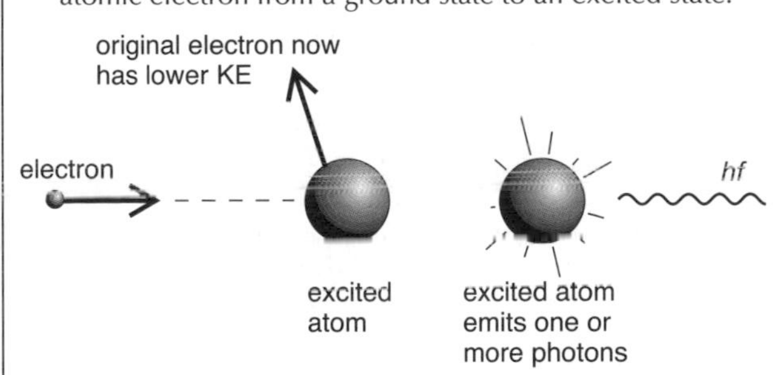

By examining the energy of an electron before and after a collision it is possible to determine the energy levels that an electron can occupy in a particular atom. The loss in energy of an electron in one collision corresponds to an energy difference between the ground state and an excited state.

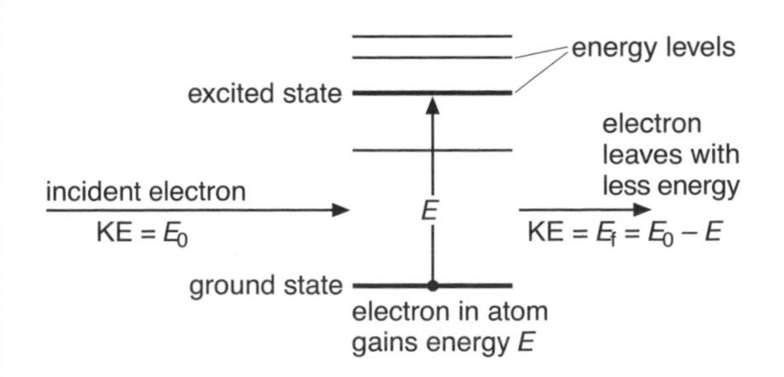

The fact that colliding electrons lose energy in well defined amounts is further evidence for the existence of electron energy levels in atoms.

J3 Radioactivity

Stability of the nucleus

If the number of neutrons ($A - Z$) in the nucleus is plotted against the number of protons (Z) for all known nuclides, the general form of the graph is like this:

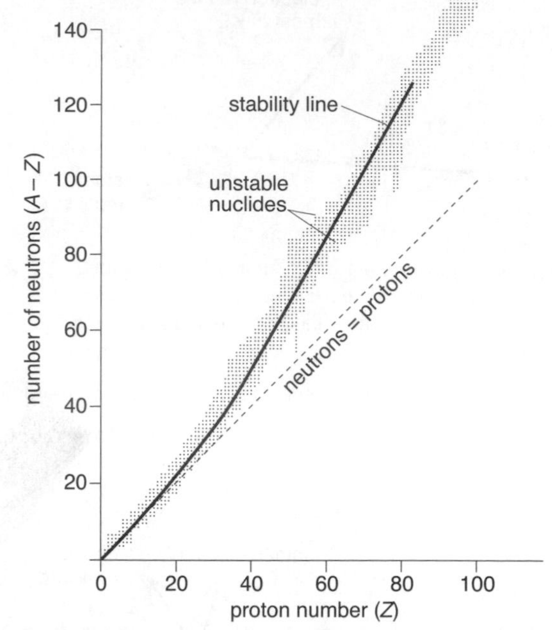

Stable nuclides These have nuclei which are stable. They occur along the solid line.

- Light nuclides (those at the lower end of the graph) have about equal numbers of protons and neutrons.
- The heaviest nuclides have about 50% more neutrons than protons.
- The most stable nuclides tend to have an even number of protons and an even number of neutrons. This is because each group of 2 protons and 2 neutrons in the nucleus makes an especially stable combination.

Unstable nuclides These occur either side of the solid line. They have unstable nuclei which, in time, *disintegrate* (break up) – usually by emitting an ***alpha*** particle or ***beta*** particle and maybe ***gamma*** radiation as well (see also J4). The disintegration is called ***radioactive decay***.

Element	Symbol	Z
hydrogen	H	1
helium	He	2
lithium	Li	3
beryllium	Be	4
boron	B	5
carbon	C	6
nitrogen	N	7
oxygen	O	8
fluorine	F	9

Element	Symbol	Z
neon	N	10
sodium	Na	11
iron	Fe	26
cobalt	Co	27
nickel	Ni	28
protactinium	Pa	91
uranium	U	92

Alpha (α) decay

An alpha (α) particle consists of 2 protons and 2 neutrons, so it has a charge of $+2e$. It is identical to a nucleus of helium-4 and may be represented by either of these symbols:

$${}^{4}_{2}\alpha \qquad {}^{4}_{2}\mathrm{He}$$

If an atom emits an α particle, its proton number is decreased, so it becomes the atom of a different element. For example, an atom of radium-226 emits an α particle to become an atom of radon-222, as shown by this equation:

$${}^{226}_{88}\mathrm{Ra} \rightarrow {}^{222}_{86}\mathrm{Rn} + {}^{4}_{2}\alpha$$

radium-226   radon-222   α particle

Note:

- Radon-222 and the α particle are the ***decay products***.
- The nucleon numbers on both sides of the equation balance ($226 = 222 + 4$) because the total number of protons and neutrons is conserved (unchanged).
- The proton numbers balance ($88 = 86 + 2$) because the total amount of positive charge is conserved.
- Alpha decay tends to occur in heavy nuclides which are below the stability line (see above graph) because it produces a nuclide which is closer to the line.

Beta (β) decay

β^- decay This is the most common form of beta decay. The main emitted particle is an electron. It has a charge of $-1e$ and may be represented by either of these symbols:

$${}^{0}_{-1}\beta \qquad {}^{0}_{-1}\mathrm{e}$$

Note:

- The beta particle is not a nucleon, so it is assigned a nucleon number of 0.
- The 'proton number' is –1 because the beta particle has an equal but opposite charge to that of a proton.

During β^- decay, a neutron is converted into a proton, an electron, and an almost undetectable particle with no charge and near-zero mass called an antineutrino. The electron is emitted, along with the antineutrino. For example, an atom of boron-12 emits a β^- particle to become an atom of carbon-12, as described by this equation:

$${}^{12}_{5}\mathrm{B} \rightarrow {}^{12}_{6}\mathrm{C} + {}^{0}_{-1}\mathrm{e} + {}^{0}_{0}\bar{\nu}$$

boron-12   carbon-12 β^- particle   antineutrino

Note:

- The nucleon numbers balance on both sides of the equation. So do the proton numbers.
- β^- decay tends to occur in nuclides above the stability line (see graph at top of page).

β^+ decay Here, the main emitted particle is a ***positron***, with the same mass as an electron, but a charge of $+1e$. It is the ***antiparticle*** of an electron. For example, an atom of nitrogen-12 emits a β^+ particle to become an atom of carbon-12, as described by the following equation:

$${}^{12}_{7}\mathrm{N} \rightarrow {}^{12}_{6}\mathrm{C} + {}^{0}_{+1}\mathrm{e} + {}^{0}_{0}\nu$$

boron-12   carbon-12 β^+ particle   neutrino

Other types of decay

Electron capture

A proton in the nucleus captures an orbiting electron and becomes a neutron. The energy lost by the electron is emitted as an X-ray.

Transmutation

One element changes into another. This can occur when atoms are bombarded by other particles. For example, if a high-energy α particle strikes and is absorbed by a nucleus of nitrogen-14, the new nucleus immediately decays to form a nucleus of oxygen-17 and a proton. This is an example of a ***nuclear reaction***. It can be described by the following equation:

$${}^{14}_{7}\mathrm{N} + {}^{4}_{2}\alpha \rightarrow {}^{17}_{8}\mathrm{O} + {}^{1}_{1}\mathrm{p}$$

nitrogen-14   α particle   oxygen-17   proton

Note:

- The nucleon numbers balance on both sides of the equation. So do the proton numbers.

J4 Radioactive decay

Properties of alpha particles, beta particles, and gamma radiation

Type	α	β	γ
nature	2 p + 2 n	e	electromagnetic (see C1 and C2)
charge	+ 2*e*	− 1*e*	no charge
speed (typical) (*c* = speed of light)	0.1*c*	up to 0.9*c*	*c*
energy (typical)	10 MeV	0.03 to 3 MeV	1 MeV
ionizing effect: ion pairs per mm in air	~10^5	~10^3	~1
penetration (typical)	stopped by: 30-50 mm air a sheet of paper	stopped by: 3-5 mm aluminium	intensity halved by 100 mm lead
effect of magnetic field (*B* out of paper) *not to scale*	+	slow fast −	(undeflected)

Note:

- α, β, and γ emissions all cause ***ionization*** – they remove electrons from atoms (or molecules) in their path. The removed electron (–) and the charged atom or molecule (+) remaining are called an ***ion pair***. The ionized material can conduct electricity.
- α particles interact the most with atoms in their path, so they are the most ionizing and the least penetrating.
- Unlike α particles, β particles are emitted from their source at a range of speeds.
- For γ radiation emitted from a point source in air, intensity ∝ 1/(distance from source)^2 (see D2).
- γ radiation is not stopped by an absorber, but its intensity is reduced.
- α and β particles are deflected by magnetic fields, as predicted by Fleming's left-hand rule (see H7 and H11). They are also deflected by electric fields.

Detecting alpha, beta, and gamma emissions

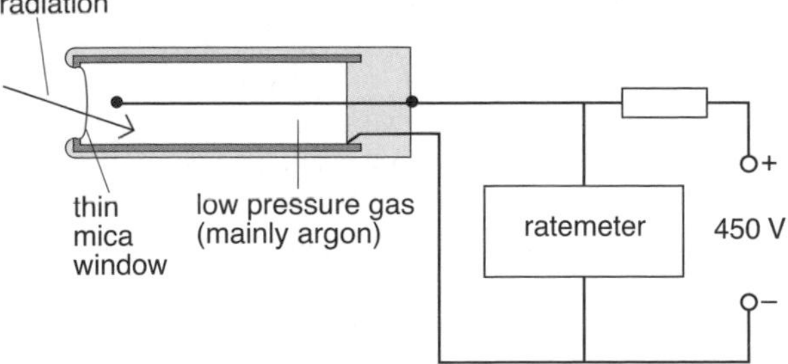

Detectors In the ***G–M tube*** (Geiger–Müller tube) above, a high voltage is maintained across the gas. When, say, a beta particle enters through the thin window, it ionizes the gas and makes it conduct. This causes a pulse of current in the circuit. The ***ratemeter*** registers the ***count rate*** (average number of pulses per second).

In a ***solid state detector***, incoming radioactive emissions ionize the semiconductor material in a diode (see F1 and F2).

Telling α, β and γ apart To tell one type of radiation from another, absorbing materials of different thicknesses can be placed between the source and the detector. For example, a thick aluminium plate will stop α and β particles, but not γ radiation.

Background radiation

Note: although alpha and beta emissions are particles, they are often referred to as radiation along with gamma radiation. This can be confusing.

There is a low level of ionizing radiation all around us. Its sources include radioactive materials present in rocks and soil, and cosmic radiation (high-energy particles from space). This is called ***background radiation***. It includes all the ionizing radiation present (except ultraviolet photons):

- α, β, and other high-energy particles
- γ and X-rays (photons with higher energy than ultraviolet).

In radioactivity experiments, allowance must be made for the background radiation at that location.

Note: in cosmology (see L5), *background radiation* refers to the microwave background radiation that is all around us as a result of the Big Bang. So if you are asked about background radiation you must take into account the context of the question. Radiation from mobile phones and other communication signals are also all around us, but they are in the microwave or radio wave part of the spectrum, so are not sources of background radiation.

Activity

The ***activity*** of a radioactive source is the number of disintegrations occurring within it per unit time.

The SI unit of activity is the ***becquerel*** (**Bq**):

1 becquerel = 1 disintegration s^{-1} (1 Bq = 1 s^{-1})

The activity of a typical laboratory source is ~ 10^4 Bq.

Each disintegration produces an α or β particle and, in many cases, γ radiation as well. The γ radiation is emitted as a 'packet' of wave energy called a ***photon*** (see E1). Particles and photons cause pulses in a detector, so the count rate is a measure of the activity of the source.

Note:

- The activity of a source is unaffected by chemical changes or physical conditions such as temperature. However, it does decrease with time (see next page).

Irradiation and contamination

If the body was ***irradiated*** with α, β, and γ emissions:

- α particles would not penetrate the skin (they could cause skin cancer, but would not affect the internal organs)
- β radiation would penetrate the whole body and could cause damage to any organ; however, most of the gamma radiation would pass straight through the body without interacting with it.

If the body was ***contaminated*** with an α-, β-, or γ-emitting material (for example, by swallowing or breathing in dust or radon gas):

- α particles are the most ionizing and would cause most damage (for example, lung cancer if breathed in).

For more details of radiation dose and effect on the body see M5.

The decay law

Unstable nuclei disintegrate spontaneously and at random. However, the more undecayed nuclei there are, the more frequently disintegrations are likely to occur. For any particular radioactive nuclide, on average

activity ∝ number of undecayed nuclei

If N is the number of undecayed atoms after time t, then the activity is the rate of change of N with t. So, in calculus notation, the activity $A = -dN/dt$. (The minus sign indicates that a *decrease* in N gives *positive* activity.)

With a suitable constant, λ, the above proportion can be rewritten as an equation:

$$-\frac{dN}{dt} = \lambda N \qquad (1)$$

λ is called the ***radioactive decay constant***. Each radioactive nuclide has its own characteristic value. (Note that the symbol λ is also used for wavelength.)

By applying calculus to equation (1), a link between N and t can be obtained:

$$N = N_0 e^{-\lambda t} \qquad (e = 2.718) \qquad (2)$$

where N_0 is the initial number of undecayed nuclei.

This is known as the ***radioactive decay law***.

A graph of N against t has the form shown below. The graph is an *exponential* curve (see also H3).

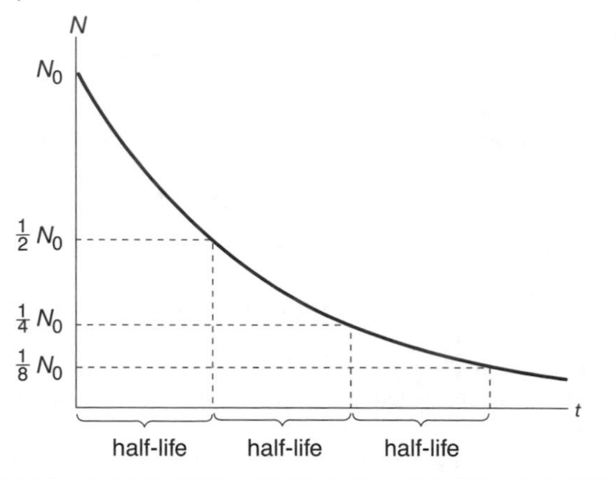

Half-life

There are two alternative definitions for this term.

The ***half-life*** of a radioactive nuclide is:

A the average time taken for the number of undecayed nuclei to halve in value,

B the average time taken for the activity to halve.

Version A is illustrated in the graph above. One feature of the exponential curve is that the half-life is the same from whichever point you start.

In equation (2), the half life, $t_{\frac{1}{2}}$, is the value of t for which $N = N_0/2$. Substituting this in the equation, taking logs, and rearranging gives

$$t_{\frac{1}{2}} = \frac{0.693}{\lambda} \qquad (0.693 = \ln 2)$$

Combining equations (1) and (2) gives $-dN/dt = \lambda N_0 e^{-\lambda t}$. So the activity ($-dN/dt$) also decreases exponentially with time, and a graph of *activity* against *time* has the same general form as the graph above. Version B of the half-life definition follows from this.

Half-lives for some nuclides

Nuclide	Half-life
potassium-40	1.3×10^9 years
plutonium-239	24 400 years
carbon-14	5730 years
strontium-90	28 years
magnesium-28	21 hours
radon-224	55 seconds

How Science Works

Using radioisotopes

Elements are a mixture of isotopes. The radioactive ones are called radioisotopes. Some can be produced artificially by transmutations in a nuclear reactor. They have many uses.

The **benefits** of using radioisotopes are illustrated by the many uses, some of which are given in the following section. It is possible to image things without cutting into them – this applies to the human body and to structures such as bridges and aeroplane wings.

The **risks** arise because of the ionization effects on living cells. Cells can be killed, or the DNA can be damaged and mutate causing cancerous growth of the cell.

There are many uses for which the benefit outweighs the risk, especially if suitable precautions are taken to minimize the risk.

When deciding which isotope to use, or explaining why a particular isotope has been chosen, include the following factors in your choice.

The type of emission chosen will depend on the required

- range
- penetration
- ionizing effect.

The particular isotope chosen will depend on

- whether it is an α-, β- or γ-emitter (see above)
- whether a long or short half-life is required
- the chemical effects – for example, how the isotope is to be delivered and whether it is toxic.

J5 Uses of radioisotopes

Examples of radioisotopes

Testing for cracks γ rays have the same properties as short-wavelength X-rays, so they can be used to photograph metals to reveal cracks. A γ source is compact and does not need an electrical power source like an X-ray tube.

Cancer treatment γ rays can penetrate deep into the body and kill living cells. So a highly concentrated beam from a cobalt-60 source can be used to kill cancer cells in a tumour. Treatment like this is called ***radiotherapy***.

Carbon dating Living organisms are partly made from carbon which is recycled through their bodies and the atmosphere as they obtain food and respire. A tiny proportion is radioactive carbon-14 (half-life 5730 years). This is continually forming in the upper atmosphere as nitrogen-14 is bombarded by cosmic radiation. When an organism dies, no new carbon is taken in, so the proportion of carbon-14 is gradually reduced by radioactive decay. By measuring the activity, the age of the remains can be estimated to within 100 years. This method can be used to date organic materials such as wood and cloth.

Tracers Radioisotopes can be detected in very small (and safe) quantities. This means that they can be used as tracers – their movements can be tracked. Examples include:

- tracking a plant's uptake of fertilizer from roots to leaves by adding a tracer to the soil water,
- detecting leaks in underground pipes by adding a tracer to the fluid in the pipe.

Dating rocks When rocks are formed, some radioisotopes become trapped. As decay continues, the proportion of radioisotope (e.g. potassium-40) decreases, while that of its decay product (e.g. argon-40) increases. The age of the rock can be estimated from the proportions. The decay of uranium-238 to lead-206, and of uranium-235 to lead-207, is also used for dating rocks.

Smoke detectors These contain a tiny α source which ionizes the air in a small chamber so that it conducts a current. Smoke particles entering the chamber attract ions and reduce the current. This is sensed by a circuit which triggers the alarm.

Choosing a radioisotope

Imaging techniques require good range and penetration, so they use γ-emitters. For crack detection in metal, and killing cancer cells, cobalt-60 has a half-life of 5.3 years and has a high enough activity without needing replacement too often.

For imaging the body, the radioisotope technetium-99m is useful. It is a γ-emitter with a half-life of 6 hours and will have effectively disappeared from the body in a couple of days (see M5).

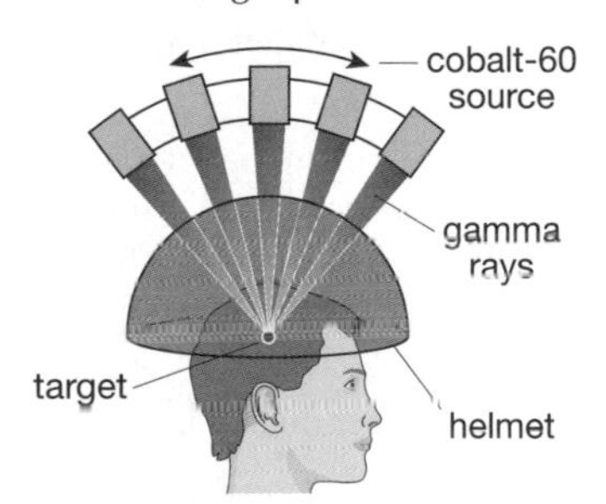

Dating techniques require very long half-lives and use the remaining naturally occurring radioisotopes that are present in the material to be dated. Carbon-14 has a half-life of 5730 years.

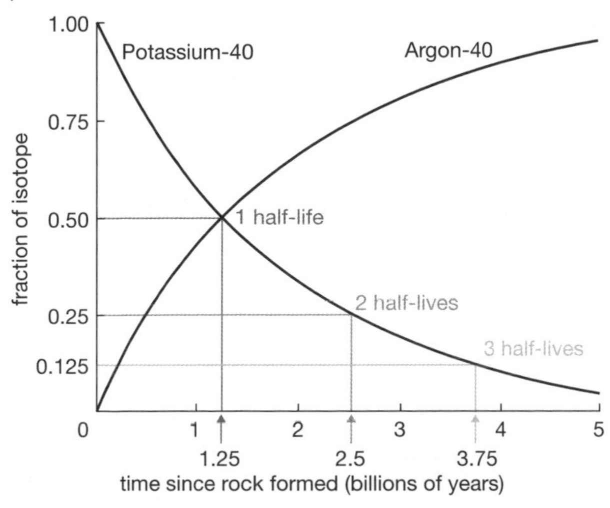

A smoke detector may use americium-241, which has a half-life of 432 years and is an α-emitter. The particles have short range – from the source to the detector – and are stopped by smoke. The activity does not change very much over a long period of time, so the rate of α emission will stay fairly constant, and the smoke detector should continue to work for years.

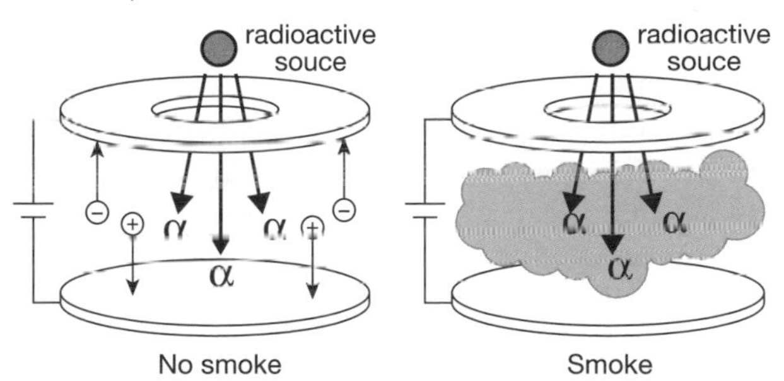

For imaging the thyroid gland, which takes up iodine, the radioisotope iodine-123 (γ-emitter with a half-life of 13 h) may be used. For treating thyroid cancer, iodine-131 (β- and γ-emitter with a half-life of 8 days) may be used.

To find a leak in a pipe, to check if packages are full, or to measure the thickness of materials, β-emitters are often the best choice. This is because α particles would always be absorbed before reaching the detector, and γ radiation would always pass straight through and be detected.

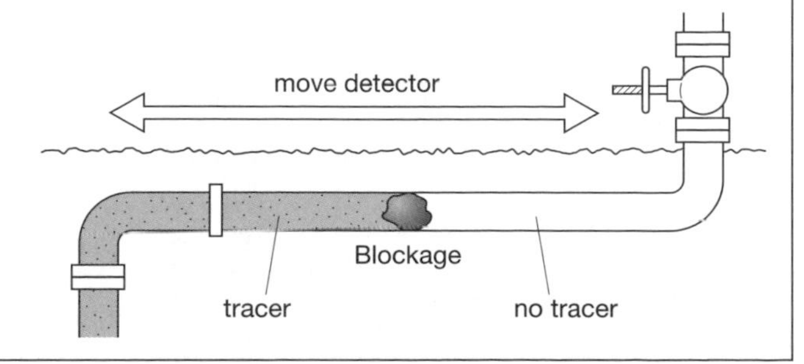

J6 Nuclear energy

Atomic mass

The ***unified atomic mass unit* (u)** is used for measuring the masses of atomic particles. It is very close to the mass of one proton (or neutron). However, for practical reasons, it is defined as follows:

$$1\ \text{u} = \frac{\text{mass of carbon-12 atom}}{12}$$

Converting into kg, 1 u = 1.66×10^{-27} kg.

Energy and mass

One conclusion from Einstein's ***theory of relativity*** is that energy has mass. If an object gains energy, it gains mass. If it loses energy, it loses mass. The change of energy ΔE is linked to the change of mass Δm by this equation:

$$\Delta E = \Delta mc^2$$

where c is the speed of light: 3×10^8 m s^{-1}

c^2 is so high that energy gained or lost by everyday objects produces no detectable change in their mass. However, the energy changes in nuclear reactions produce mass changes which are measurable. For example, when a fast α particle is stopped, its mass decreases by about 0.2%. The mass of an object when it is at rest is called its ***rest mass***.

With nuclear particles, energy is often measured in MeV (the electronvolt, eV, is defined in H11):

$$1\ \text{MeV} = 1.60 \times 10^{-13}\ \text{J}$$

From data on mass changes, scientists can calculate the energy changes taking place. With nuclear particles, mass is usually measured in u (see J1). By converting 1 u into kg and applying $\Delta E = \Delta mc^2$, it is possible to show that

1 u change in mass	is equivalent to	931 MeV change in energy

Mass defect

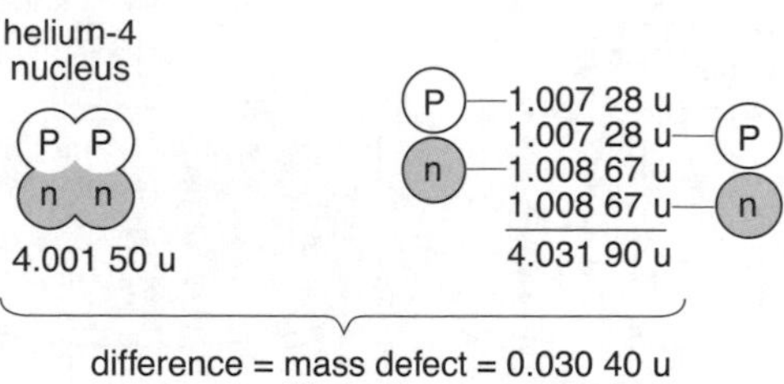

difference = mass defect = 0.030 40 u

mass defect per nucleon = $\frac{0.030\ 40}{4}$ = 0.007 60 u

A helium-4 nucleus is made up of 4 nucleons (2 protons and 2 neutrons). The calculation above shows that the nucleus has less mass than its four nucleons would have as free particles. The nucleus has a ***mass defect*** of 0.030 40 u.

The reason for the mass defect is as follows. In the nucleus, the nucleons are bound together by a strong nuclear force. As work must be done to separate them, they must have less potential energy when bound than they would have as free particles. Therefore, they must have less mass.

All nuclides have a mass defect (apart from hydrogen-1 whose nucleus is a single proton). For example:

	Mass defect	Mass defect per nucleon
hydrogen-2	0.002 40 u	0.001 20 u
iron-56	0.528 75 u	0.009 44 u
lead-208	1.757 84 u	0.008 45 u
uranium-238	1.935 38 u	0.008 13 u

Binding energy

The ***binding energy*** of a nucleus is the energy equivalent of its mass defect. So it is the energy needed to split the nucleus into separate nucleons. For example, a helium-4 nucleus has a mass defect of 0.030 40 u. As 1 u is equivalent to 931 MeV, 0.030 40 u is equivalent to 28.3 MeV. So the binding energy of the nucleus is 28.3 MeV.

Note:
- The term 'binding energy' is rather misleading. 'Unbinding energy' would be better. 28.3 MeV is the energy needed to 'unbind' the nucleons in helium-4.

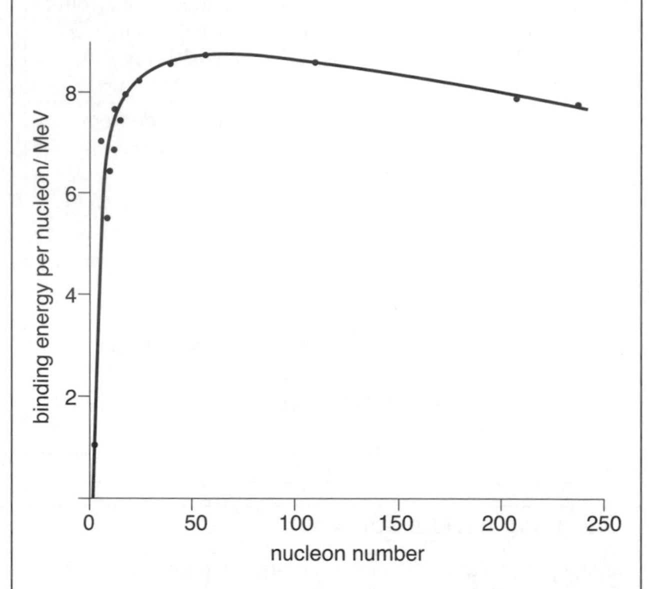

The stability of a nucleus depends on the ***binding energy per nucleon***. The graph above shows how this varies with nucleon number. The line gives the general trend; points for some individual nuclides have also been included.

Note:
- Nuclei near the 'hump' of the graph are the most stable, because they need most 'unbinding energy' per nucleon.
- A graph of *mass defect* against *nucleon number* has the same general form as the graph above.

If nucleons become rearranged so that they have a *higher* binding energy per nucleon, there is an *output* of energy.

Radioactive decay Unstable nuclei decay to form more more stable products, so energy is released. In α decay, for example, this is mostly as the kinetic energy of an α particle. When the α particle collides with atoms, it loses KE and they speed up. So radioactive decay produces heat.

Nuclear reactions The ***fission*** and ***fusion*** reactions on the next page give out energy. During fission, heavy nuclei split to form nuclei nearer the 'hump' of the graph. During fusion, light, nuclei *fuse* (join) to form heavier ones.

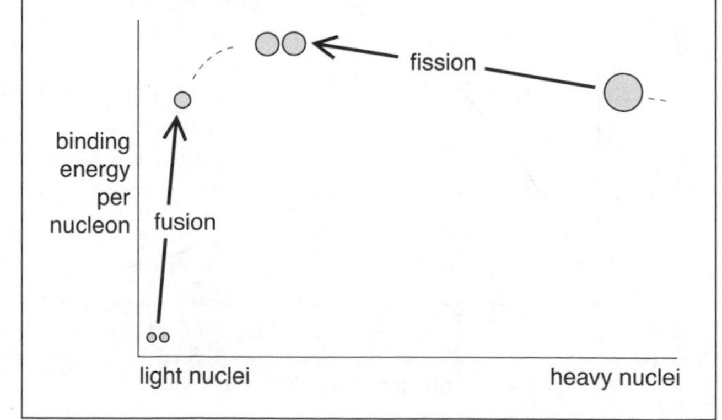

J7 Fission and fusion

Nuclear fission

During ***nuclear fission***, a heavy nucleus (e.g. of uranium or plutonium) splits to form two nuclei of roughly the same mass, plus several neutrons. Rarely, fission happens spontaneously. More usually, it occurs when a neutron hits and is captured by the nucleus. For example, here is a typical fission reaction for uranium-235:

$$^{235}_{92}\text{U} + ^{1}_{0}\text{n} \rightarrow ^{144}_{56}\text{Ba} + ^{90}_{36}\text{Kr} + 2^{1}_{0}\text{n}$$

The reaction releases energy, mostly as KE of the heavier decay products (see also I2). So fission is a source of heat.

Note:

- The energy released per atom by fission (about 200 MeV) is about 50 million times greater than that per atom from a chemical reaction such as burning.

Chain reaction The fission reaction above is started by one neutron. It gives off neutrons which may cause further fission and so on in a ***chain reaction*** (see diagram at right):

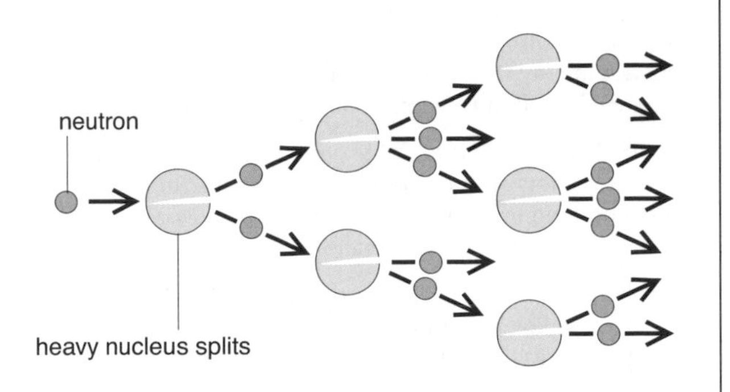

Uncontrolled chain reactions are used in nuclear weapons. Controlled chain reactions take place in ***nuclear reactors*** (see right) and release energy at a steady rate. The most commonly used fissionable material is uranium-235.

To maintain a chain reaction, a minimum of one neutron from each fission must cause further fission. However, to achieve this, these problems must be overcome:

- If the fission material is less than a certain ***critical mass***, too many neutrons escape without hitting nuclei.
- The fission of uranium-235 produces medium-speed neutrons. But slow neutrons are better at causing fission.
- Less than 1% of natural uranium is uranium-235. Over 99% is uranium-238, which absorbs medium-speed neutrons without fission taking place.

Thermal reactors

In a nuclear power station, the heat source is usually a ***thermal reactor***. (Otherwise, the layout is as for a fuel-burning station.) In the reactor, there is a steady release of heat as fission of uranium-235 takes place. It is known as a *thermal* reactor because the neutrons are slowed to speeds associated with thermal motion.

Nuclear fuel This is uranium dioxide in which the natural uranium has been enriched with extra uranium-235. 1 kg of this fuel gives as much energy as about 25 tonnes of coal.

Moderator This is a material which slows down the medium-speed neutrons produced by fission. Some reactors use graphite as a moderator. The PWR uses water as a moderator.

Control rods These are raised or lowered to control the rate of fission. They contain boron, which absorbs neutrons.

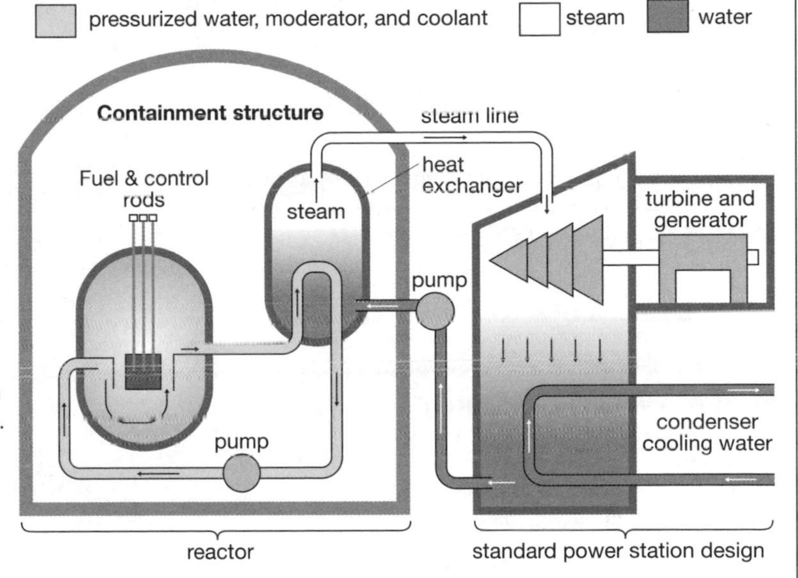

Pressurised water reactor (PWR)

Coolant (e.g. water or carbon dioxide gas)
This carries heat from the reactor to the heat exchanger. In the PWR, this is pressurized water.

How Science Works

Benefits and risks of thermal reactors

Benefits of thermal reactors include

- high energy output for the amount of fuel used
- no carbon dioxide produced in the reactor (although some may be produced in mining and preparing the fuel)
- many useful isotopes made by bombarding stable isotopes with neutrons in the core of a reactor.

Risks of thermal reactors include

- release of radioactive material during operation
- release of radioactive material due to a fault
- the long half-life of radioactive waste, which means it must be kept safely away from the biosphere (living things) for 10 000 to 1 000 000 years.

To reduce the risks as much as possible, the reactors have thick concrete shielding to absorb radiation. There is a heat exchanger so that the coolant does not mix with the water in the power-generating plant. If radioactive dust or gas escaped, this would contaminate the environment and be dangerous to living things, so the reactor is enclosed and the level of radioactivity in the environment is monitored.

The design of the reactor is 'fail safe'. This means that, for example, the control rods can drop into the reactor to stop the reaction, and they do not rely on electric motors, in case the power switches off in an emergency.

Nuclear waste

How Science Works

There are three types of radioactive waste.

Low-level waste: for example, laboratory equipment and protective clothing. It can be sealed in drums and buried in trenches in clay rocks (so that if the drums corrode the radioactivity will not seep into the water supply).

Intermediate-level waste: for example, fuel cladding from nuclear reactors. The drums of waste can be put in concrete casks and then in reinforced concrete trenches. Clay above and below prevents contamination of water.

High-level waste: for example, spent fuel. This has to be stored in cooling water ponds until the level of activity drops enough for it to be sent for reprocessing. Liquid waste is vitrified (sealed in glass). The high-level waste is stored in steel canisters containing concrete. It can be kept in a safe guarded compound, or buried in deep underground sites.

The problem is whether the site will stay geologically stable for 1 000 000 years.

Do the benefits of thermal reactors outweigh the risks? To make this decision, the Government consults scientists for advice.

For some years no additional nuclear power stations were built because it was thought to be too risky. Now, to reduce our dependence on fossil fuel (because of global warming), there are plans to build more nuclear power stations. Newer designs are said to be safer, and we already buy electricity from France that is generated using nuclear reactors. Global warming could be more devastating than a nuclear reactor disaster, and the power cuts that would result from not having enough power-generating plants would be a risk to our health and well-being too. However, the problems of waste storage have not changed.

Nuclear fusion

Reactors using ***nuclear fusion*** are many years away. Current research is based on the fusion of hydrogen-2 (called ***deuterium***) and hydrogen-3 (called ***tritium***):

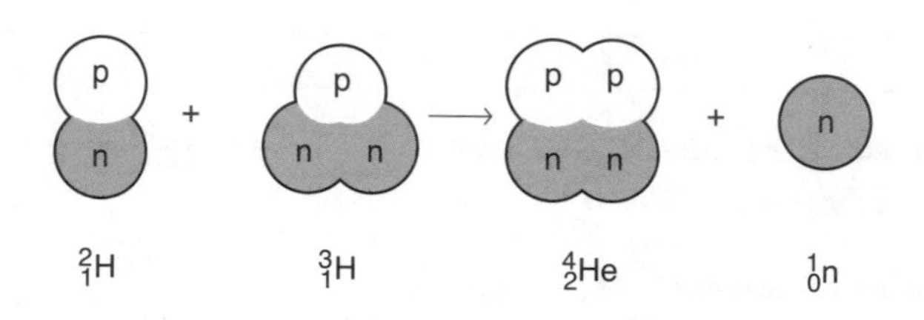

Although the energy release per fusion is less than 10% of that from a fission reaction, fusion is the better energy source if the processes are compared *per kg* of material.

Fusion is much more difficult to achieve than fission because the hydrogen nuclei repel each other.

- For the nuclei to collide at a high enough speed for fusion, the gas has to be heated to 10^8 K or more so that the atoms lose their electrons and the gas becomes a ***plasma***.
- No ordinary container can hold such a hot material and keep it compressed. Scientists are experimenting with magnetic fields to trap the nuclei in a ring-shaped container (called a tokamak).

The advantages of a fusion reactor will be:

- Fuels will be readily obtainable. For example, deuterium can be extracted from sea-water.
- The main waste product, helium, is not radioactive.
- Fusion reactors have built-in safety. If the system fails, fusion stops.

The Sun gets its energy from the fusion of hydrogen, though using a different reaction from that on the left. Its huge size and gravity maintain the conditions needed.

Progress in fusion research

How Science Works

In the UK, fusion research takes place at JET (Joint European Torus). They have shown that fusion is possible, but so far it requires more input energy than output energy.

A new larger tokamak is needed to make progress. The International Thermonuclear Experimental Reactor (ITER) is a joint international research and development project that aims to demonstrate the scientific and technical feasibility of fusion power. It involves the European Union, Japan, China, India, Korea, Russia, and the USA. ITER will be constructed at Cadarache in France.

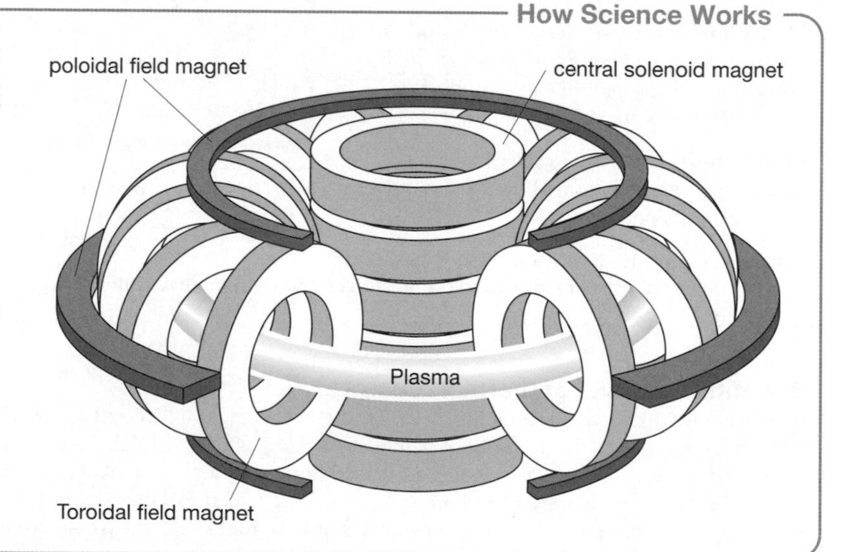

K1 Particle physics – 1

Detectors

In experiments using accelerators (see H12), detectors are needed to reveal the paths of the particles produced.

Bubble chamber This is filled with liquid hydrogen whose pressure is suddenly reduced so that it is ready to vaporize. Charged particles entering the chamber ionize the hydrogen. This triggers vaporization, so that a trail of bubbles is formed along the track of each particle.

Drift chamber This is a gas-filled chamber containing, typically, thousands of parallel wires. Incoming particles cause a trail of ionization in the gas. Their track is worked out electronically by timing how long it takes ionization electrons to drift to the nearest sense wires. A computer processes the signals and displays the tracks graphically.

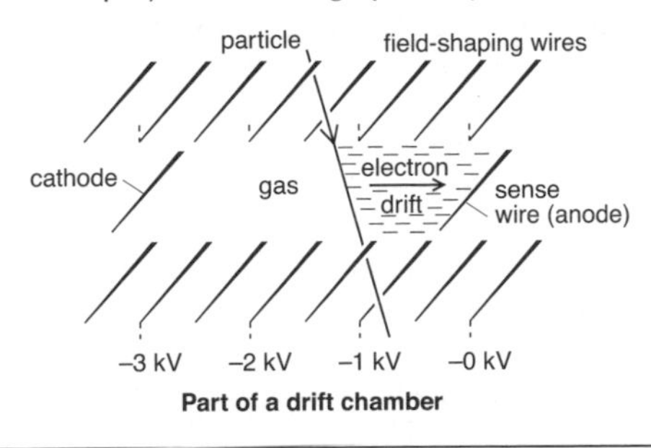

Part of a drift chamber

How Science Works

Excited states of the nucleus

After a particle is emitted or absorbed by a nucleus, it may be unstable and decay again, but it may also be in an ***excited state.***

To return to its ground state, it emits a gamma photon (gamma radiation).

Example

Technetium-99m, a radioisotope used in medicine (see M5) is produced by beta decay of molybdenum-99 (which has a half-life of 66 hours).

$^{99}_{42}\text{Mo} \rightarrow {}^{99}_{43}\text{mTc} + {}^{0}_{-1}\text{e} + {}^{0}_{0}\overline{\nu}$

The 'm' denotes the excited state, which has a half-life of 6 hours and decays by gamma emission.

$^{99}_{43}\text{mTc} \rightarrow {}^{99}_{43}\text{Tc} + \gamma$

A sample of the radioisotope molybdenum-99 is prepared in a reactor and delivered to a hospital. It is used to provide samples of Technetium-99m as required for about a week.

Identifying particles

A particle's properties can be identified from

- the direction of curvature of its track in a magnetic field
- the radius of curvature of the track
- the change in radius of the path as it loses energy
- the length of the track
- the density of the track.

Although uncharged particles do not leave tracks their properties can be deduced from the tracks of charged particles they interact with.

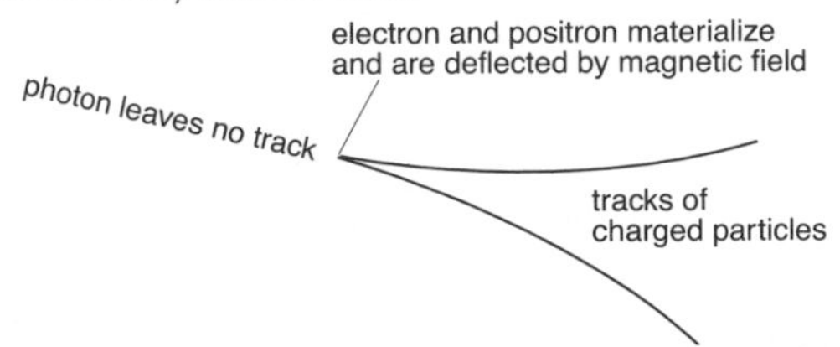

Direction of curvature

When a charged particle moves perpendicularly to a magnetic field it moves in a circular path. The direction of curvature of a track depends on the field direction and the sign of the charge on the particle and can be deduced using the 'left-hand rule'.

some tracks of particles in a bubble chamber with magnetic field into the paper

(no tracks) γ e^+ e^- Ξ^0 Λ π^- p^+

Degree of curvature

The radius of curvature r of a particle is directly proportional to the momentum of the particle. The momentum p is given by the equation

$$p = Bqr$$

where B is the magnetic flux density.

Most subatomic particles have a charge of $+1.6 \times 10^{-19}$ or -1.6×10^{-19} C so $p \propto r$.

For a given particle, kinetic energy $= p^2/2m$, so energetic particles therefore have high momentum and high radius of curvature of their tracks. This means that more energetic particles have straighter tracks.

Change in curvature

A particle that loses energy quickly in a bubble chamber will change curvature quickly. An electron loses energy very quickly and its path is seen to spiral inwards as the radius of curvature of the path decreases. The radius of curvature of a proton, however, would hardly change at all.

Track length

The track length of a particle ends either because it collides with another particle or because it decays into other particles. It is possible to deduce from conservation laws whether subsequent tracks are due to a collision or a decay. The length of the track of a particle that decays provides information about its lifetime. Longer tracks mean that the particle has a long life and is therefore more stable.

Density of the track

A dense track means that the particle produces a lot of ionization and some deductions can be made about the particle's properties. For example a fast-moving electron produces less ionization than a proton of similar energy. However, a fast-moving proton would also produce very little ionization, so although it is possible to draw some conclusions using the density of the track other evidence is needed.

Matter and antimatter

Most matter particles, such as the proton, electron, and neutron, have corresponding ***antiparticles***. These have the same rest mass as the particles but opposite charge and spin.

Apart from the antiparticle of the electron e^-, which is the positron e^+, antiparticles are given the same symbol as the particle but with a bar over the top.

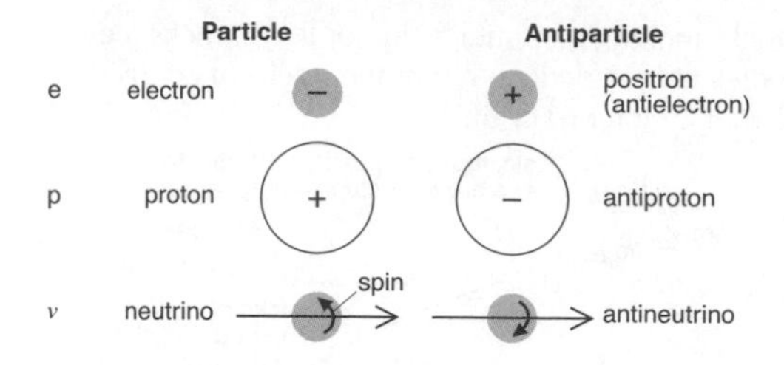

When a particle and its antiparticle meet, in most cases, they ***annihilate*** each other and their mass is converted into energy as given by $E = mc^2$. For example, the annihilation of an electron and positron may produce a pair of gamma photons.

Note:

- There are far more particles than antiparticles in the Universe, so annihilation is extremely rare.

Pair production

Energy can also be converted into mass. For example, if a gamma photon has at least 1.02 MeV of energy, it may, when passing close to a nucleus, convert into an electron-positron pair (total rest mass 1.02 MeV/c^2). In high-energy collisions, heavier particles (and antiparticles) may materialize from the energy supplied.

The strong nuclear force

In the nucleus, the nucleons (neutrons and protons) are bound together by the strong nuclear force. The strong force:

- is strong enough to overcome the Coulomb repulsion between protons, otherwise they would fly apart,
- has a short range, ~10^{-15}m, and does not extend beyond neighbouring nucleons,
- becomes a repulsion at very short range, otherwise the nucleus would collapse.

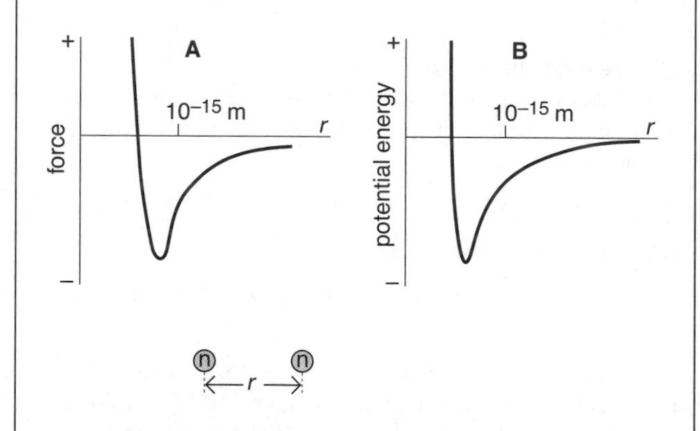

Graph A shows how the strong force varies with separation, for two neutrons. (An *attractive* force is *negative*.)

Graph B shows how the potential energy of the neutrons varies with separation. Minimum potential energy corresponds with the position of zero force in graph A.

Escaping from the nucleus: α decay

The graph below shows how the potential energy of an α particle varies along a line through the centre of a nucleus. Outside the nucleus, there is Coulomb repulsion (giving positive PE). Inside the nucleus, this is overcome by the strong force (giving negative PE). The result is a potential energy 'well' with a ***Coulomb barrier*** around it. Within this well, there are different energy levels.

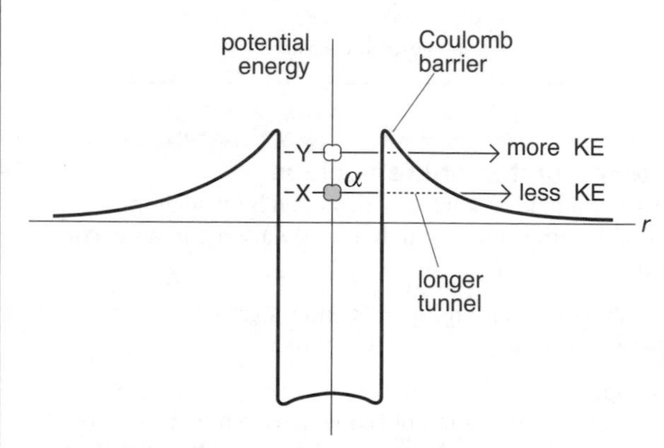

Above, an α particle has formed at X. It seems to be trapped by the Coulomb barrier. But because of the ***uncertainty principle*** (see G8) there is a chance that it may briefly 'borrow' enough energy to tunnel through the barrier and escape. This is called ***quantum mechanical tunnelling***.

An α particle at Y will escape with more KE – and is likely to escape sooner because the tunnel is shorter and easier to pass through. That is why α particles with the *highest* KEs come from nuclides with the *shortest* half-lives.

Nuclear radii

Results of experiments show that the radius of a nucleus R is proportional to $A^{1/3}$ where A is the nucleon number.

Hence $R = R_0 A^{1/3}$

where R_0 is a constant having a value of 1.2×10^{-15} m.

Nuclear density

The mass of a nucleus $= A \times$ the mass of a nucleon

The volume of a nucleus $= \frac{4}{3}\pi(1.2 \times 10^{-15} A)^3$

The mass of a nucleon is 1.7×10^{-27} kg

The density of a nucleus is therefore 2.3×10^{17} kg m^{-3}

This is enormous when compared with the density of water (1000 kg m^{-3}). The difference is due to the fact that most of the volume of atoms is empty space. Only in neutron stars where the atoms are stripped of their electrons is it possible that such large densities exist on a large scale.

Measurements of masses in particle physics

A commonly used unit of mass in particle physics is the unit MeV/c^2 or GeV/c^2. This avoids the necessity to continually convert particle energies to mass or vice versa when considering collisions and decays.

$$1 \text{ MeV} \equiv 1.6 \times 10^{-13}\text{ J}$$
$$1.6 \times 10^{-13}\text{ J} \equiv 1.6 \times 10^{-13}/(3.0 \times 10^8)^2$$
$$\equiv 1.78 \times 10^{-30}\text{ kg}$$

Using this unit the mass of a proton is 960 MeV/c^2 and that of an electron is 0.51 MeV/c^2.

K2 Particle physics – 2

Fundamental particles

Fundamental particles are particles that cannot be divided into smaller particles. In the ***standard model of particle physics*** (our model today), there are 12 fundamental particles from which matter is made (there will also be 12 corresponding antimatter particles). There are 6 types of ***lepton*** and 6 types of ***quark*** (and also 6 corresponding ***antileptons*** and 6 ***antiquarks***). The main difference between quarks and leptons is that quarks feel the strong force and leptons do not.

Leptons

Leptons have no size and, in most cases, low or no mass. There are three generations of leptons, but only the first (the electron and its neutrino) occurs in ordinary matter. Muons and muon neutrinos are produced in the upper atmosphere by cosmic rays, but the tau has so far only been seen in laboratory experiments. The neutrino, ν, produced by beta$^-$ decay is the electron-antineutrino ν_e.

	Leptons	
Charge	**⁻e**	**0**
1st generation	electron e^-	electron-neutrino ν_e
2nd generation	muon μ^-	muon-neutrino ν_μ
3rd generation	tau τ^-	tau-neutrino ν_τ

Quarks

Symmetry theory predicts that there should be 3 generations of quarks to match the 3 generations of leptons. They have a fractional charge of $+\frac{2}{3}$ e or $\frac{1}{3}$ e. The top quark is the most massive fundamental particle (almost 200 times the mass of the proton).

	quarks	
Charge	$+\frac{2}{3}$**e**	$-\frac{1}{3}$**e**
1st generation	up u	down d
2nd generation	charmed c	strange s
3rd generation	top t	bottom b

Note:

- Ordinary matter contains only the first generation of quarks. Very high energies are needed to make hadrons of other quark generations. These hadrons quickly decay into first generation particles.
- Individual quarks have never been detected.

Hadrons

Hadrons are particles made from quarks that are held together by the strong force. This force is so strong that quarks have never been found individually. There are two types of hadron:

Baryons are made of 3 quarks.

Mesons are made of a quark and an antiquark.

To describe how quarks can join together, each quark is assigned a baryon number of $\frac{1}{3}$ (each antiquark has ***baryon number*** $-\frac{1}{3}$).

All baryons have a baryon number of 1, antibaryons are –1, and all other particles, including mesons, are 0.

Similarly, the strange quark is assigned a strangeness of –1 (and the antistrange has +1) – all other quarks have strangeness 0. Strange particles have a long lifetime.

	quarks	charge	Baryon number	Strangeness	Name of particle	symbol
Baryon	uud	+1	+1	0	proton	p
	udd	0	+1	0	neutron	n
	$\overline{Uud}$	–ß1	–1	0	antiproton	$\overline{P}$
	$\overline{Udd}$	0	–1	0	antineutron	$\overline{n}$
				0	↓(more baryons)	
Meson	$\overline{uu}$	0	0	0	pion (pi zero)	π^0
	$\overline{dd}$	0	0	0	pion (pi zero)	π^0
	$\overline{ud}$	+1	0	0	pion (pi plus)	π^+
	$\overline{du}$	–1	0	0	pion (pi minus)	π^-
	$\overline{ds}$	0	0	–1	kaon	k^0
	$\overline{ds}$	0	0	+1	kaon	k^0
	$\overline{us}$	+1	0	–1	kaon	k^+
	$\overline{su}$	–1	0	+1	kaon	k^-
		0			↓more mesons	

Note: π0 has a very short lifetime, as the quarks will annihilate each other. The π^+ is the antiparticle of the π^- (so the symbol with the bar is not usually used). There is a similar pattern for other mesons.

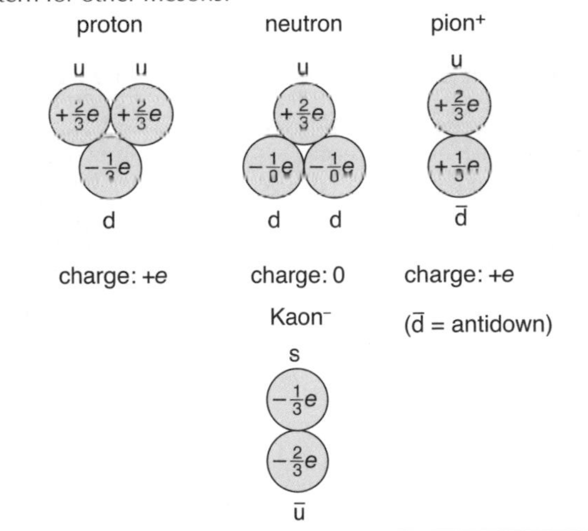

Conservation laws

There are conservation laws for *momentum* and *total energy*. However, as mass and energy are equivalent, the total energy must include the rest energy (see G6).Particles have various ***quantum numbers*** assigned to them. These are needed to represent other quantities which may be conserved during interactions. For example:

Charge In any interaction, this is conserved: it balances on both sides of the equation (see J3 for examples).

Lepton number This is +1 for a lepton, –1 for an antilepton, and 0 for any other particle. For example, a 'free' neutron decays, after about 15 minutes, like this:

neutron	→	proton	+	electron	+	antineutrino
(0)		(0)		(+1)		(–1)

The numbers (in brackets) have the same total, 0, on both sides of the equation, so lepton number is conserved. This applies in any type of interaction.

Baryon number This is +1 for a baryon, –1 for an antibaryon, and 0 for any other particle. It is conserved in all interactions.

Strangeness This is needed to account for the particular combinations of 'strange particles' (certain hadrons) produced in some collisons. It is conserved in strong and electromagnetic interactions, but not in all weak ones.

Charm relates to the likelihood of certain hadron decays.
Spin relates to a particle's angular momentum.
Topness and **bottomness** are further quantum numbers.

Fundamental forces

Force	Range/m	Relative strength	Effects, e.g.
strong	$\sim 10^{-15}$	1	Holding nucleons in nucleus
electro-magnetic	∞	$\sim 10^{-2}$	Holding electrons in atoms; holding atoms together
weak	$\sim 10^{-17}$	$\sim 10^{-5}$	ß decay; decay of unstable hadrons
gravitational	∞	$\sim 10^{-39}$	Holding matter in planets, stars, and galaxies

Particles *interact* by exerting forces on each other. There are four known types of force in the Universe (see chart above). As electric and magnetic forces are closely related, they are regarded as different varieties of one force, the electromagnetic. ***Grand unified theories (GUTs)*** seek to link the strong, weak, and electromagnetic forces. Gravitational force has yet to be linked with the others. It is insignificant on an atomic scale.

The weak force and beta decay

See also J3

The proton is stable, but all other hadrons outside the nucleus are unstable. Eventually they will decay, including the neutron. Inside the nucleus, whether protons or neutrons decay depends on whether they are needed for stability. If the nucleus is unstable, then the weak force causes one of the beta decays to occur.

$${}^1_0n \rightarrow {}^1_1p + {}^0_1e + {}^0_0\bar{\nu}_e$$

(In terms of quarks: W? emission turns a d into a u – see right).

When there is a nuclear reaction, the conservation of momentum and energy can be used to calculate the energy and momentum of the decay products. In beta decay, the beta particles are emitted with a range of possible energies. This is how the ***neutrino*** was discovered. There has to be another particle with the missing energy and momentum, otherwise the conservation laws are violated.

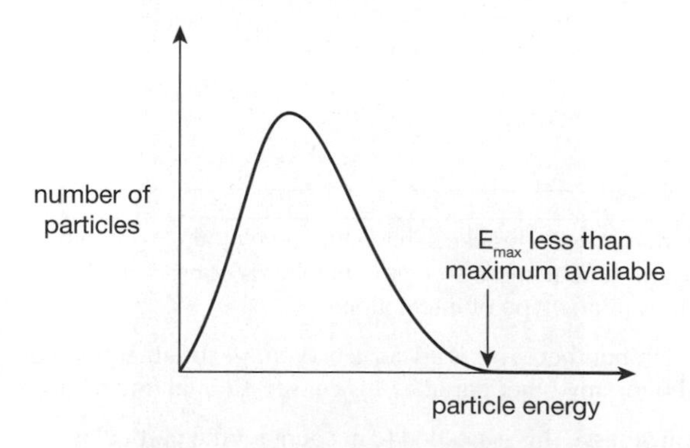

Initially thought to have a rest mass of zero, scientists are still debating this point. The experiments that do show a mass have such a small value that it could be experimental error. Some properties are consistent with mass and some are not. The best description at the present time is 'zero or negligible' rest mass.

Force carriers

Like other particles, nucleons need not be in contact to exert forces on each other. To explain how the strong force is 'carried' from one nucleon (e.g. neutron) to another, the idea of ***exchange particles*** is used:

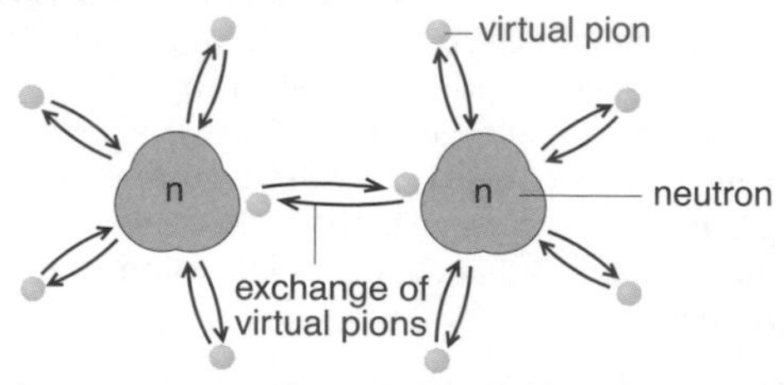

Each nucleon is continually emitting and reabsorbing ***virtual pions***, which surround it in a swarm. When close, two nucleons may exchange a pion. The momentum transfer produces the effect of a force (attractive or repulsive).

Note: The emitting nucleons lose no mass, so virtual pions are only allowed their brief existence by the uncertainty principle. To create 'real' pions, the missing mass must be supplied by the energy of a collision.

All the fundamental forces are believed to be carried by exchange particles. For example, electrons repel each other by exchanging ***virtual photons***. This process can be represented by a ***Feynman diagram*** as below. For 'real' photons to exist, energy must be supplied.

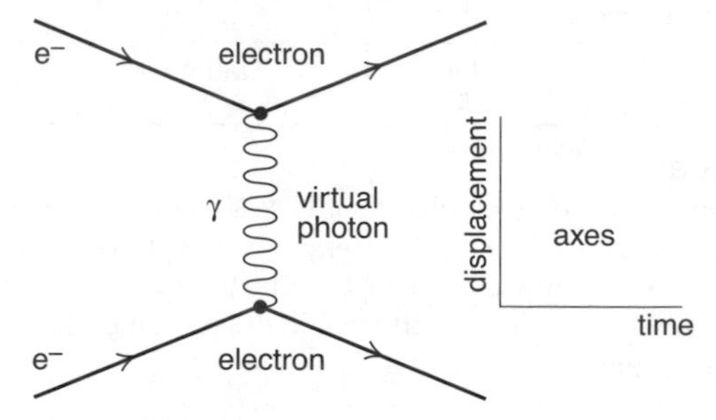

The particles that carry the fundamental forces are known as ***gauge bosons***. They are listed in the chart below.

Note:

- Quarks are bound together by ***gluons***. As nucleons and pions are made of quarks, the gluon would seem to be the basic force carrier for all strong interactions.
- The existence of the graviton is speculation only.

Force	Gauge bosons		
strong	gluon		
electromagnetic	photon		
weak	W^+	W^-	Z^0
gravitational	graviton		

Note: particle physicists are also using the LHC (see page H12) to look for the ***Higgs boson***, a particle they think may be responsible for giving particles mass.

Energy and the uncertainty principle

According to the ***uncertainty principle***, a particle's momentum and position cannot both have precise values. There is a level of uncertainty about them. One consequence of this is that the law of conservation of energy can, briefly, be *violated* (disobeyed). A particle can have more energy than it 'ought' to, by an amount ΔE, provided that this is paid back in a time Δt, where $\Delta E.\Delta t \approx h$. This has important consequences for the behaviour of particles, including quantum mechanical tunnelling in K1.

L1 Telescopes – 1

Refracting telescopes

In the telescope on the right, the ***objective*** lens focuses light from a distant object, to form a real image just on the principal focus of the ***eyepiece*** lens. The eyepiece lens then forms a virtual, magnified image of this real image. Set like this, the telescope is in ***normal adjustment.***

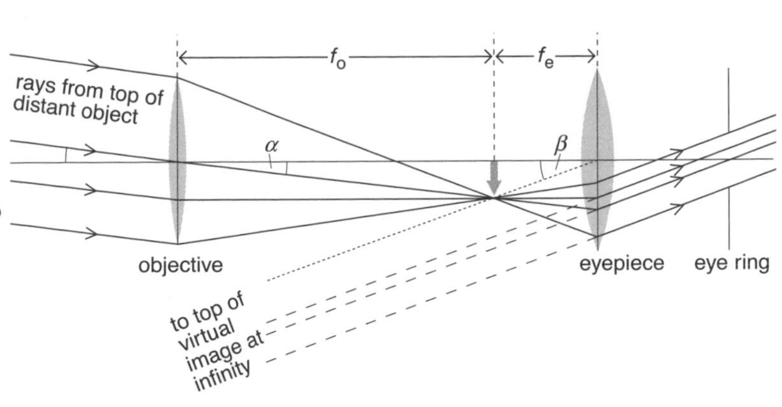

Angular magnification This is defined as the ratio of the two angles β and α in the diagram on the right:

$$\text{angular magnification, M} = \frac{\text{angle subtended by image at eye}}{\text{angle subtended by object at unaided eye}}$$

$$\text{angular magnification} = \frac{\beta}{\alpha}$$

From triangles in the diagram, it can be shown that

$$\text{angular magnification} = \frac{f_o}{f_e}$$

For example, if the focal lengths of the objective and eyepiece lenses are 100 cm and 5 cm respectively, then the angular magnification (in the above setting) is 20.

Aperture The diameter of the objective lens is called its ***aperture***. It is difficult to make wide-aperture lenses which do not give a distorted image. But if this problem is solved, a wider aperture gives

1 a brighter image,
2 a greater resolving power (see below).

Resolving power This is $1/\theta_{min}$, where θ_{min} is the smallest angle between two points such that their images can just be *resolved* (seen separately and unmerged).

Diffraction limits the resolving power because a telescope's objective lens acts rather like a wide slit (see D3). As a result, the images of two 'point' stars will each appear as tiny, circular fringe systems with intensities as below.

According to the ***Rayleigh criterion***, two images are just resolved if the central maximum of one coincides with the first minimum of the other. Increasing the aperture D gives less spread-out fringe systems and a greater resolving power ($\approx D/\lambda$).

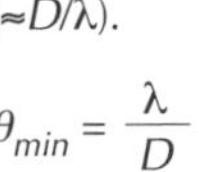

$$\theta_{min} = \frac{\lambda}{D}$$

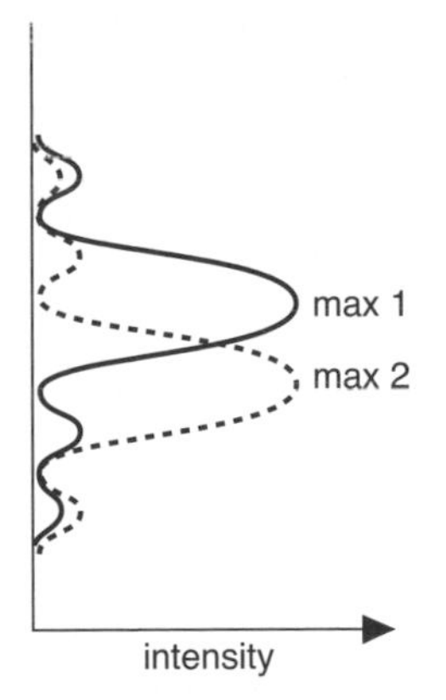

Note, in the above diagram:

- The final image is inverted.
- For the widest angle of view, the eye is placed in the ***eye ring*** position (the position where the eyepiece forms a real image of the objective lens).
- By moving the eyepiece, the real image can be formed just inside its principal focus. The final, virtual image is then closer to the eye. However, it is more straining for the eye to look at this image for a long time.

Airy's disc

When a distant point object is viewed with a telescope, the image observed is a disc of light caused by diffraction. This is called **Airy's disc**, named after the 19th century astronomer Sir George Airy.

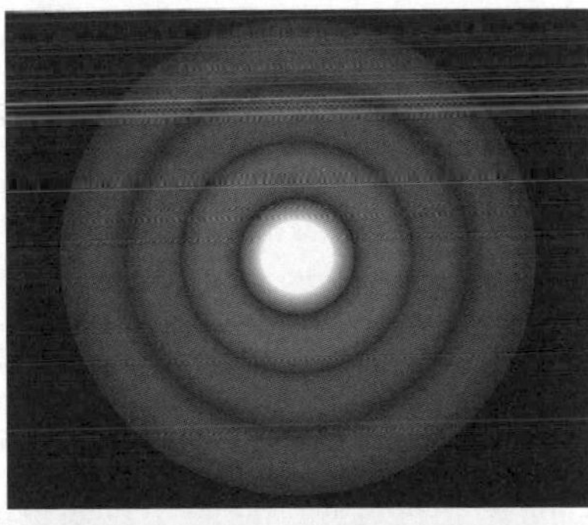

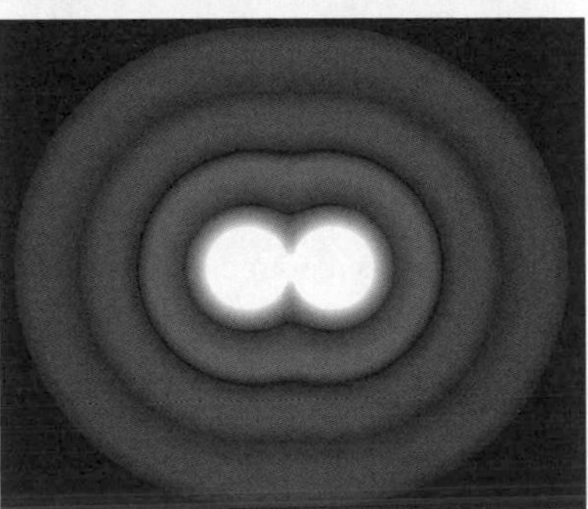

The disc becomes smaller when larger-aperture telescopes are used. When two such objects are close together the diffraction discs merge so that the images are not resolved. Larger-aperture telescopes improve resolution.

Lens and mirror equations

Concave mirrors are *converging* systems, with similar image-forming properties to convex lenses.

Convex mirrors are *diverging* systems, with similar image-forming properties to concave lenses.

Equations (1) and (2) on the opposite page apply to all thin lenses and mirrors. However, when using them, note:

- *Converging* systems have *positive* focal lengths.
- *Diverging* systems have *negative* focal lengths.

Focal point of a mirror

Light from a distant object converges toward the focus of a concave mirror.

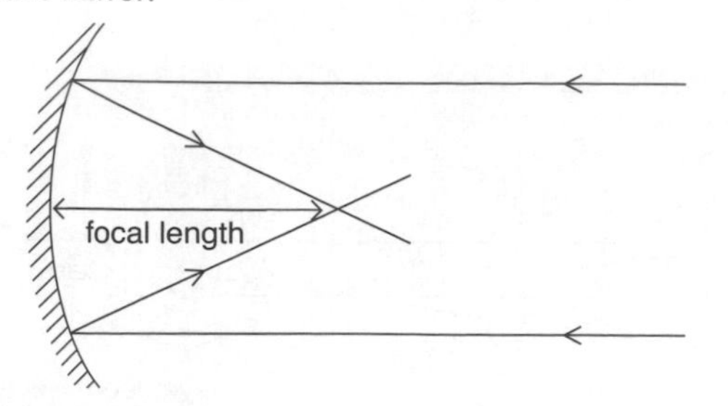

Light directed toward the focus of a spherical convex mirror is reflected as a parallel beam.

Light directed toward one focus of a hyperbolic convex mirror converges toward the other focus.

Reflecting telescopes

In large-aperture telescopes, the rays are brought to a focus by a concave mirror rather than an objective lens. With a mirror, support can be provided at the back to stop it flexing, and it is easier to reduce image distortion. Often, there is no eyepiece. The real image is formed either on a photographic plate, for later enlargement, or on a ***charge-coupled device (CCD)***, which stores it electronically.

Cassegrain arrangement

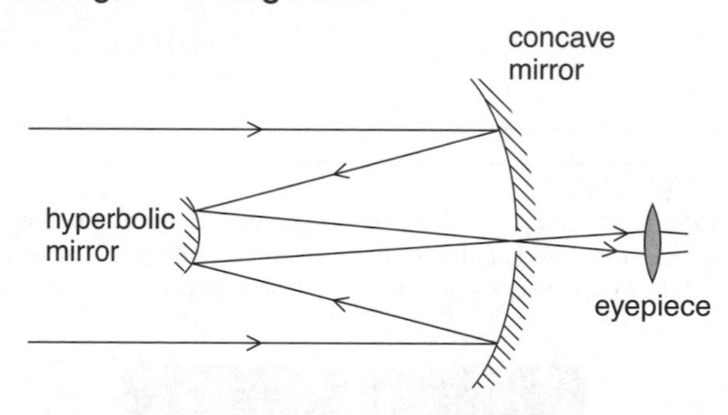

The convex mirror is hyperbolic and has one focus at the same point as that of the large concave mirror. The other focus is at the pole (centre). Light that passes through a small hole at the pole is then viewed through an eyepiece.

Merits of reflecting and refracting telescopes

As well as producing high magnification an astronomical telescope has to view faint objects so a large aperture is needed.

Small ***refracting telescopes*** are very convenient to use but to make large–aperture telescopes for astronomy a highly transparent material is needed that has few flaws to reflect the light. A large-diameter lens is heavy and can only be supported at its rim so that when mounted it suffers mechanical distortions which produce optical distortions in the images.

The large-aperture mirrors required in a ***reflecting telescope***

- are easier to construct since they can be supported anywhere behind it
- are silvered on the front surface so no refraction is involved
- be used with a wider range of wavelengths including those outside the optical region
- can be more easily constructed to eliminate aberrations than lenses.

These are amongst the reasons why large telescopes for astronomical and other purposes are usually reflecting rather than refracting.

Problems with mirrors and lenses

A mirror that is silvered on the rear surface suffers from multiple images. One is produced by partial reflection at the front surface and another by reflection from the back surface. This can be overcome by using mirrors that are silvered on the front surface.

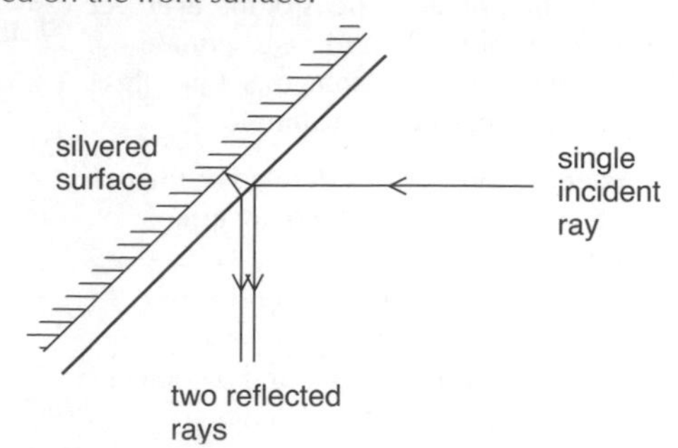

Spherical aberration

Light that is incident on a spherical lens or mirror a long way from its axis is brought to a focus at a different position from light incident closer to the axis.

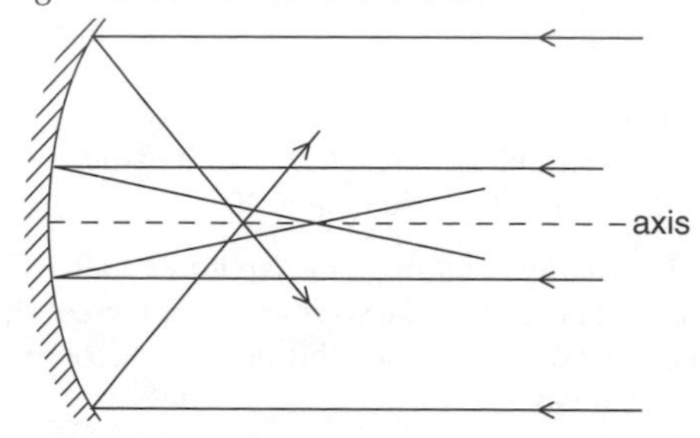

The result is a blurred image.

Spherical aberration can be reduced by restricting the lens aperture. But this reduces the brightness of the image and makes resolution worse.

On parabolic mirrors, all rays are brought to a focus at the same point and spherical aberration is eliminated.

Chromatic aberration

Light of different colours refracts differently when passing through a lens. The image of an object emitting blue light is in a different position to that formed by one in the same position that is emitting red light.

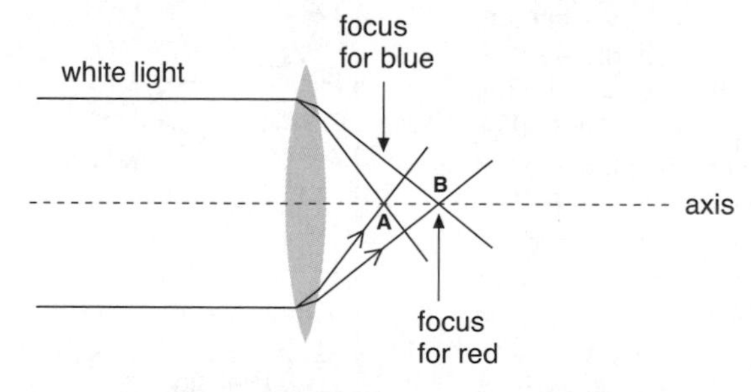

A screen placed at **A** would show an image with a blue edge and a screen at **B** would show an image with a red edge.

White objects emit all colours and there are therefore many images formed of an object, all in different positions.
The result is a blurred image with coloured edges.

Problems can be reduced by using combinations of converging and diverging lenses made from materials of different refractive indices.

Because mirrors do not depend on refraction there is no chromatic aberration with mirrors that are silvered on the front surface.

L2 Telescopes – 2

Radio telescopes

Radio telescopes are reflecting telescopes. They are used to form images of distant objects due to the emission of microwaves or radio waves rather than visible light. The telescope scans the sky and builds up an image formed by the variation in intensity of radio waves coming from different directions.

Most have a large concave dish to reflect incoming radio waves towards the antenna. Microwaves need a smooth metal reflector. With longer wavelengths, wire mesh can be used, provided the mesh size is less than about $\lambda/20$. A computer-generated 'radio image' is built up by ***scanning*** the source line by line.

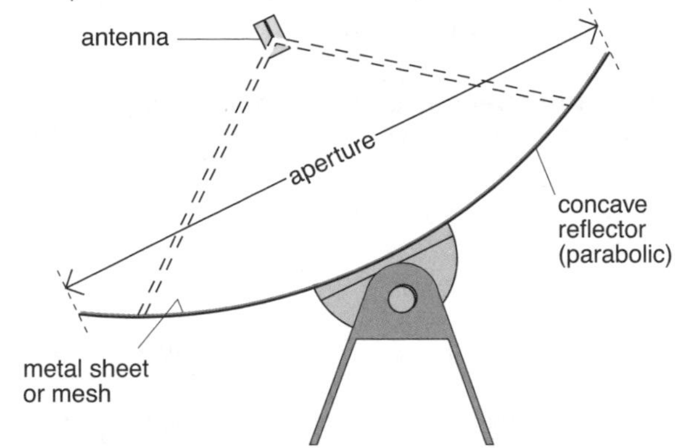

Radio telescopes need large diameters

- to collect sufficient energy to form an image
- to produce adequate resolution.

Radio telescopes have to deal with much longer wavelengths than optical instruments, so they need much wider apertures to give the same resolving power. Effective apertures of many kilometres can be achieved by linking several telescopes electronically.

Using the Rayleigh criterion, a radio telescope forming an image using 30 cm waves would need a diameter of about 3 km to produce the same resolution as that of a naked eye forming an optical image.

Charge-coupled device (CCD)

This is used in many optical (e.g. light) telescopes instead of a photographic plate. It is more sensitive, and the image on it can be processed electronically. It consists of a silicon chip divided into an array of tiny photodiodes, each contributing one picture element *(a **pixel**)* to the whole picture. When photons fall on a photodiode they release photoelectrons from the surface. The number of electrons liberated is proportional to the intensity of the light (see D5). Charge builds up in the ***pixel***, which can be thought of as a bucket, or ***potential well***, that is filling up with electrons. When the exposure is complete, the electron pattern is identical to the image formed on the CCD.

$n \propto I$

A measure of how many incident photons are converted to photoelectrons is given by the ***quantum efficiency*** of the pixel, and is greater than 70%.

The charge is then processed, read off pixel by pixel, row by row, to give a digital signal that can be amplified and processed by a computer for display on a screen.

Collectors and detecting radiation

Information about planets, stars, and galaxies is obtained by analysing the electromagnetic radiation they emit. Depending on the source, this can range from radio waves to γ radiation. Some form of telescope is used as the ***collector***. This is linked to a ***detector***.

Electromagnetic radiation	Telescope: collector, e.g.	detector, e.g.
radio	concave metal dish	antenna
infrared	concave mirror	solid state detector
light ultraviolet	concave mirror (see C5)	photographic plate or CCD
X-rays γ-rays	concave metal dish	solid state detector

Siting telescopes Most incoming radiation is blocked by the Earth's atmosphere. That which does pass through includes light, parts of the radio spectrum, and some infrared and ultraviolet. Radio waves can pass through interstellar dust, which blocks light from galactic centres.

Telescopes are sited as follows:

- Radio telescopes are usually ground-based.
- Optical telescopes are mounted as high in the atmosphere as possible (e.g. on mountain tops) or above it. This is to reduce image quality problems caused by atmospheric refraction and 'light pollution' from cities.
- Infrared, ultraviolet, X, and γ radiation telescopes are placed in high-altitude balloons or orbiting satellites. Satellite-based instruments include:
  HST, the *Hubble Space Telescope* (optical),
  COBE, the *Cosmic Background Explorer* (microwave),
  IRAS, the *Infrared Astronomical Satellite*.

Magnitude

The ***apparent magnitude***, *m*, of a star is a measure of its observed brightness. On the scale of *m* values, 0, 1, 2, 3 etc represents an order of *decreasing* observed brightness (see L5).

- By definition, a star for which $m = 1$ appears 100 times brighter than a star for which $m = 6$.
- The brightest stars in the sky have negative values of *m*.
- A star's apparent magnitude can be deduced by analysing its image on a photographic plate or CCD.

The ***absolute magnitude***, *M*, of a star is the apparent magnitude it would have if it were 10 parsecs away.

- Absolute magnitude is directly related to luminosity and does not depend on the star's distance from the Earth.

Using the inverse square law for intensity (see L4), it can be shown that a star's distance *d* (in pc), apparent magnitude *m*, and absolute magnitude *M* are linked like this:

$$m - M = 5 \log \frac{d}{10} \quad (1)$$

For methods of estimating distances to stars, see H2.

Star	Apparent magnitude (m)	Absolute magnitude (M)	Distance/ pc
Sirius	–1.5	1.4	2.7
Rigel	0.1	–7.2	290
Deneb	1.3	–7.2	490
Proxima C	11.1	15.5	1.3

L3 Measuring distance and speed

The Doppler effect

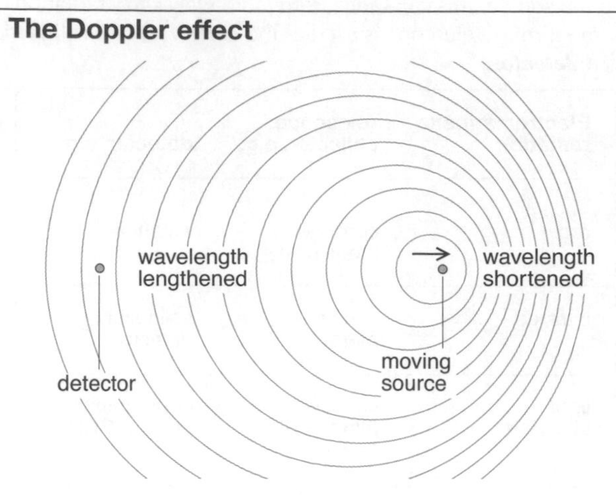

If a wave source is receding (moving away) from a detector, the waves reaching the detector are more spaced out, so their measured wavelength is increased and their frequency reduced. This is an example of the ***Doppler effect***. It causes the change of pitch which you hear when an ambulance rushes past with its siren sounding.

$$\frac{\Delta f}{f} = -\frac{v}{c} \text{ and } \frac{\Delta\lambda}{\lambda} = \frac{v}{c} = z$$

f and λ are the emitted frequency and wavelength. Δf and $\Delta\lambda$ are the observed changes – both defined as *increases*.

How Science Works

Red shift

Information about a star's temperature composition and motion can be found by analysing its spectrum (see D1).

Star motion can be fast enough to cause a detectable Doppler shift in light waves. If a star is moving *away* from the Earth, its spectral lines are shifted towards the *red* end of the spectrum. If v is the relative velocity of recession, and v is small compared with the speed of light, c.

- The Sun's rotation causes a broadening of its spectral lines, because light from the receding side is red-shifted while that from the approaching side is blue-shifted. This effect can be used to work out the speed of rotation.

Answering questions

Re-read your answers and make sure that the examiner can understand what you mean.

When frequency increases, wavelength decreases, and vice versa. So writing: 'The Doppler effect is the change of wavelength and frequency when a source moves. If the source moves towards the observer it increases.' is not clear enough; the first sentence is OK, but the second could refer to frequency or wavelength so will gain no marks.

When you describe beams of travelling waves, some words can be used for distance or for time. Words such as 'longer' and 'shorter', for example, may be unclear.

'It will arrive at point B faster' is unlikely to gain an explanation mark, because 'It will arrive at point B faster ***because it has taken a shorter route***' is quite different from 'It will arrive at point B faster ***because it has a greater velocity***', and the examiner cannot tell whether the student understands the situation or not.

Measuring distance

Distance can be measured by sending out a pulse of waves and measuring the time it takes for the echo to return.

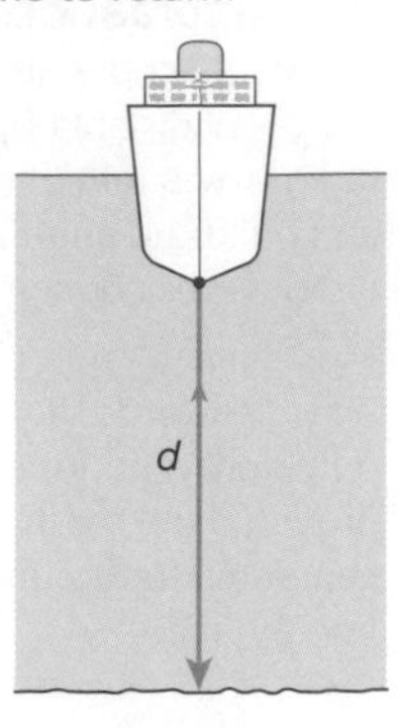

Applications of this are ultrasonic or infrared range finders, ships' sonar, and radar. This is also how ultrasound scanners build up images (see M3).

If you know the speed of the waves, v, the distance, d, is $d = v\,t/2$.

It is important to remember to divide by 2, as the pulse has travelled there and back, twice the distance, in the time measured.

If d is known, this method can be used to work out the speed of the waves.

Measuring speed (pulse–echo method)

A pulse–echo method can be used to work out the speed of an object approaching or moving away.

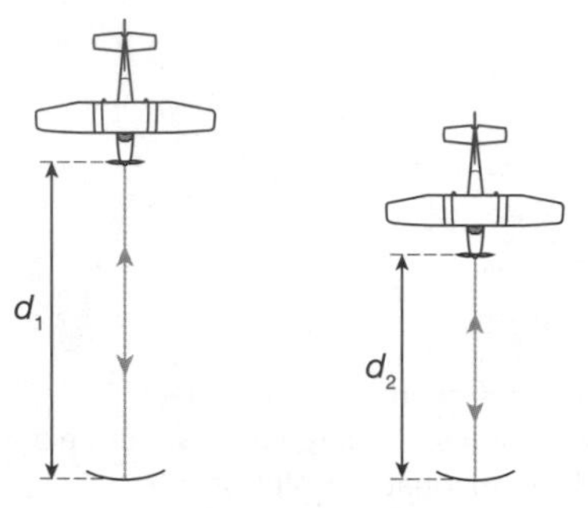

This is done by making two measurements of the distance (see above) at a fixed time apart. If the two distances are d_1 and d_2, the average speed is

$v = (d_1 - d_2)/t$

v will be positive if the object is approaching and negative if it is moving away.

How Science Works

Blood flow

When a beam of ultrasound is sent into the body, any motion within the body causes a Doppler shift in the reflected ultrasound. This can be used to check the heart beat or blood flow in an unborn baby. Continuous, rather than pulsed, ultrasound is used. Any difference between the outgoing and returning frequencies is heard as a tone or displayed on a screen.

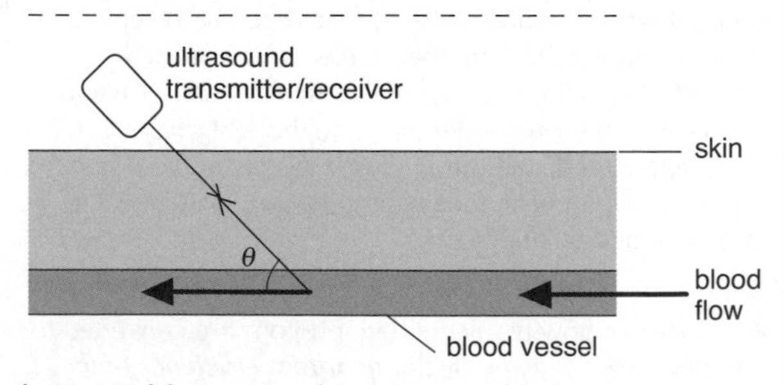

The original frequency of the ultrasound is f.
The change in frequency due to the Doppler effect is Δf.
The wave speed in the medium is c.
The speed of the blood v is given by

$$v = \frac{c \times \Delta f}{2f\cos\theta}$$

The volume flow rate of blood through the blood vessel is Av where A is the area of cross-section of the blood vessel.

L4 Astrophysics – 1

Solar System, stars, and galaxies

The Earth is one of many ***planets*** in orbit around the Sun. The Sun, planets, and other objects in orbit are together known as the ***Solar System***.

Most of the planets move in near-circular orbits. Many have smaller ***moons*** orbiting them (see H5 for orbital equations and laws). ***Comets*** are small, icy objects in highly elliptical orbits around the Sun. Planets, moons, and comets are only visible because they reflect the Sun's light.

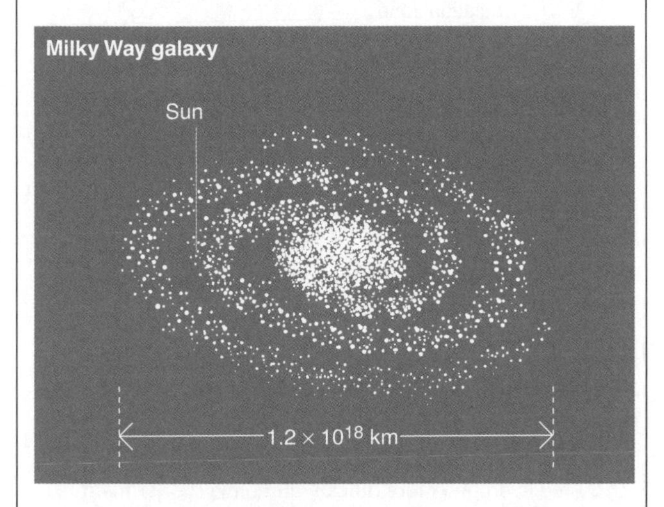

The Sun is one star in a huge star system called a ***galaxy***. Our galaxy contains about 10^{11} stars, as well as interstellar matter (thinly-spread gas and dust between the stars). Our galaxy, called the ***Milky Way***, is slowly rotating, with a period of more than 10^8 years. It is held together by gravitational attraction. It is just one of many billions of galaxies in the known ***Universe*** (see L5).

Normal galaxies emit mostly light. However, about 10% of galaxies have active centres which emit strongly in other parts of the electromagnetic spectrum as well.

Distance units

In astrophysics, the following distance units are used.

Light-year (ly) This is the distance travelled (in a vacuum) by light in one year. 1 light-year = 9.47×10^{15} m.

Astronomical unit (AU) This is the mean radius of the Earth's orbit around the Sun. 1 AU = 1.50×10^{11} m.

Parsec (pc) This is the distance at which the mean radius of the Earth's orbit has an angular displacement of one arc second (1/3600 degree). (See also *Parallax* below).

1pc = 3.26 ly = 2.06×10^5 AU = 3.09×10^{16} m

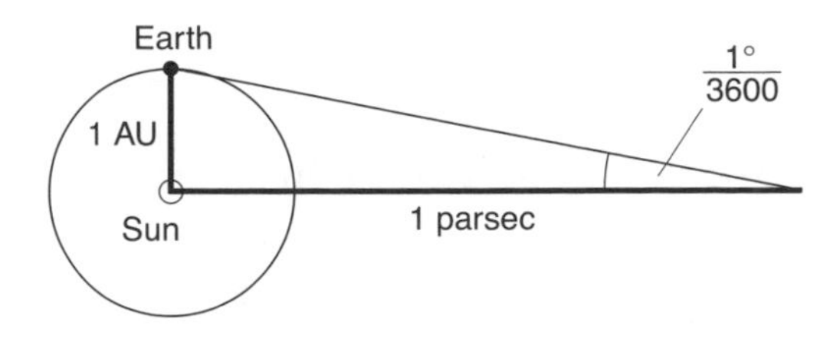

diameter of Earth = 1.3×10^4 km
diameter of Sun = 1.4×10^6 km
radius of Earth's orbit = 1.5×10^8 km = 1 AU

diameter of Solar System = 50 AU
distance to nearest star (*Proxima Centauri*) = 2.7×10^5 AU = 4.2 ly = 1.3 pc

diameter of galaxy (*Milky Way*) = 1.3×10^5 ly = 40 kpc
distance to neighbouring galaxy (*Andromeda*) = 2.2×10^6 ly = 0.7 Mpc

Luminosity and observed brightness

The ***luminosity*** of a star is the rate at which it radiates energy. The unit of luminosity is the watt (W).

The observed brightness of a star depends on its luminosity *and* on its distance from Earth. A very luminous star can appear dim if it is far enough away, because the intensity of its radiation obeys an inverse square law (see below).

Estimating distances to stars and galaxies

These are the main methods used:

Parallax As the Earth orbits the Sun, nearby stars appear to move agains the background of very distant stars. The nearer the star, the greater its apparent movement. This effect is called ***parallax***.

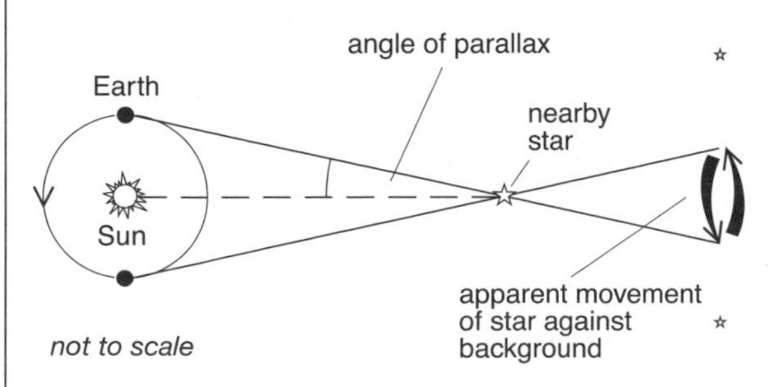

By measuring the angle of parallax, the distance can be calculated using trigonometry. The method is suitable for distances up to ~100 pc. (The pc – parsec, see above – was defined to aid calculations involving parallax.)

Inverse square law The distance of a star or galaxy is worked out by comparing its apparent and absolute magnitudes. This method is suitable for very distant objects. However, it requires 'standard' sources (sometimes called 'standard' candles') of known luminosity, for example

- stars whose positions on the Hertzsprung–Russell diagram (see page 124) luminosity can be worked out by spectral analysis
- cepheid variables, which are stars with a brightness that varies regularly with a period that can be used to calculate the luminosity. This can then be compared to the observed brightness.
- supernovae (see L7).

The Hubble law Galaxies have a red shift which is proportional to their distance from the Earth (see H3). This can be used to estimate the distance.

Classifying stars

Stars can be classified according to their spectra. The main spectral classes are: O, B, A, F, G, K, M. This represents an order from high to low temperature (see diagram below).

Here are details of some of the spectral classes:

- O-type stars are the hottest and appear blue-white. Helium lines are prominent in their absorption spectra.
- A-type stars appear white. Hydrogen lines are prominent in their absorption spectra.
- G-type stars, like the Sun, appear yellow-white. There are many metallic lines in their absorption spectra.
- M-type stars are the coolest, and appear red. Banding in their absorption spectra indicates the presence of molecules.

The Hertzsprung–Russell (H–R) diagram

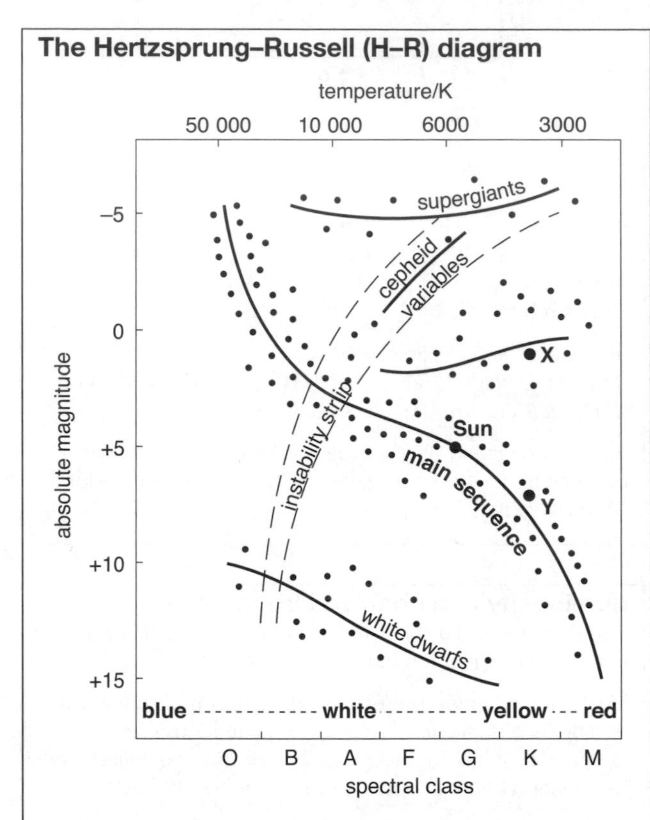

This is a diagram in which the absolute magnitudes of stars are plotted against their spectral classes. A simplified version is shown above.

Note:

- Star X is the same spectral class as star Y, but higher up the diagram, i.e. X is at the same temperature as Y, but radiates more light. So X is larger than Y.
- Terms such as ***giant*** and ***dwarf*** indicate star size.
- The points on an H–R diagram occur in zones. Most stars, including the Sun, belong to the ***main sequence***.

Birth of a star

Stars form in huge clouds of gas (mainly hydrogen) and dust called ***nebulae.*** The Sun formed in a nebula about 5×10^9 years ago. The process took about 5×10^7 years:

Gravity pulled more and more nebular matter into a concentrated clump called a ***protostar***. The loss of gravitational PE caused a rise in core temperature which triggered the fusion of hydrogen and the release of energy (see J6). Thermal activity stopped further gravitational collapse. The Sun had become a main sequence star. (Its planets had formed in an orbiting disc of nebular matter.)

Life and death of a star

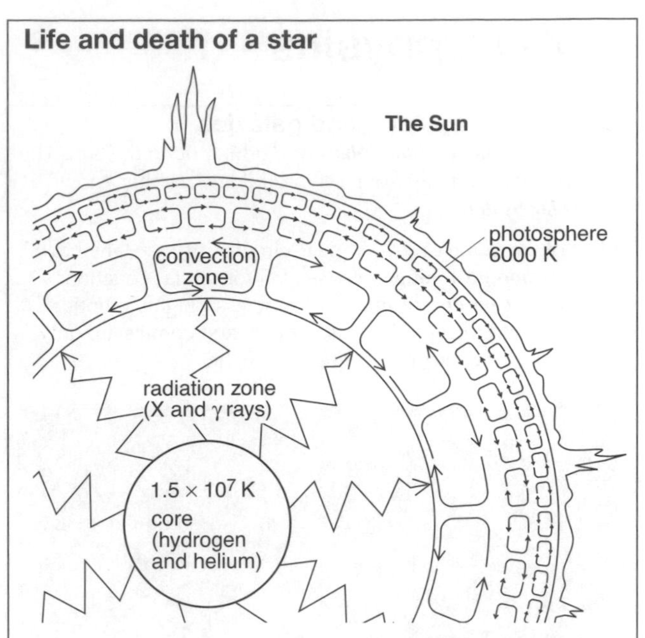

The Sun gets most of its energy from the ***proton–proton chain***, a multi-stage fusion process which converts hydrogen-1 into helium-4. Hotter, more massive stars use the ***CNO cycle***. This also changes hydrogen-1 into helium-4, but involves carbon, nitrogen, and oxygen nuclei.

The Sun is about half way through its life on the main sequence (about 10^{10} years). Hotter, more massive stars consume hydrogen more quickly and have shorter main sequence lives.

When all its hydrogen has been converted into helium, the Sun will take the path shown on the H–R diagram below.

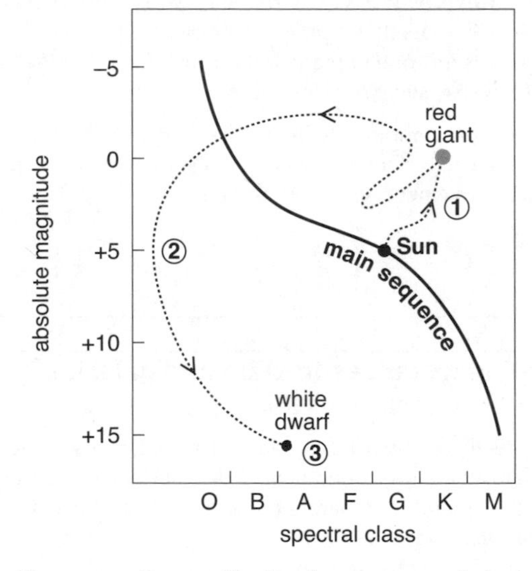

1 The core collapses. The Sun becomes a ***red giant*** as its outer layers expand and cool (and engulf the Earth). With the core temperature rising to over 10^8 K, energy is released by the fusion of helium into carbon.

2 After further changes, the outer layers expand and drift off into space. The core and inner layers become a ***white dwarf*** whose core is so dense that the normal atomic structure breaks down. The electrons form a ***degenerate electron gas*** whose pressure stops further collapse.

3 Fusion ceases. The white dwarf cools and fades for ever.

Note:

- Stars less massive than the Sun end up as white dwarfs, without going through the giant stages.
- Massive stars become giants or supergiants, then end up as ***neutron stars*** or ***black holes*** (L7).

L5 Cosmology

The structure of the Universe

The study of the Universe, its origins, and evolution is called ***cosmology***.

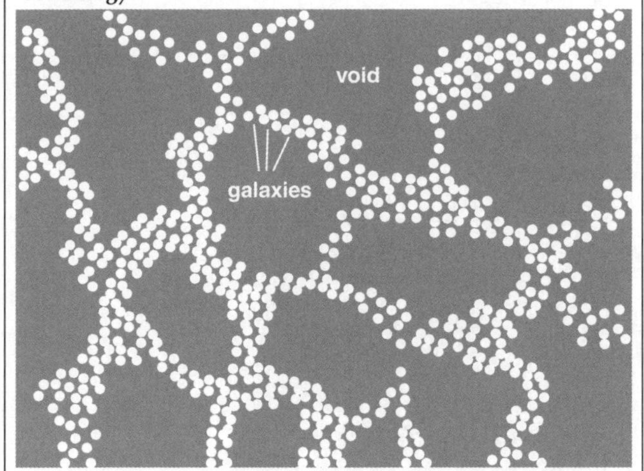

The Universe contains billions of galaxies. Their average separation is ~ 10^6 light-years. Together, they form a network of long, clumpy filaments with huge voids (spaces) in between. Despite their local irregularities, the galaxies are, on a large scale, evenly distributed in all directions.

The motion of galaxies indicates that they are surrounded by massive amounts of thinly spread, invisible material. This is called ***dark matter*** (see L6). Its nature is not yet known.

Hubble's law

Measurements of Doppler red shifts (see L3) indicate that, in general, the galaxies are receding from each other. The further away the galaxy, the greater is its red shift and, therefore, the greater its recession velocity.

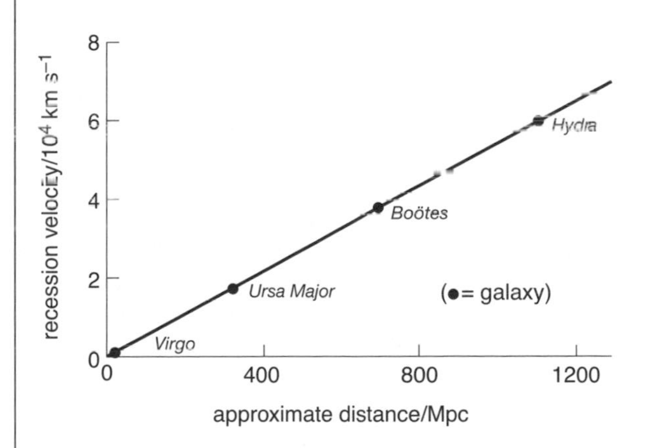

According to ***Hubble's law***, the distance d of a galaxy and its recession velocity v are linked by this equation:

$$v = H_0 d \quad (1)$$

H_0 is called the ***Hubble constant***. Large distances are difficult to estimate accurately, so the value of H_0 has a high uncertainty. However, it is thought to lie in the range 50–100 km s^{-1} Mpc^{-1} (1.6–3.2×10^{-18} s^{-1}). Its value is important for several reasons:

- It enables the distances of the most remote galaxies to be estimated from their red shifts.
- The age of the Universe can be estimated from it (see above right). (H_0 has dimensions of 1/time.)
- The fate of the Universe depends on it (see next page).

The expanding Universe

The most generally accepted explanation of galactic red shifts is that the Universe is expanding. At zero time, all its matter and energy was together in a highly concentrated state.

Estimating the age of the Universe If a galaxy is d from our own, and has a steady recession velocity v, then separation of the galaxies must have occurred at a time d/v ago. This represents the approximate age of the Universe. From equation (1) $d/v = 1/H_0$, so

$$\text{age of the Universe} \approx 1/H_0$$

This gives an age in the range 1–2×10^{10} years (10–20 billion years).

Note:

- The above calculation assumes constant v. However, recent observations suggest that the Universe's rate of expansion may actually be increasing, although the reason for this is not yet clear.

Olbers' paradox In the 17th century, it was pointed out that, if the stars continued out to infinity, the night sky should be white, not dark – because light must be coming from every possible direction in the sky. This became known as Olbers' paradox.

Two reasons for the dark night sky have been suggested:

- In an expanding Universe, red-shifted wavelengths mean reduced photon energies (see E1), so the intensity of the light from distant stars is reduced.
- There is a limit to our observable Universe. If, say, the Universe is 15 billion years old, then we have yet to receive light from stars more than 15 billion light-years away. So everything beyond that distance looks dark.

The cosmological principle This says that, apart from small-scale irregularities, the Universe should appear the same from all points within it (i.e. the distribution of galaxies and their recession velocities should appear the same from all points).

The hot Big Bang theory

According to this theory, sometimes called the ***standard model***, the Universe (and time) began about 10–20 billion years ago when a single, hot 'superatom' erupted in a burst of energy called the ***Big Bang***. As expansion and cooling took place, particles and antiparticles formed. Further cooling meant that combinations were possible, so nuclei and then atoms formed – and eventually galaxies (see next page).

Fundamental forces In the instant after the big bang, the fundamental forces (see K2) existed as one superforce. But within 10^{-11} s, they had separated from each other.

Cosmic background There is a steady background radiation which comes from every direction in space. It peaks in the microwave region, and corresponds to the radiation from a black body (see I7) at 2.7 K. It is thought to be the red-shifted remnant of radiation from the big bang. Its presence is predicted by the Big Bang theory.

Inflation The standard model cannot satisfactorily explain why, on a large scale, the Universe and its microwave background radiation are so uniform. Mathematically, it is possible to overcome this difficulty by assuming that the early Universe went through a brief period of very rapid inflation, when its volume increased by a factor 10^{50}.

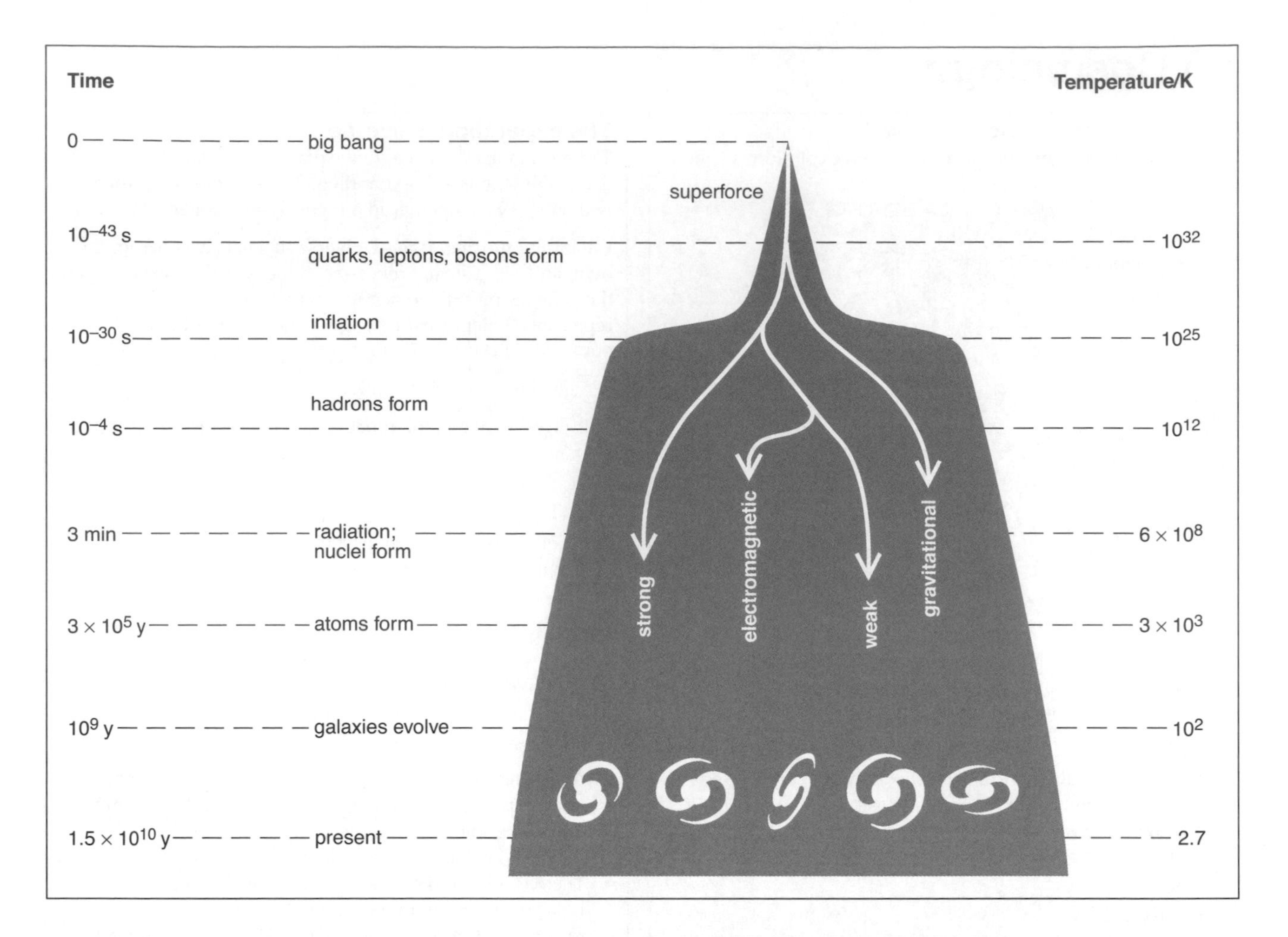

The fate of the Universe

Gravity is affecting the expansion of the Universe. The fate of the Universe depends on how its average density ρ compares with a certain **critical density** ρ_0:

- If $\rho < \rho_0$ the expansion continues indefinitely.
- If $\rho = \rho_0$ the expansion continues, but the rate falls to zero after infinite time.
- If $\rho > \rho_0$ the expansion reaches a maximum, and is followed by contraction.

The average density of the Universe is thought to be close to the critical density.

Linking ρ_0 and H_0 The critical density depends on the value of the Hubble constant. A higher H_0 means a higher recession velocity per unit separation. So a higher density is needed to stop the expansion. It can be shown that

$$\rho_0 = \frac{3H_0^2}{8\pi G}$$ where G is the gravitational constant (see H4)

This gives ρ_0 in the range 5–20 × 10^{-27} kg m^{-3}.

Models of the Universe

The Big Bang was not an explosion into existing space. Space itself started to expand. The galaxies are separating because the space between them is increasing.

Space has three dimensions of distance (represented by x, y, and z co-ordinates) and one of time. According to Einstein's theory of general relativity, gravity causes a curvature of space-time. If gravity is sufficiently strong, it may produce a 'closed' Universe, as shown below.

To visualize the expansion of the Universe, it is simpler to use models with only two of the distance dimensions. Imagine that the Universe is on an expanding, elastic surface. Three possible models are shown below. In each case, the galaxies move apart as the surface stretches. From any position on the surface, each galaxy recedes at a velocity that is proportional to its distance away.

Note:

- It is the value of the critical density, and therefore of the Hubble constant, which decides whether we live in an open, flat, or closed Universe.

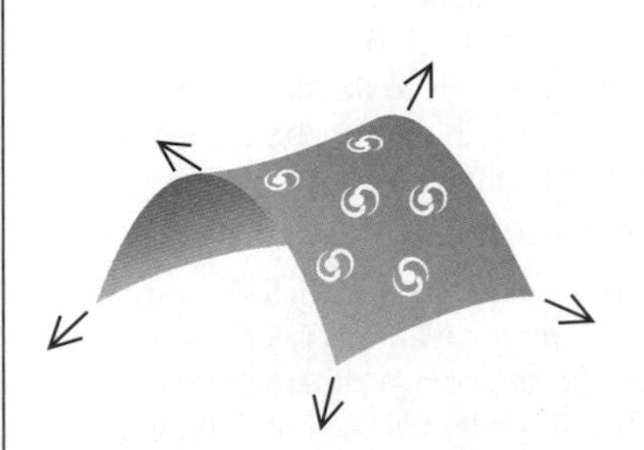

Open Universe ($\rho < \rho_0$) The surface is infinite and unbounded.

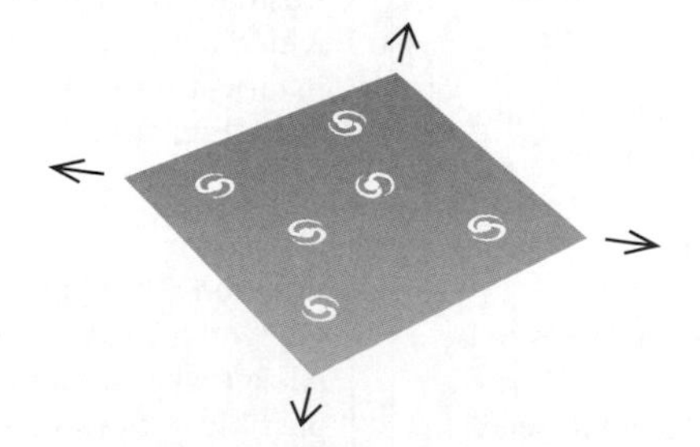

Flat Universe ($\rho = \rho_0$) The surface is infinite and unbounded.

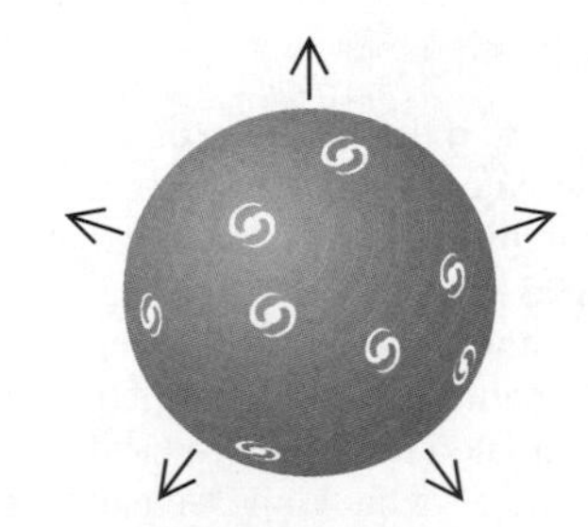

Closed Universe ($\rho > \rho_0$) The surface is finite and bounded.

L6 Dark matter

How Science Works

Why dark matter?

In the early 1930s Swiss physicist Fritz Zwicky calculated the velocities of galaxies and found eight that were moving 400 times faster than theory predicted. He suggested that 'dark matter' that could not be seen increased the gravity in the galaxy.

There have been other anomalies, such as the orbital speed of galaxies in clusters – which is too high – and the formation of galaxies after the Big Bang – which occurred far too quickly, based on calculations involving visible matter.

A summary of the suggested explanations:

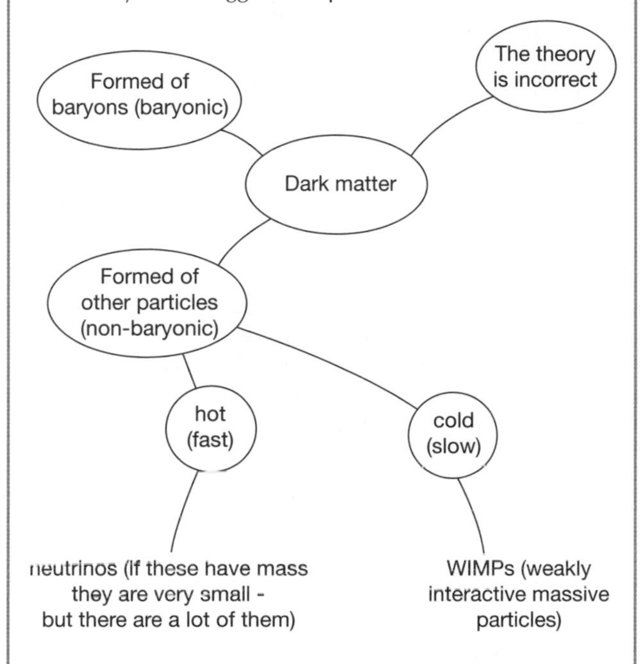

Baryonic matter consists of protons and neutrons. Suggestions for baryonic matter that could account for the dark matter in the universe include: black holes, neutron stars, brown dwarfs, and even planets.

The Wilkinson Microwave Anisotropy Probe (WMAP) was launched by NASA in 2001. It calculated the age and size of the Universe more accurately than before and also showed that the Universe is composed of 4% ordinary matter, 23% of an unknown type of dark matter, and 73% dark energy (error ± 5%.)

Experiments to detect neutrinos include the Super-Kamiokande in Japan.

The (so far theoretical) WIMPs feel only the weak force and gravity – they are a little like neutrinos but with large mass. Experiments to try to detect them include the Boulby Salt mine experiment in the UK (under the North Sea), the Soudan mine experiment in the USA, and the (completed) DAMA-NaI experiment in Italy.

How Science Works

Dark energy

Calculations show that about 5 billion years ago the expansion of the Universe speeded up. To account for this it is suggested that there is dark energy in the Universe, which we cannot detect.

How Science Works

Theories of dark matter and dark energy

The development of the Big Bang theory is a good example of the tentative nature of scientific knowledge. It is very much 'a work in progress'. All over the world, over many years, scientists working in cosmology and astrophysics, and also those working in nuclear and particle physics, contribute observations and ideas.

Theories are used to make predictions and then experiments are carried out to try and confirm or reject the theory. At the same time observations are made and then calculations done to see if they 'fit' with the current theory. There is no suggestion that the theory should be kept at all costs. If a new, better explanation comes along, it will be accepted. As time goes on, however, and lots of evidence appears to fit with the theory, scientists will look for modifications to a theory that seems to work well, rather than starting all over again with a new theory.

This is also a good example of scientists validating new knowledge and ensuring integrity. They publish the details of their experiments and their results and read the publications of other scientists. There are several different types of experiments used to look for the same thing, so the work can be checked. Scientists meet at conferences to discuss results – both positive and negative – and the conference proceedings are published, too.

L7 Astrophysics - 2

More stellar objects

Supernovae When a massive star enters its giant phase, its core becomes so hot that carbon is fused into heavier elements. If the star exceeds about 8 solar (Sun's) masses, iron is produced. As this is at the top of the binding energy curve (see J6), fusion no longer supplies energy. The core collapses, causing a shock wave which blows away the star's outer layers in a huge explosion called a ***supernova***. For a few days, this is millions of times brighter than a star. Elements ejected from supernovae eventually 'seed' the nebulae in which new stars and planets will form. Supernovae can be used as standard candles (see L4) to determine distance.

Neutron stars If the core of a supernova exceeds about 1.4 solar masses, the degenerate electron gas cannot resist gravitational collapse. Electrons and protons are pushed together to form neutrons. The result is a neutron star – essentially a giant nucleus about 10–30 km across.

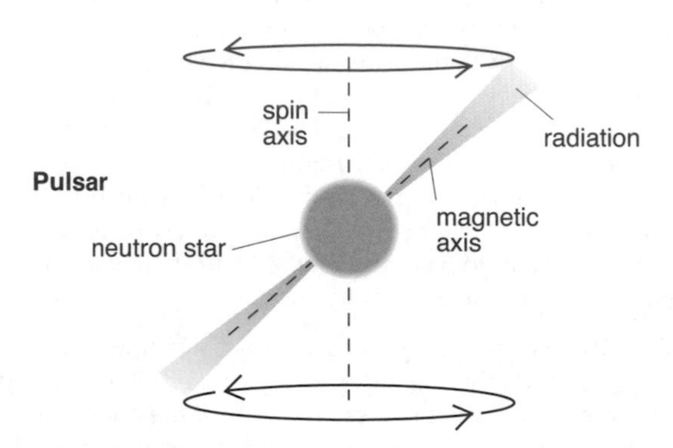

Pulsars These emit radio, light, or X-ray pulses at up to 500 times per second. They are believed to be rapidly spinning neutron stars. Pulses are detected because the star sends out two narrow radiation beams which rotate with it, rather like the beams of light from a lighthouse.

Binary stars These are two stars which rotate about a common centre of mass. If they are close, gravity may pull material from one to the other. If one is a neutron star or black hole, material falling into it will give off X-rays.

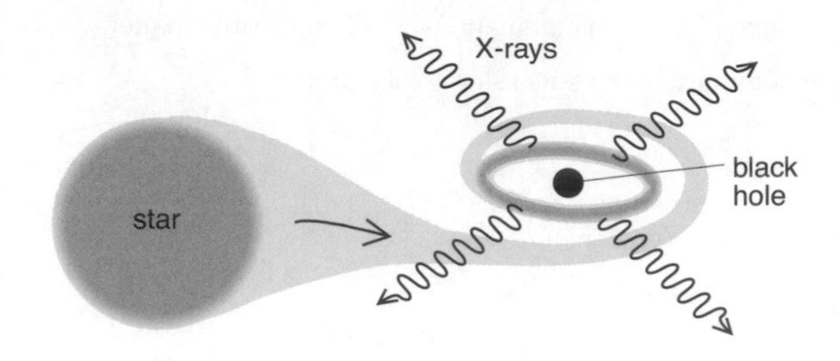

X-ray binary system

Quasars These have red shifts which suggest that they are the most distant objects in the Universe (see *Hubble's law* in L5). If they really are distant, they radiate as much energy as some galaxies, but have only the volume of a solar system. Each may be the active centre of a galaxy where nebular matter surrounds a supermassive black hole.

Note:

- Some scientists argue that quasars are closer, less luminous objects whose red shifts have some other cause.

Black holes

If the core of a supernova exceeds about 2.5 solar masses, even the neutrons formed cannot resist gravitational collapse. The core shrinks to become a black hole . For a black hole, the escape velocity (see H4) is greater than the speed of light, and no particles or radiation can escape.

There is thought to be a supermassive black hole at the centre of a galaxy.

Event horizon

Because no light escapes from a black hole, we cannot see anything occurring inside it. A black hole is surrounded by an imaginary sphere, which is the boundary between things we can see outside and the invisible inside. This boundary is called the ***event horizon***.

The radius of the event horizon is called the ***Swarzschild radius***, *R*, and is found for a black hole of mass *M* by setting the escape velocity equal to the speed of light, *c*.

$$R = \frac{2\,GM}{c_2}$$

More about classifying stars

See also L4

Spectral class	Intrinsic colour	Temperature (K)	Prominent absorption lines
O	blue	25 000–50 000	He+, He, H
B	blue	11 000–25 000	He, H
A	blue-white	7500–11 000	H (strongest) ionized metals
F	white	6000–7500	ionized metals
G	yellow-white	5000–6000	ionized and neutral metals
K	orange	3500–5000	neutral metals
M	red	<3500	neutral atoms, TiO

M1 Medical physics – 1

The human eye

Read D4 first.

In the eye, the cornea and lens form a real image on the ***retina***. The focus is adjusted by changing the shape of the lens – a process called ***accommodation***. This enables a normal eye to form a clear image of any object between its ***near point*** (about 25 cm away) and infinity. The amount of light reaching the retina is controlled by the ***iris***.

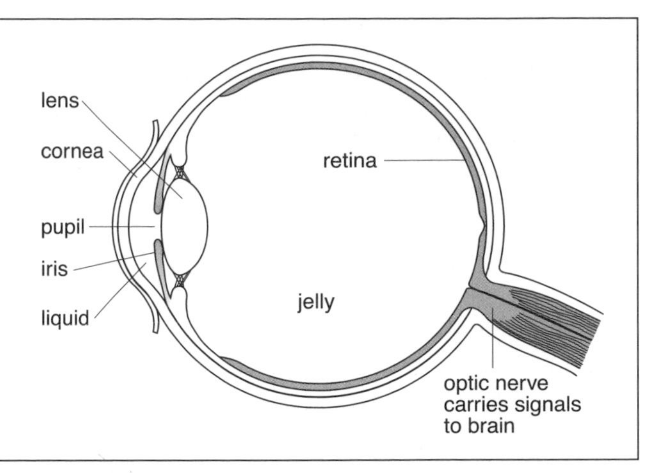

The retina contains millions of light-sensitive cells which send signals to the brain. ***Rods*** are the most sensitive, but do not respond to colour. Different ***cones*** respond to the *red, green,* and *blue* regions of the spectrum.

Myopia (short sight) A short-sighted eye cannot accommodate for distant objects. The rays are brought to a focus before they reach the retina. The defect is corrected by a *concave* spectacle lens.

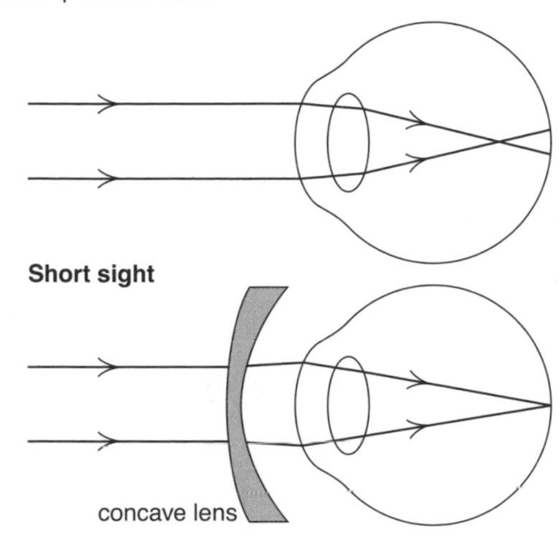

Hypermetropia (long sight) A long-sighted eye cannot accommodate for close objects. The rays are still not focused by the time they reach the retina. The defect is corrected by a *convex* spectacle lens.

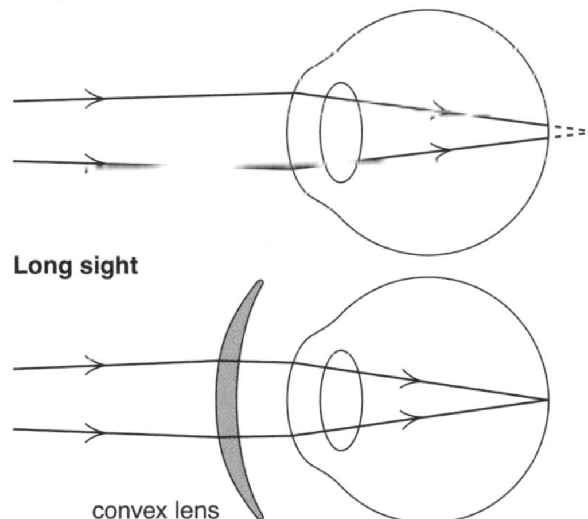

Lens power This is defined as $1/f$. With f in metres, the unit is the ***dioptre*** (D). (Note: *diverging* lenses have *negative* powers). The power of the eye's lens system varies. It is lowest when distant objects are being viewed.

The powers of close lenses can be added algebraically. For example, if a normal eye must reduce its power to 50 D to see distant objects, but a short-sighted eye cannot get beneath 54 D, then the defect is corrected by a lens of power –4 D, i.e. a concave lens with $f = \frac{1}{4} = -0.25$ m.

Astigmatism This effect occurs when the curve of the cornea is not perfectly spherical. So, for example, vertical lines might be seen in clearer focus than horizontal ones.

Scotopic vision

This is the term that describes vision in low light or night-time conditions. Rods are not sensitive to colour but have their peak response in the green part (wavelength ≈ 500 nm) of the visible spectrum. They are very sensitive to light and are the principal receptors in dim light.

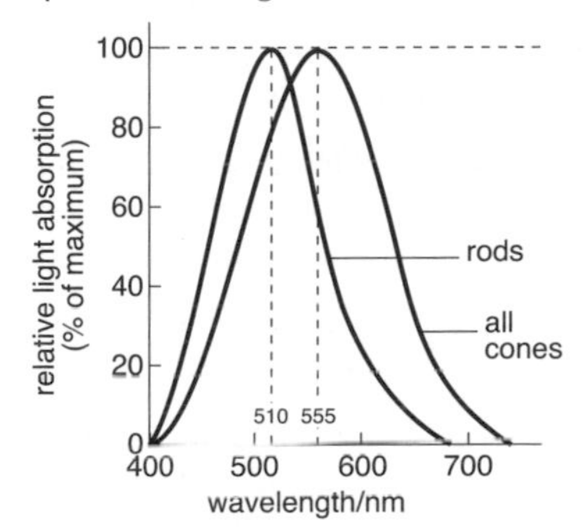

The frequency responses of the rods are shown in the diagram above.

Photoptic vision

Although they are not as sensitive as rods, cones are able to differentiate between different colours. There are three types of cone that are sensitive to different colours. The sensitivity of the cones is illustrated in the graph below.

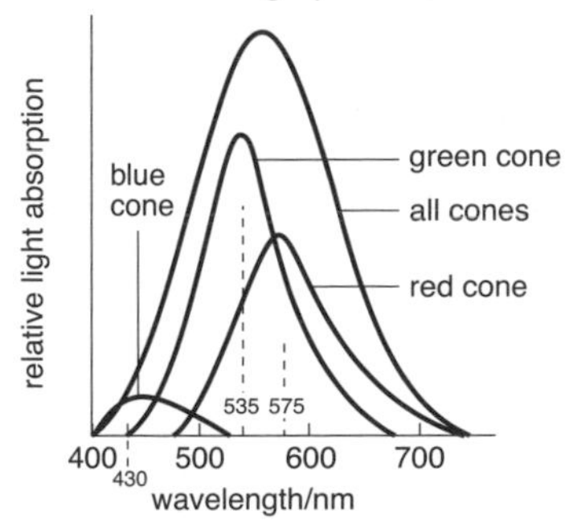

They are used in bright light or daytime conditions.

Persistence of vision

When a rod or cone has been stimulated by light it takes about 0.2 s for the image to disappear. Successive flashes merge or 'fuse' to produce a continuous image if they arrive frequently enough. In dim light 5 Hz is adequate to produce a continuous image.

In high-intensity conditions images need to arrive more frequently. In cinemas 24 frames per second are projected and in TV 60 images per second are transmitted.

The heart

The diagram shows schematically the action of a heart.

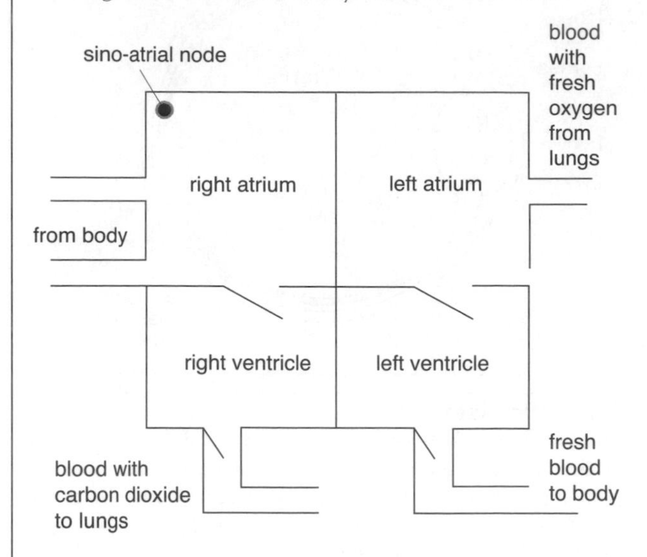

The heart is essentially a pump. There are two separate paths through which blood flows. The valves are one-way valves so that blood can only flow one way.

In path 1, blood is pumped to the body via arteries so that oxygen reaches the body tissue. At the same time the blood picks up carbon dioxide returning to the heart through veins.

In path 2 the blood is pumped to the lungs where it releases the carbon dioxide and picks up a fresh supply of oxygen.

The ***sino-atrial node*** is the pacemaker, a group of cells that generate electric signals, or action potentials, that trigger the heart muscle to contract. The contraction spreads from this region throughout the heart.

Action potentials

Signals in the form of voltage pulses are passed round the body by the nervous system. The signals are sent by living cells called ***neurons***. The signals travel along ***axons***. The signals may stimulate routine tasks such as the heartbeat. Some nerves work as a result of an outside stimulus, such as when you touch something very hot. Some neurons send signals to the brain, others directly to muscles causing them to contract in response to an emergency.

When not transmitting the resting potential of the neuron is about –70 mV. This is the difference between the voltages inside and outside the neuron.

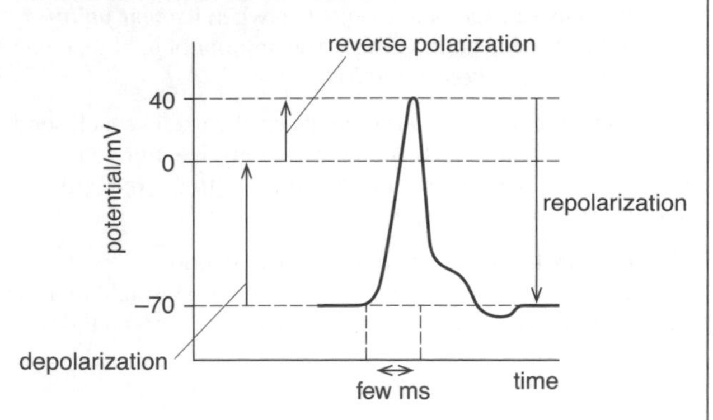

When the neuron receives a stimulus above a threshold level a voltage pulse travels along the axon to the brain or directly to the muscle causing the body to act in reaction to the stimulus.

Electrocardiograms (ECGs)

Body fluids conduct electrical signals, so the small potential differences generated by action potentials can be detected by electrodes on the surface of the skin. An ECG is a recording of the action potentials in the heart that have been detected and amplified. The graph shows the voltage variation for a single heartbeat.

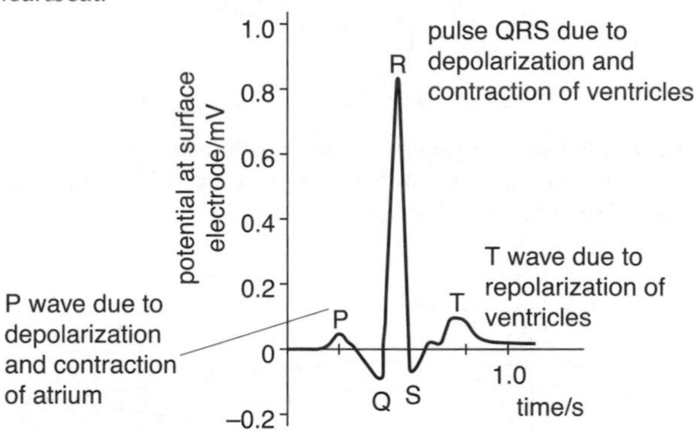

Signals detected at different parts of the body are indicative of different aspects of the body's function, so they can be used to diagnose health problems, particularly those directly related to the heart itself.

M2 Medical physics – 2

Human hearing

Read D1 first.

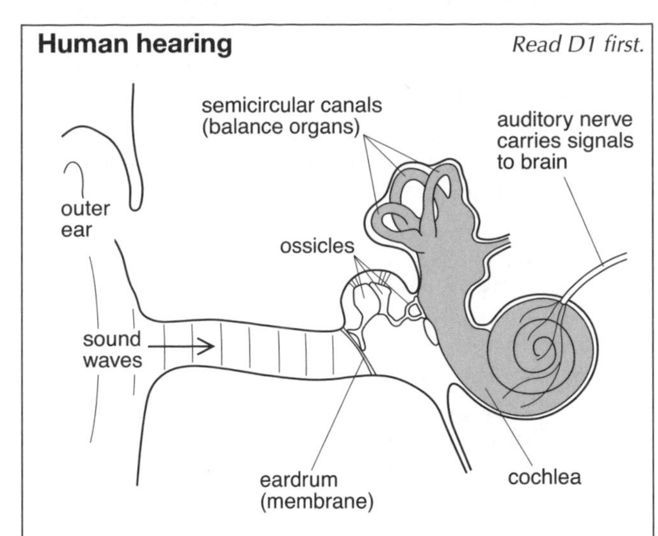

Sound waves entering the ear set up vibrations in the ***ear drum***. These are transmitted to the ***cochlea*** by ***ossicles*** (small bones) which act as levers and magnify the pressure changes. In the cochlea, sensory cells respond to different frequencies and send signals to the brain.

Frequency response The ear detects frequencies in the range 10 Hz to 20 kHz. It is most sensitive around 2 kHz.

Intensity levels The lowest intensity of sound which the ear can detect is known as the ***threshold intensity***, I_0. It is taken as 10^{-12} W m^{-2}.

The ear responds to sounds according to the *ratio* of their intensities. For example, each doubling of intensity gives the same sensation of increased loudness. For this reason, the ***intensity level*** of a sound is defined as below, where dB stands for ***decibel***, and I is the intensity of the sound:

intensity level in dB = $10 \log_{10}(I/I_0)$

so threshold = 0 dB

a 3 dB increase = double the intensity

For two sounds of intensities I_1 and I_2,

difference in intensity level in dB = $10 \log_{10}(I_2/I_1)$

Note:

- Any two sounds with the same intensity ratio I_2/I_1 have the same difference in intensity level.

dBA scale This is a dB scale, adjusted to allow for the ear's different sensitivity to different frequencies. It is used to measure noise levels. Typical values are:

hearing threshold	0 dBA	legal noise limit	90 dBA
conversation	50 dBA	near disco speakers	120 dBA
busy street	70 dBA	pain threshold	140 dBA

Hearing defects

Defects in hearing are caused by

- sound signals failing to reach the cochlea
- damage to the nerves that send signals from the cochlea to the brain.

Causes of defects

- natural ageing that gradually reduces the ability to hear high frequencies
- exposure to excessive noise such as that caused by an explosion or machinery
- an illness
- an accident
- prolonged exposure to excessively loud sounds such as that due to machines or excessively loud music. This can produce a continuous ringing sensation in the ear known as 'tinnitus'. As a health and safety measure those who work in noisy environments are required to wear ear defenders.

Using light

Read D1 first.

Endoscope This is a flexible tube used for looking inside the body. The light to illuminate the objects is transmitted along a bundle of optical fibres. The light from the objects, which forms the image at the far end, is carried to the viewing end by a *coherent* bundle of optical fibres, i.e. fibres whose positions alongside each other are the same throughout their length, so an accurate image is formed.

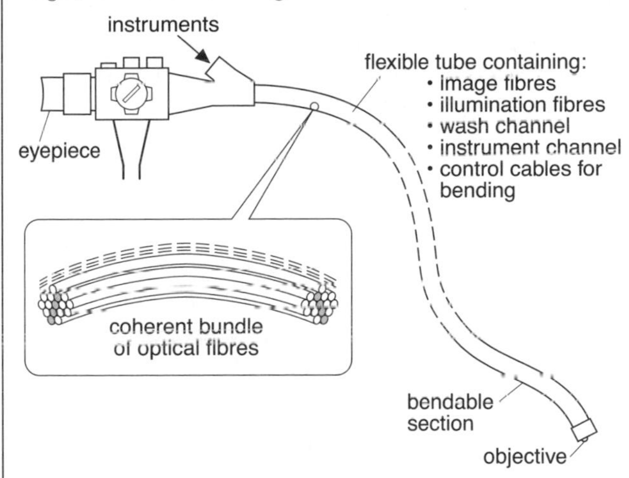

Laser This gives a fine beam of high-intensity light. It can be used like a scalpel to cut tissue, but its heating effect also seals small blood vessels at the same time. The type of laser chosen depends on the tissue – the aim being to select a wavelength which gives maximum energy absorption. In some cases, the light can be carried to the required point by an optical fibre, i.e. through an endoscope.

M3 Medical imaging – 1

Ultrasound

Read L3 first

Ultrasonic sound, or ***ultrasound***, has frequencies above the upper limit of human hearing, i.e. above 20 kHz. If a pulse of ultrasound is sent into the body, it is partially reflected at the boundaries between different layers of tissue, so their positions can be worked out from the time delays of the echoes received. The frequency used (in the 2–10 MHz range) depends on the depth of tissue. Increasing the frequency gives better resolution but poorer penetration.

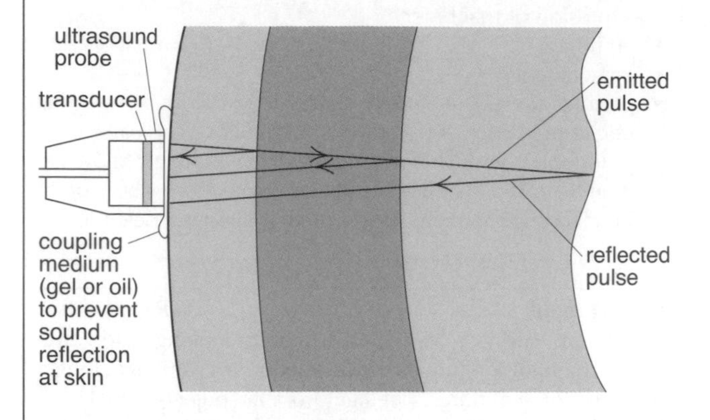

Probe This contains a ***transducer***, which sends and receives the ultrasound pulses. The ultrasound is produced using the ***piezoelectric effect***: a high frequency alternating voltage is applied across a slice of crystalline ceramic so that it vibrates at its resonant frequency and emits sound waves. When the reflected waves return, they cause vibrations in the slice, which generates a small alternating voltage.

Signals from the probe can be displayed on a oscilloscope or processed by a computer.

A-scan The reflected pulses are displayed as peaks on an oscilloscope, i.e. as **A**mplitude changes. Positions along the time axis are a measure of distances into the body.

B-scan The reflected pulses are displayed as spots whose **B**rightness is a measure of their amplitude. In a ***two-dimensional B-scan***, the probe is moved around the body in order to build up a complete cross-sectional image.

- Though its resolution is poorer, ultrasound imaging is safer than X-ray imaging and cheaper than MRI.

Destructive ultrasound Focused ultrasound, at an intensity above 10^7 W m^{-2}, can be used to break up kidney stones and gallstones, so surgery is not required.

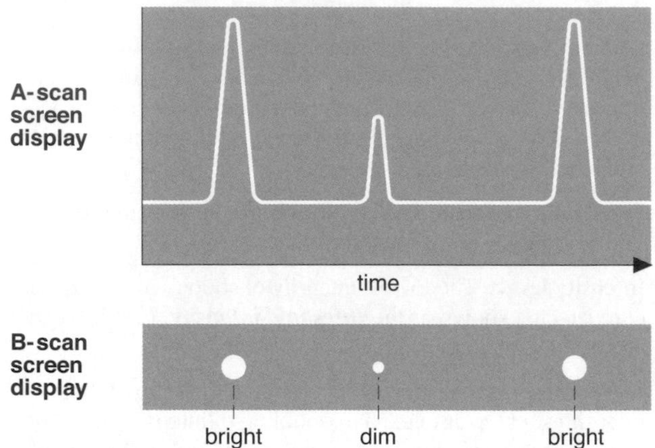

Acoustic impedance

A transmitting medium opposes the transmission of a sound pressure wave. The acoustic impedance is a measure of the resistance of a medium to the passage of sound waves.

The ***specific impedance*** of the medium $Z = \rho c$
where ρ is the density of the medium
and c is the speed of sound in the medium.

Intensity reflection coefficient α

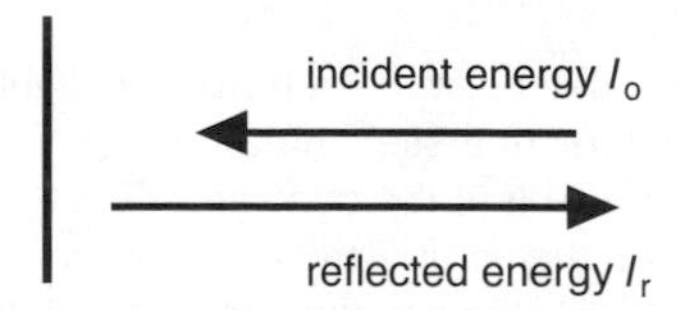

When ultrasound reaches an interface between two different media the fraction of the incident energy that is reflected is called the ***intensity reflection coefficient*** α_r and is given by

$$\alpha_r = \frac{I_r}{I_0} = \left(\frac{Z_2 - Z_1}{Z_2 + Z_1}\right)^2$$

Using ultrasound

Typical specific impedances are:

air	430 kg m^{-2} s^{-1}
fat	1.38×10^6 kg m^{-2} s^{-1}
blood	1.59×10^6 kg m^{-2} s^{-1}
bone	about 6×10^6 kg m^{-2} s^{-1}

At a muscle fat interface the reflection coefficient is 0.10 so 10% of the incident energy is reflected. Between fat and bone the figure rises to 60%. Even when only 1% of the energy is transmitted the echoes can be detected and can produce useful information.

Coupling medium

When sound is incident between air and the skin the reflection coefficient is almost 1 so that when ultrasound reaches the skin nearly all the incident sound is reflected. The coupling medium of gel or oil between the ultrasound probe and the skin reduces the amount of energy that is reflected. This is called impedance matching.

M4 Medical imaging – 2

X-rays

Read E1 and H11 first.

In the X-ray tube above, electrons gain high KE before striking a metal target. About 1% of the KE is converted into X-ray photons; the rest is released as heat. The anode is rotated rapidly to prevent 'hot spots'.

Quality The X-ray beam contains a range of wavelengths. The spectral spread is known as the ***quality***. The longer, least penetrating wavelengths are called *soft* X-rays; the shorter, more penetrating ones are *hard*. The beam can be hardened by using filters to absorb the longer wavelengths.

Tube current and voltage Increasing the current increases the intensity of the beam. Increasing the voltage increases the intensity *and* reduces the peak wavelength, i.e. it produces more photons, which are more penetrating.

Attenuation As X-rays pass through a material, their energy is gradually absorbed and their intensity reduced. This is called ***attenuation***. It is in addition to any reduction in intensity due to beam divergence.

If I_0 is the intensity of the incident beam (do not confuse with I_0 as used in ultrasound), and I is the intensity at a distance x into the material, then, for a non-diverging beam,

$$I = I_0 e^{-\mu x} \qquad (1)$$

μ is the ***total linear attenuation coefficient***. Its value depends on the absorbing material and on the photon energy. X-rays are more attenuated by bone than by soft tissue. The difference is greatest for photons of around 30 keV energy.

The half-thickness $x_{\frac{1}{2}}$ of an absorber is the value of x for which $I/I_0 = \frac{1}{2}$. From equation (1): $x_{\frac{1}{2}} = \log_e 2/\mu$

The ***mass attenuation coefficient*** μ_m (= μ/ρ) is often used in place of μ. This is the attenuation per unit mass and is constant for a given atomic number and photon energy.

Image contrast enhancement Contrast between soft tissue is poor so to yield useful information about the digestive tract patients are given a ***barium meal***. The relatively dense barium absorbs X-rays. The meal, in the form of a drink, is taken just before the X-ray is taken and this improves the contrast between the digestive system and the rest of the body.

X-ray photographs X-rays affect photographic film, but cannot be focused, so the photos are 'shadow' pictures of the absorbing areas. For a sharp image, the X-rays need to be emitted as if from a point source. To reduce the risk of cell damage, exposure times are normally less than 0.2 s.

CAT scanning (computed axial tomography) A cross-section of the body is scanned by rotating a fan of narrow monochromatic X-ray beams around it. The intensity reduction caused by each 'slice' is measured by a detector. Then a computer carries out a mathematical analysis on the total data and uses it to construct an image of the cross-section on the screen.

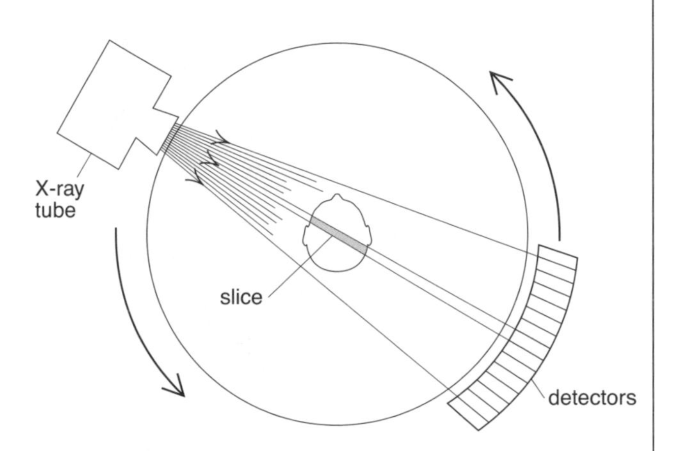

Radiotherapy Very hard X-rays (photon energies around 10 MeV) can be used to destroy cancer cells deep in the body. However, to reduce damage to the surrounding tissue, the beam is rotated around the patient so that only the target area stays in the beam all the time. (Note: gamma rays from a cobalt-60 source can be used instead of X-rays.)

How Science Works

Advantages and disadvantages of medical imaging

X-rays are ionizing radiation, so there is a risk to the patient during an X-ray examination. Staff have to leave the room when the X-ray machines are working, or use a lead screen for protection, because they are working with the ionizing radiation every day. CAT scans involve a higher dose of radiation, because they are taking lots of images, but the beam is very fine, and the detectors very sensitive, so modern scanners use a lower dose than older scanners. Ultrasound and MRI have the advantage of not using ionizing radiation. However, CAT scanners are cheaper than MRI scanners, and much quicker, so many more patients can be scanned per day.

Staff using ionizing radiation may wear film badges at work, so that these can be developed later to check the amount of exposure they have received.

MRI (magnetic resonance imaging)

This produces images of the body by scanning, but without the risks of X-rays. It uses the fact that different tissues contain different concentrations of hydrogen atoms.

About 60% of the human body consists of hydrogen atoms. The hydrogen nucleus - the proton - spins on its axis, giving it a ***magnetic moment.*** In a magnetic field a proton lines up parallel to the field and experiences a torque (see B4). The torque makes the spinning proton ***precess*** about the direction of the magnetic field (in a similar way to a spinning gyroscope precessing about the direction of the gravitational field).

There are two possible states. Most protons line up in the direction of the field (the lower energy state) but some line up against the field (the higher energy state).

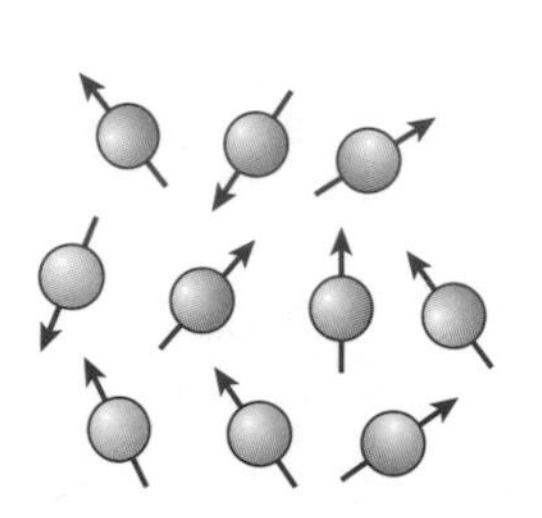

a) random protons

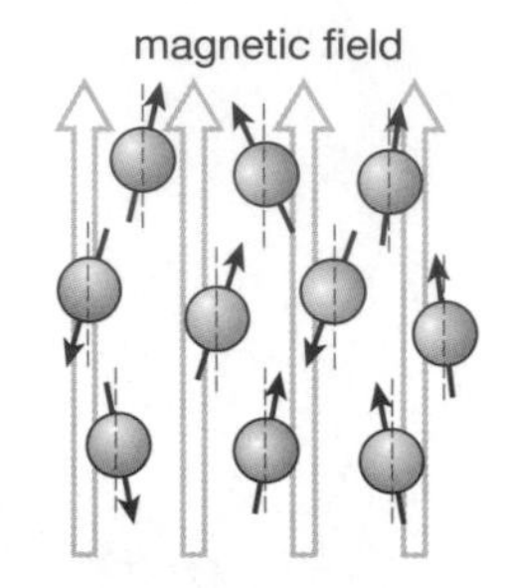

b) protons precessing around B field direction

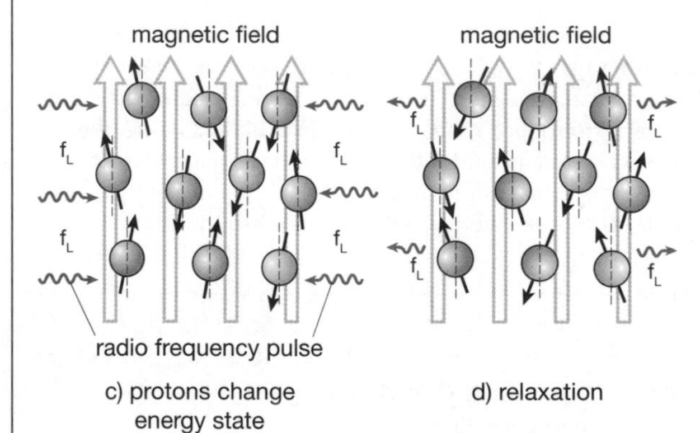

c) protons change energy state

d) relaxation

The frequency of precession of the proton is called the ***Larmor frequency***, f_L, measured in Hz. It depends on the magnetic field B:

$f_L = 4.25 \times 10^7\ B$

The Larmor frequency is in the radio frequency range. When a pulse of electromagnetic radiation at this frequency is directed at protons, some absorb the energy and move to the higher energy state. At other frequencies this does not happen – it is a ***resonance*** effect that occurs at frequency f_L.

After the pulse the protons relax back to the lower energy state, emitting radio frequency waves that can be detected, amplified, and recorded. This is why a pulse, and not continuous radio wave, must be used. The key to MRI is that these ***relaxation times*** are different for different tissue (for example, muscle, brain, and tumour tissue).

Coils are used to produce an additional, accurately calibrated, magnetic field so that every point in the body being scanned is in a different, known, magnetic field. This means that every point has a different, known, value of f_L, so that the part of the body the radio waves have come from is known from the frequency detected, f_L. A map of the tissue type in the body is built up by varying the radio frequency f_L.

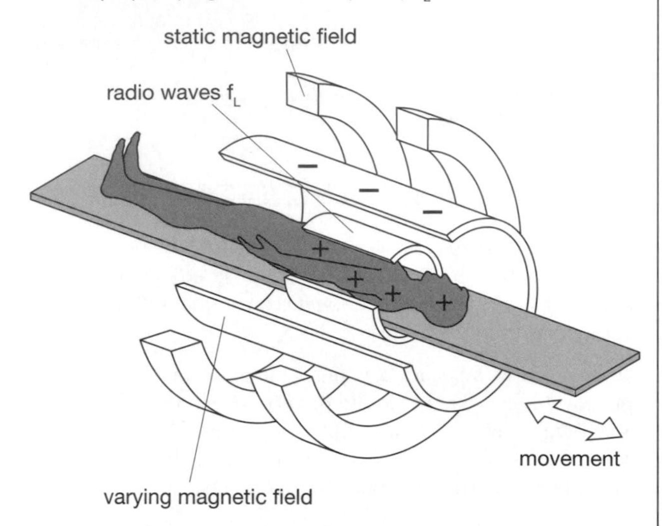

The MRI machine uses superconducting electromagnets, cooled by liquid helium, to produce a very high magnetic field. Any ferromagnetic objects are strongly attracted to it. Staff and patients must be very careful not to take ferromagnetic metal objects into the room.

Using magnetic resonance techniques it is also possible to detect and measure chemicals that are in the body without having to take samples. This is called **magnetic resonance spectroscopy (MRS)**. By monitoring chemicals at the site of a disease doctors are able to conduct research to determine whether a particular treatment is effective.

How Science Works

Advantages and disadvantages of MRI

Advantages: MRI scans do not use ionizing radiation. They give good contrast between different types of tissue, and high quality images. They can be used for imaging the brain through the skull.

Disadvantages: MRI scanners are expensive and slow. No metal objects can be scanned, including metal pins and implants inside patients.

M5 Nuclear medicine

Radiation risks

Read J4 first.

Ionizing radiations such as X and γ rays can cause cell damage which may lead to cancer or genetic changes in sex cells. The amount of damage depends on the amount of energy absorbed. This rises with increased exposure time.

Absorbed dose This is the energy absorbed per unit mass of tissue. The SI unit is the J kg^{-1}, called the ***gray*** (Gy).

Effective dose For the same amount of energy delivered, α particles cause more biological damage than X or γ rays. To allow for this, the ***effective dose*** is defined as the absorbed dose × Q, where Q is a ***quality factor***. The SI unit of effective dose is the J kg^{-1}, known in this case as the ***sievert*** (Sv).

Radiation	Quality factor	Effective dose of 1 gray in sieverts
α	20	20
β	1	1
γ	1	1
neutrons	10	10
X-rays	1	1

In the UK the average effective dose per person per year is 2.5 mSv.

Background radiation in the UK

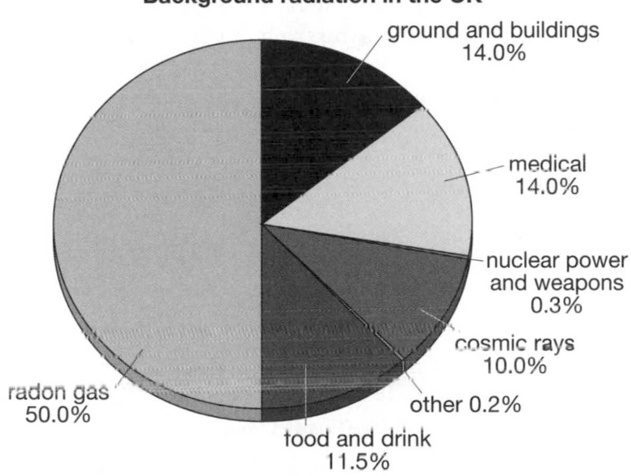

Note:

- At low levels, radiation is ***stochastic*** (random) in its effects. There is no minimum safe level. The risk of damage increases with the dose equivalent absorbed.

Physical, biological, and effective half-lives

When a radioisotope is in the body, its *effective* half-life T_E is less than its *physical* ('real') half-life T_P because biological processes gradually remove it from the body. For example, if you consume a glass of water, after about 7 to 10 days half of it will have left your body. This is the *biological* half-life of water in the body. Strontium-90 is dangerous because it has a very long biological half-life, as it is similar to calcium and gets stored in the bones. Phosphorous-32 is used for bone scans. Although it has a long biological half-life, its physical half-life is very short. An advantage of Technetium-99 is that is has short biological and physical half-lives (1 day and 6 hours, respectively). If T_B is the biological half-life, i.e. the time for half the original radioactive material to be removed:

$$1/T_E = 1/T_P + 1/T_B \qquad (T_P = T_{\frac{1}{2}} \text{ in J4})$$

Using tracers

Read J4 and J5 first.

Tracers used in medical diagnosis include the γ-emitting radioisotopes iodine-123 (half-life 13 h) and technetium-99^m (half-life 6 h). Small amounts can be carried in the bloodstream to various sites in the body.

Gamma camera γ photons from the tracer strike the sodium iodide disc, causing flashes of light whose intensity is amplified by the photomultiplier tubes. Signals from these are processed electronically, and an image built up on a CRO screen. The collimator improves image quality by only letting through γ rays travelling at right angles to the disc.

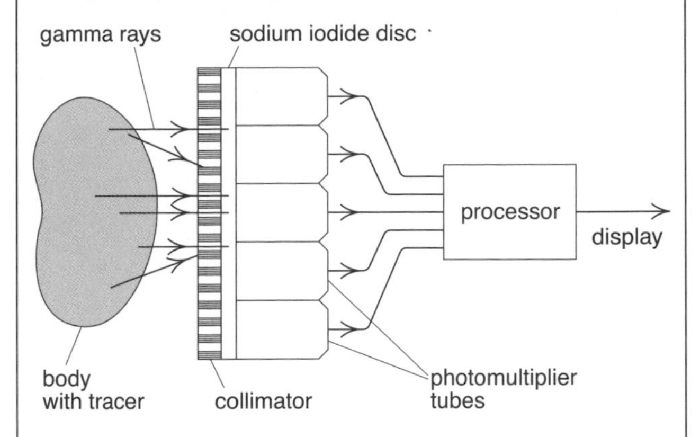

Checking blood flow in the lungs With technetium-99^m tracer present in the bloodstream, a gamma camera will reveal which parts of the lungs do not contain the tracer and, therefore, have blocked blood vessels.

Checking thyroid function The thyroid has a natural uptake of iodine. With iodine-123 tracer present in the bloodstream, the uptake can be checked by measuring the activity in the thyroid.

Physiological effects of radiation

Damage to cells may affect the individual by producing

- death due to cancer
- radiation sickness (nausea, loss of hair, tiredness)
- damage to genes that are transmitted to offspring and may produce mutations or other harmful effects.

The ions produced by radiation react inside the cell and interfere with the operation of the cell. For example, the ionization could cause a molecule to break apart if a bonding electron is lost. When a large number are destroyed by high doses the body may not be able to replace them quickly enough, causing health problems. The effect depends on the size of the dose.

Radiation may damage the DNA. This is serious because a cell may only have one copy of this so it is irreplaceable. The cell may die or become defective. The defective cell may then divide and produce more defective cells. This rapid production of defective cells is cancer.

Risk associated with radiation diagnosis or treatment

When deciding whether to treat a patient using radiation it is necessary to consider whether the risk associated with the use of radiation outweighs the risk of not using radiation for the treatment or diagnosis of a disease.

Maximum permitted level (MPL)

For those who work with radiation there is a maximum permitted dose of 50 mSv per year for the whole body or reproductive organs and 750 mSv per year for the hands, feet, etc. A maximum of 5 mSv per year is permitted for those who do not work with radiation.

N1 Rotation

Equivalent quantities

Read B14 first

Motion can be *translational* (e.g. linear) or *rotational*. Quantities used in measuring linear motion all have their rotational equivalents, as shown below.

Linear quantity	Symbol	Rotational quantity	Symbol
displacement	s	angular displacement	θ
velocity	v	angular velocity	ω
acceleration	a	angular acceleration	α
force	F	torque	T
mass	m	moment of inertia	I

Note:
- In physics, the symbol T may stand for torque, period or tension. Above, it represents torque.
- Moment of inertia is the property of a body which resists angular acceleration. It is explained below.
- The equivalent quantities above do not have the same units.

Moment of inertia

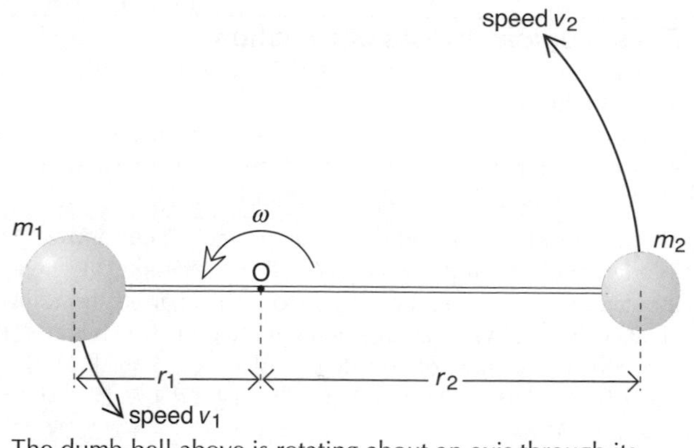

The dumb-bell above is rotating about an axis through its centre of mass O. (The bar of the dumb-bell has no mass.) All parts of the dumb-bell have the same angular velocity ω. The rotational KE of the system is the sum of the linear KEs of the two masses. So

$$\text{rotational KE} = \tfrac{1}{2}m_1v_1^2 + \tfrac{1}{2}m_2v_2^2$$

But $v_1 = \omega r_1$ and $v_2 = \omega r_2$. So the above equation can be rewritten:

$$\text{rotational KE} = \tfrac{1}{2}(m_1r_1^2 + m_2r_2^2)\omega^2$$

Comparing this with the equation for rotational KE in the right-hand table at the top of the page shows that

$$\text{moment of inertia } I = m_1r_1^2 + m_2r_2^2$$

This principle applies to all objects – for example, the one on the right. The moment of inertia is the sum of the mr^2 terms for all the particles in an object. Expressed mathematically:

$$\text{moment of inertia } I = \Sigma(mr^2)$$

Equivalent equations

The equations used when dealing with linear motion all have their equivalents in rotational motion, as shown below.

Linear equation		Rotational quantity	
velocity	$v = \frac{\theta}{t}$	angular velocity	$\omega = \frac{\theta}{t}$
acceleration	$a = \frac{v}{t}$	angular acceleration	$\alpha = \frac{\omega}{t}$
force	$F = ma$	torque	$T = I\alpha$
work done	$W = Fs$	work done	$W = T\theta$
KE	$E = \frac{1}{2}mv^2$	KE	$E = \frac{1}{2}I\omega^2$
power	$P = Fv$	power	$P = T\omega$
momentum	$p = mv$	angular momentum	$L = I\omega$

Note:
- The equations above are in simplified, non-calculus form. They assume uniform changes, and uniform forces and torques.
- Any rotational equation can be found by taking the linear equation and replacing the symbols with their rotational equivalents.
- Whether the motion is linear or rotational, work and energy are measured in joules. (Similarly power is measured in watts.)

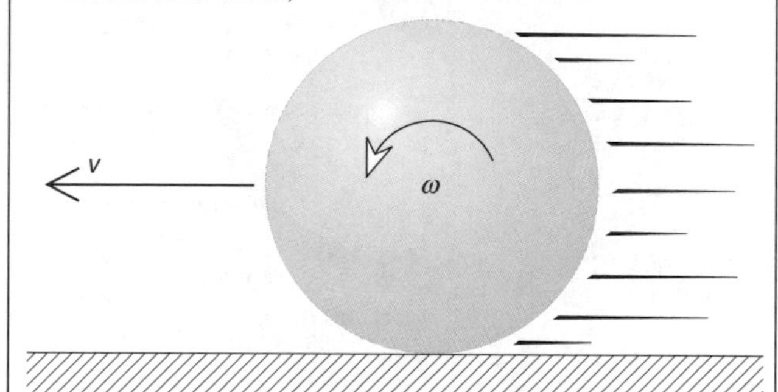

The rolling ball above has both linear *and* rotational motion. In this case,

$$\text{total KE} = \tfrac{1}{2}mv^2 + \tfrac{1}{2}I\omega^2$$

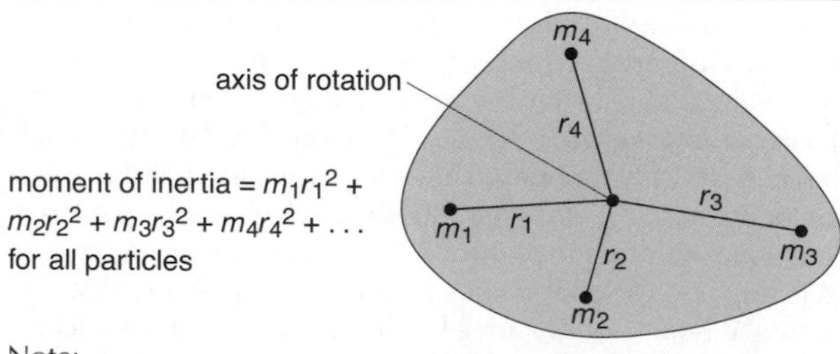

moment of inertia = $m_1r_1^2 + m_2r_2^2 + m_3r_3^2 + m_4r_4^2 + \ldots$ for all particles

Note:
- An object's moment of inertia depends on which axis it is being rotated about. The more spread out the mass, the higher the moment of inertia.
- From this, it follows that objects with the same mass can have different moments of inertia.

O1 Turning points in physics

Waves and particles

Corpuscles or waves? Newton suggested that light might be made up of high-speed 'corpuscles' (particles). Huygens (in 1680) proposed an alternative wave theory which satisfactorily accounted for reflection and refraction.

The wave theory could not explain how 'empty' space could transmit waves. So it was argued that space must be filled with an invisible medium, called the ***ether***.

The corpuscular theory predicted that light should speed up when passing from air to water. The wave theory predicted the opposite. In 1862, the speed of light in water was measured and found to be less than in air.

Electromagnetic waves Maxwell's electromagnetic theory (1864) predicted that oscillating charges should emit waves which would travel through an electromagnetic field at a speed of $1/\sqrt{\varepsilon_0\mu_0}$ (= speed of light) (see below). Later (in 1888), Hertz demonstrated the existence of radio waves.

Note:

- Maxwell's theory is an example of a ***classical*** theory – one that does not deal with quantum effects.

Black body radiation The graph shows the energy distribution in a black body's spectrum (see I7). The dotted line is the distribution predicted by classical theory. In classical theory, oscillating particles are assumed to emit waves continuously, over a continuous range of energies. But at higher frequencies, there is a huge discrepancy between the predicted and experimental results. This is known as the ***ultraviolet catastrophe***.

Quantum theory This was proposed by Planck (in 1900). It solved the black body spectrum problem by restricting energy changes to multiples of hf. Later, it led to the concepts of the ***photon*** and ***wave–particle duality***.

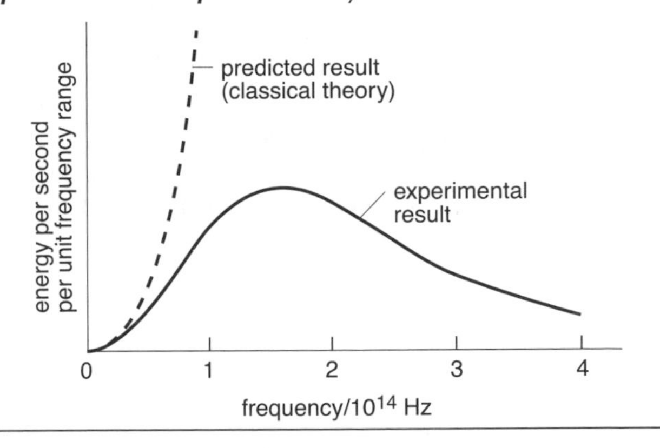

The special theory of relativity

Looking for the ether As the Earth moves through space, the ether should flow past it. In 1887, Michelson and Morley tried to detect this flow with an ***interferometer***:

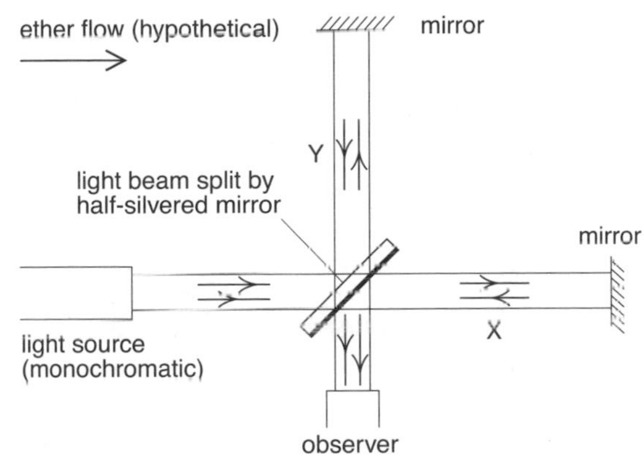

Beams X and Y recombine to form an interference pattern (see C3). If there is an ether flow as above, X's travel time should be slightly longer than normal. But when the apparatus is rotated 90°, Y's travel time should be slightly longer than normal, so the interference pattern should shift. No shift was detected. All experiments indicated that the measured velocity of light in a vacuum is *invariant* (always the same).

Postulates Einstein rejected the idea of absolute motion through space: motion could only be relative to the observer's ***frame of reference***. He developed his ***special theory of relativity*** (1905) from the following two postulates (assumptions):

1. Physical laws (e.g. the laws of motion) are the same in all *inertial* (unaccelerated) frames of reference.
2. The speed of light in a vacuum has the same measured value in all inertial frames of reference.

Deductions From these postulates, Einstein deduced that length and time measurements could not be absolute. They must depend on the relative motion of the observer. Some results of his mathematical analysis are given on the right.

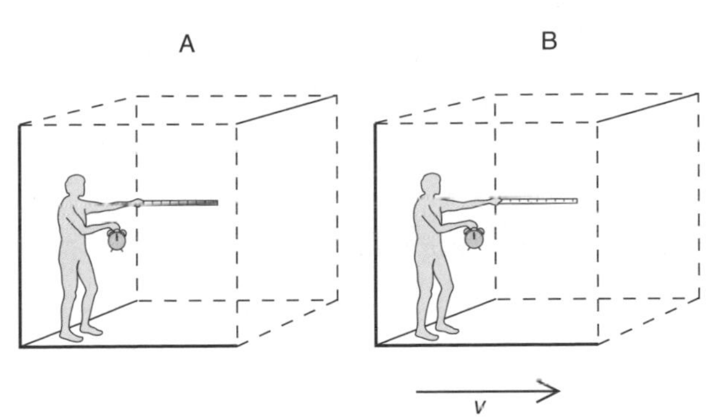

Above, frame B has a velocity v relative to frame A.

- An event in B takes time t_0 when measured by an observer in the same frame. This is its ***proper time***. But to an observer in A, the same event takes time t, where:

$$t = \frac{t_0}{\sqrt{1 - v^2/c^2}}$$

(c = speed of light in a vacuum)

Note: $t > t_0$. To the observer in A, a clock in B runs slow: there is ***time dilation*** (enlarging).

- If an object in frame B has a ***proper length*** l_0 (in the v direction), its length as observed from A is given by

$$l = l_0\sqrt{1 - v^2/c^2}$$

Note: $l < l_0$. To the observer in A, the object in B is shortened: there is ***length contraction***.

- The object's mass as observed from A also depends on v:

$$m = \frac{m_0}{\sqrt{1 - v^2/c^2}}$$

Note: $m > m_0$. As v increases, m increases. When $v = c$, m is infinite, and so is the KE. This effectively makes the speed of light a universal speed limit.

- Observed increases in mass and energy are linked by the equation $\Delta E = \Delta mc^2$ (see J6).

Discovering the electron

Read E8 first.

Cathode rays 19th century experiments with electric discharges through gases at very low pressure (as below) indicated that invisible rays were travelling from the cathode to the anode in the discharge tube. These were called ***cathode rays***. Perrin (in 1895) demonstrated that they were negatively (–) charged.

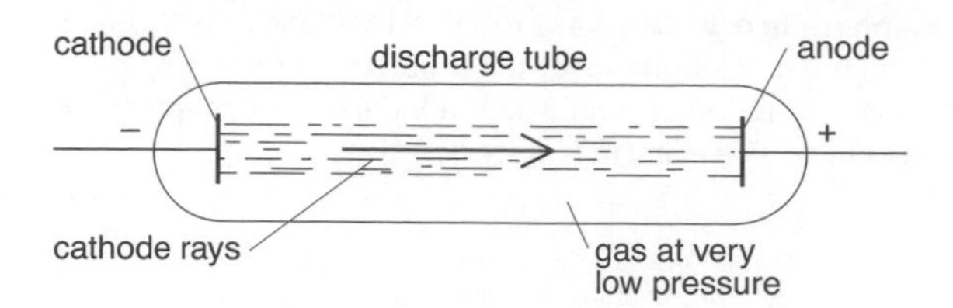

Specific charge (charge/mass) J. J. Thomson suggested that cathode rays were particles, which he called ***electrons***. In 1897, he measured their specific charge (e/m_e). A modern version of his experiment is shown below:

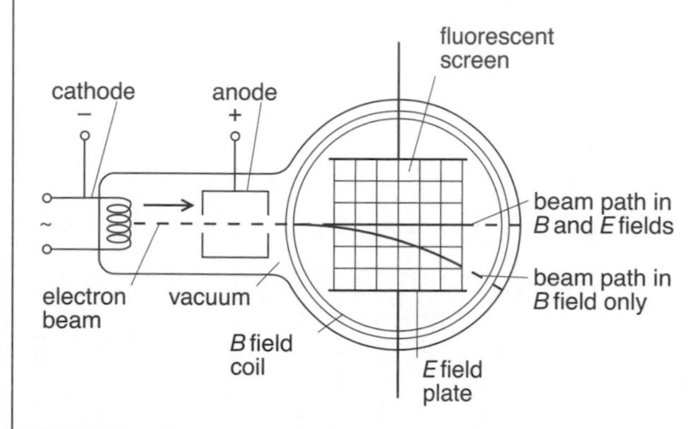

The electron beam is deflected by a magnetic field (B), then restored to a straight path using an electric field (E).

When the magnetic field alone is acting

$$Bev = m_e v^2/r$$

So $e/m_e = v/rB$ (quantities defined as in H11)

With r and B measured, e/m_e can be calculated if v is known. When the electric and magnetic forces on the electron beam are equal, $Bev = Ee$. So $v = E/B$.

Electronic charge *e* Millikan (in 1909) observed oil droplets as they fell at terminal velocity through air, then measured the *change* in terminal velocity (Δv) which occured when a droplet gained charge (q) and a vertical electric field (E) was applied. In this case, electric force = Eq.

So $Eq = 6\pi\eta a\Delta v$ (see B9)

giving $q = 6\pi\eta a\Delta v / E$

(a was deduced from terminal velocity measurements when E was zero). Millikan took hundreds of measurements and found that q was always a multiple of a basic charge e.

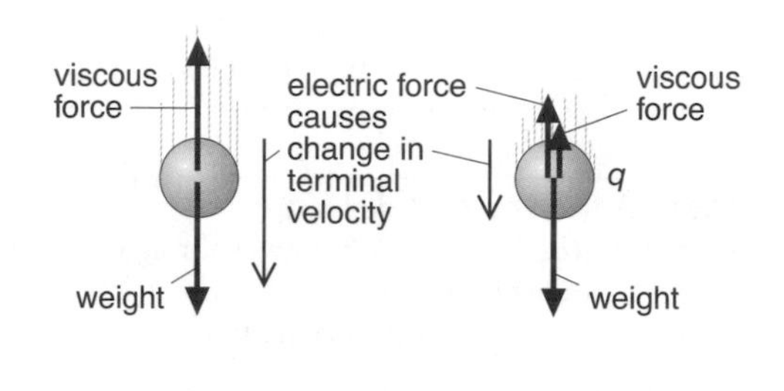

Scanning tunnelling microscope (STM)

Read E2 first

To examine the surface of a metal, a sharp metal tip – so sharp that the end is a single atom – is scanned across the sample at height d above the surface (d is typically about 1 nm).

A small voltage (usually 0.1–2 V) is applied between the tip and the sample. In a classical physics model there is no current between the sample and the tip until the two touch. In the quantum mechanical model, just before the two touch there is a tiny current (called the ***tunnelling current***) as there is a small probability that electrons will disappear (tunnel) from the tip and reappear in the sample. The tunnelling current can be experimentally measured between the tip and the sample, so this is good evidence for the quantum theory.

The size of the tunnelling current increases exponentially with decreasing distance d, so the height of the tip can be adjusted to keep the current, and therefore d, constant. As the tip is scanned across the sample, and moves up and down, it draws out the contours of the individual atoms on the surface.

The scanning of the tip across the surface and its movement up and down is performed using piezoelectric crystals, which change size as a pd is applied to them.

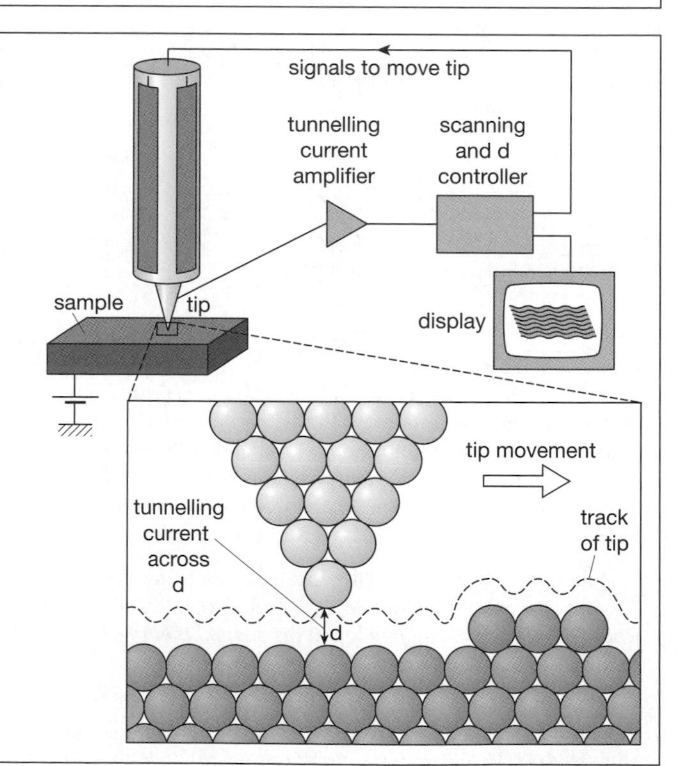

P1 Energy and the environment – 1

Energy for the Earth

The Earth's prime source of energy is ***solar radiation*** (radiation from the Sun). Its effects include:

- maintaining temperatures on Earth
- maintaining winds, ocean currents, and weather systems
- maintaining plant and animal life
- maintaining atmospheric composition
- millions of years ago, supplying the energy now stored in fossil fuels (e.g. oil, natural gas, and coal).

Non-solar energy

Heat from inside Earth

Energy in nuclear fuels

Tidal energy (due to Moon's gravitational pull)

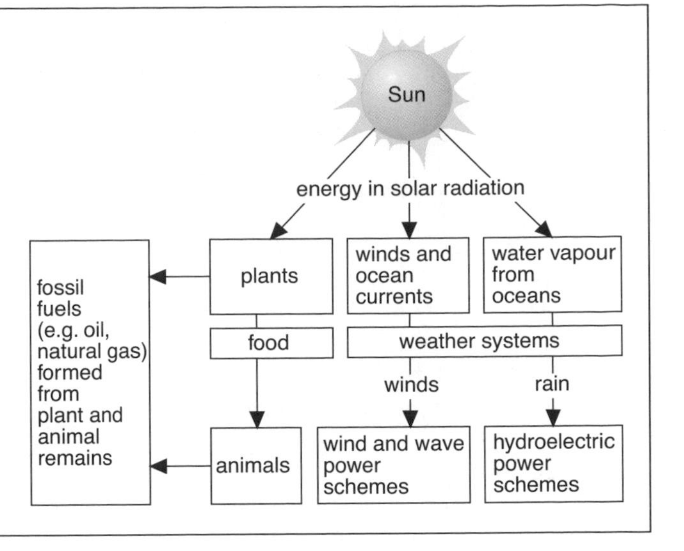

Maintaining atmospheric composition

Animals take in oxygen. They use it to 'burn up' their food and obtain energy. This process is called ***respiration*** and it produces carbon dioxide and water. (Burning fossil fuels also produces carbon dioxide and water.)

Plants take in carbon dioxide and water. With these, they make food using energy from sunlight. This process is called ***photosynthesis*** and it gives out oxygen.

Note:

- Plants also use oxygen for respiration. But overall, they make more oxygen than they consume.
- For photosynthesis, plants absorb light at the red and blue ends of the visible spectrum. They transmit or reflect green light, which is why they look green.

Together, plants and animals maintain the amounts of oxygen and carbon dioxide in the atmosphere. By various chemical processes, some nitrogen is also incorporated in the bodies of living organisms, but this is recycled.

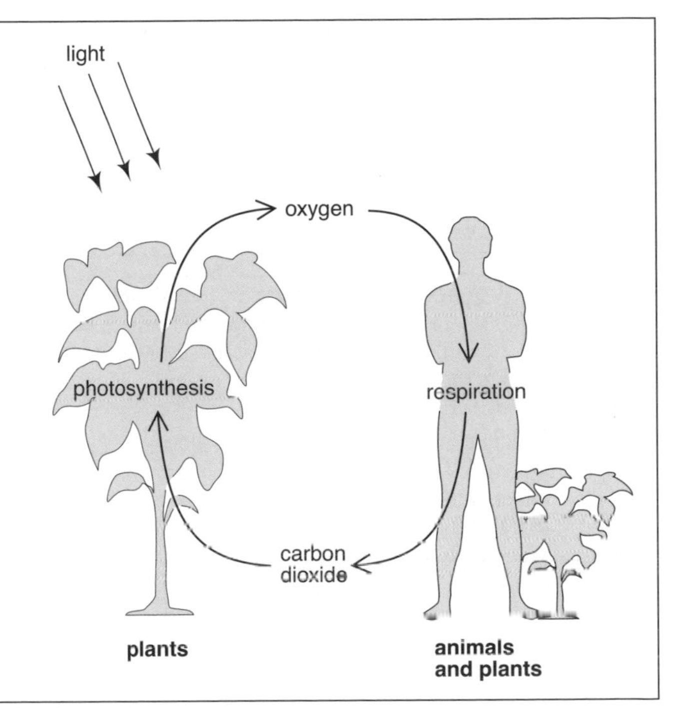

Water balance

Overall, the rate at which water evaporates from the oceans and land regions is equal to the rate at which water is returned by precipitation (rain, snow, and hail). On average, there is a ***water balance***. However, at any given time, each region may have an imbalance – in which case, water is either going into storage or coming out of it.

The oceans are the main water storage system. Others include the polar ice caps, soil, plants, and porous rocks. Underground water-bearing rocks are called ***aquifers***. Water can stay locked away in them for thousands of years.

Melting icebergs do not raise the sea level (when floating they displace their own weight of water – see above right). Melting ice from land will contribute to a rise in sea level.

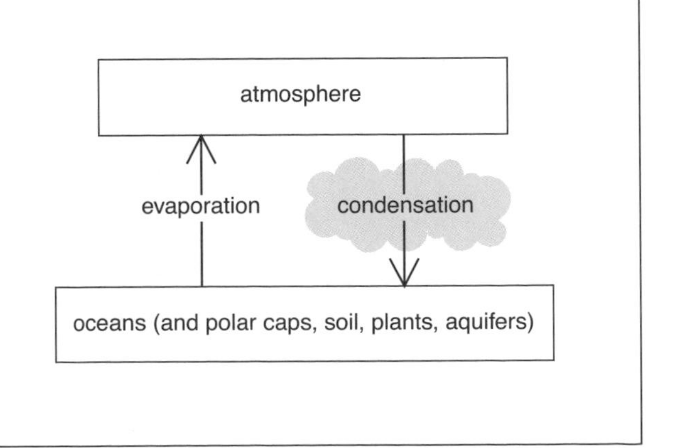

Flotation

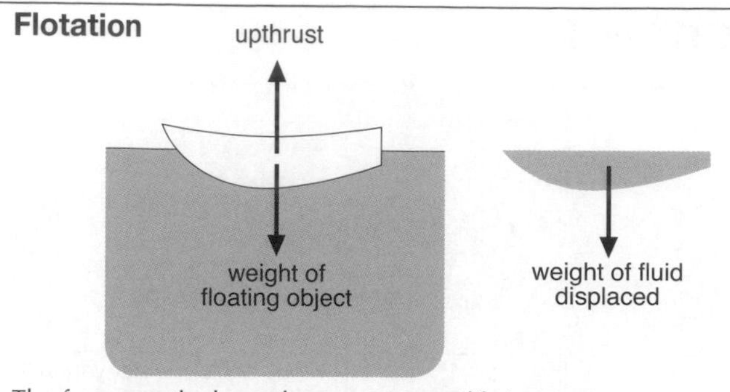

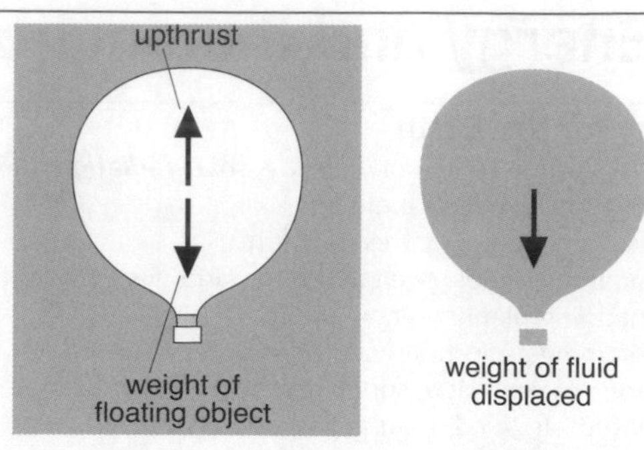

The forces on the boat above are in equilibrium, so, weight of boat = upthrust on boat. But, by Archimedes' principle (see B9), upthrust = weight of fluid displaced. So

weight of floating object = weight of fluid displaced

This is known as the ***principle of flotation***.

A given volume of hot air weighs less than the same volume of cold air because it has a lower density.

A hot-air balloon like the one above just floats in the cold air around it when the total weight of hot air, fabric, and load is equal to the weight of cold air displaced.

Solar constant

This is the amount of solar energy arriving at the Earth's outer atmosphere per second per square metre (at right angles to the radiation). It is equal to 1.35 kW m^{-2}.

- About 25% of the incoming solar radiation is reflected back into space by the atmosphere and clouds. It does not reach the Earth's surface.
- The amount of solar energy per second striking each square metre of the surface, depends on the time of day, season, atmospheric conditions, and latitude. For example, the radiation reaching region B on the right is spread over a larger area than that reaching A.

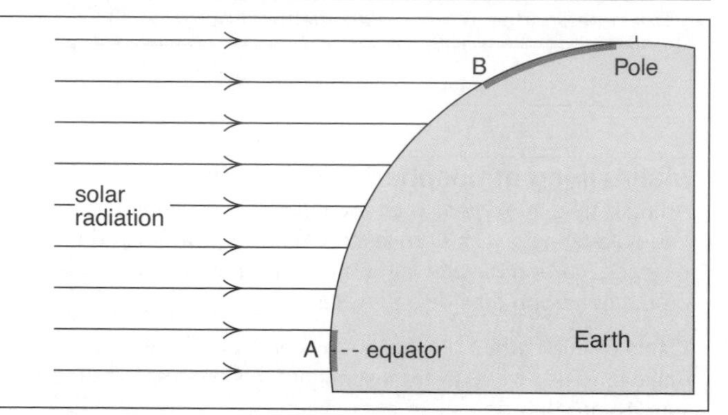

The natural greenhouse effect

Read I7 first.

The Earth gains energy from solar radiation, and loses it by emitting radiation into space. But overall, the *rates* of energy gain and loss are the same. This is another example of dynamic equilibrium.

If the Earth had no atmosphere, its average surface temperature would be about −18 °C. However, the 'heat trapping' effect of the atmosphere, called the ***greenhouse effect***, means that dynamic equilibrium occurs at about 15 °C. The raised temperature is caused like this:

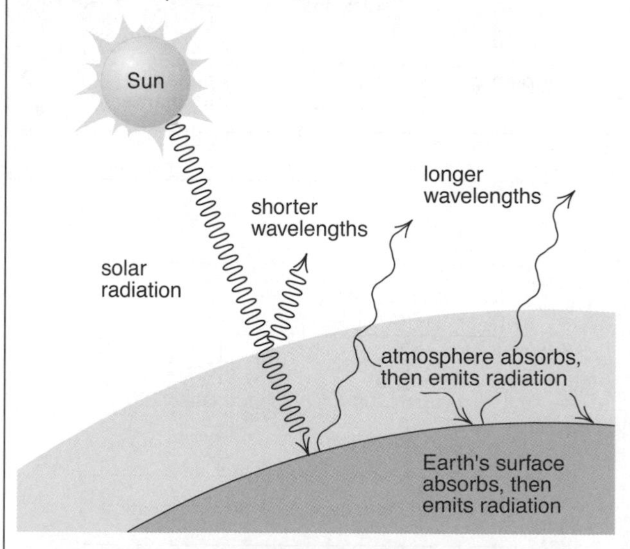

Visible and short-infrared radiation from the Sun pass easily through the atmosphere and warm the Earth's surface. However, being cooler than the Sun, the surface radiates back at longer infrared wavelengths (see I7). Some of these wavelengths are absorbed by molecules of water vapour and carbon dioxide in the atmosphere. These then emit infrared in all directions - including downwards. So the atmosphere and surface are both warmed, and dynamic equilibrium is reached at a higher temperature.

Global warming

By burning fossil fuels, industrial societies are putting carbon dioxide into the atmosphere at a faster rate than plants can absorb it. This is adding to the greenhouse effect and may be causing ***global warming***. In the past 100 years, the average global temperature has risen by almost 1°C.

- Though often called the 'greenhouse effect', global warming is an *addition* to the natural greenhouse effect.
- Extra carbon dioxide may not be the only cause of global warming. Global temperatures have always fluctuated.
- Other 'greenhouse gases', include methane (from paddy fields, animal waste, and oil and gas fields), and CFCs.

The increase in carbon dioxide in the atmosphere is less than half the extra emitted. Possible reasons are:

- With extra carbon dioxide in the air, plant growth is increased, so more of the gas is absorbed.
- Some carbon dioxide dissolves in the oceans.

Future effects of global warming These cannot be predicted with any certainty. Melting of polar ice may cause a rise in sea level. Evaporation from the oceans will increase, so some regions may get more rain. But shifts in climate may mean that some regions are drier.

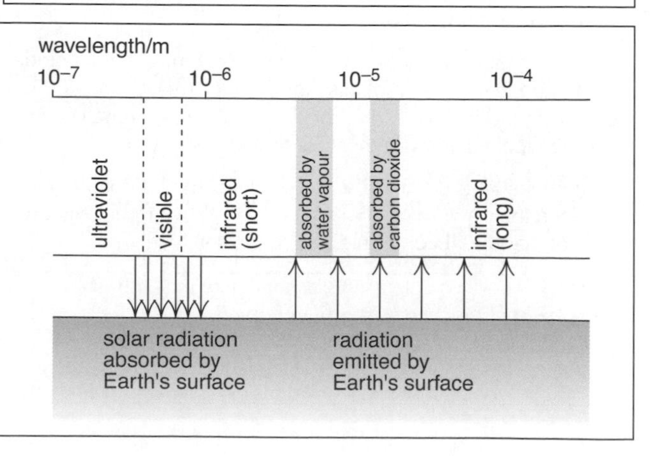

P2 Energy and the environment – 2

Using fossil fuels

Industrial societies get over 90% of their energy by burning oil, natural gas, and coal. However, these fossil fuels took millions of years to form in the ground. They are effectively ***non-renewable*** and supplies are ***finite*** (limited).

Resources of fossil fuels are the total quantities known to exist in the Earth.

Reserves are the quantities which could now be extracted commercially. They are less than resources.

Reserves (2007): energy/10^{20} J		
	World	UK
oil	71	0.21
natural gas	67	0.15
coal	180	0.043

Future consumption This will decide how long reserves will last. It is difficult to predict but will be affected by:

- the increase in world population
- economic growth in developed countries
- industrialization of developing countries
- efficiency of fuel use
- levels of building insulation.

Problems from fossil fuels Major problems include:

- atmospheric pollution, and risk of global warming
- risk of marine pollution from oil
- economic debt in developing countries which import oil.

Pumped storage

The electricity supply industry has to respond instantly to changes in demand for mains power. A ***pumped storage scheme*** aids this process by acting as an energy store. When demand is high, water flows from the top reservoir and turns the turbines which drive the generators. When the demand is low, the generators are used as motors. Power is taken from stations with spare generating capacity, and water is pumped back up to the top reservoir.

In an area of high rainfall or snow, the scheme may contribute to energy resources, but the purpose is to store surplus energy for later use. A disadvantage is that there are few suitable sites and they will often be in areas of outstanding natural beauty. Flooding valleys may be necessary, for either the top or bottom reservoir.

Calculating power output The following calculation assumes that gravitational PE is converted into electrical energy with an efficiency of 100%. If a mass m of water flows from the top reservoir in time t and loses a height h,

electrical energy output = loss of PE = mgh

Dividing by t gives

electrical power output = mgh/t

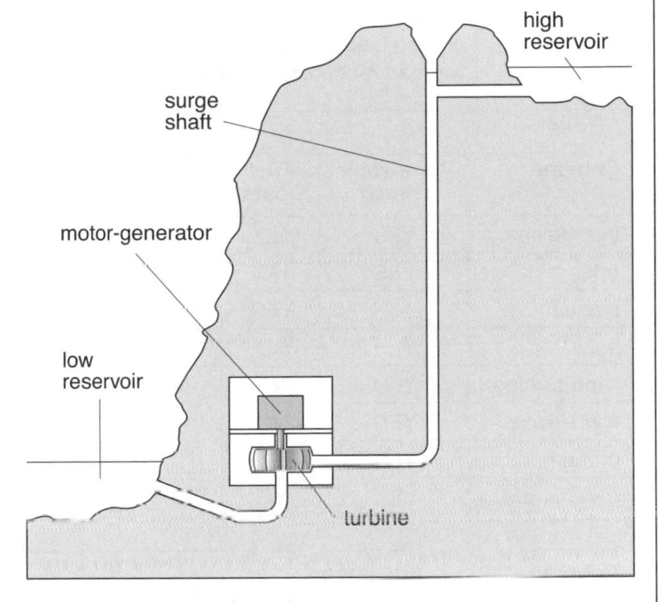

Note:

- In practice, the efficiency is, typically, about 75%.

Power stations: efficiency of fuel use

As with all heat engines, the efficiency of a fuel-burning power station is limited by the second law of thermodynamics (see I5). With additional energy losses (e.g. frictional), the efficiency is reduced to about 40%. So more heat is produced (60%) than electrical energy.

Above, you can see what happens to each 1000 J of energy released by burning fuel in a typical power station. A diagram like this is called a ***Sankey diagram***. Amounts of energy are represented by the widths of the arrows.

Combined heat and power (CHP) In a fuel-burning power station, waste heat produces slightly warmed cooling water. By running the power station at *reduced* generating efficiency, it is possible to supply the local district with water hot enough for heating systems. In a CHP scheme like this, the *overall* efficiency of fuel use is greatly improved.

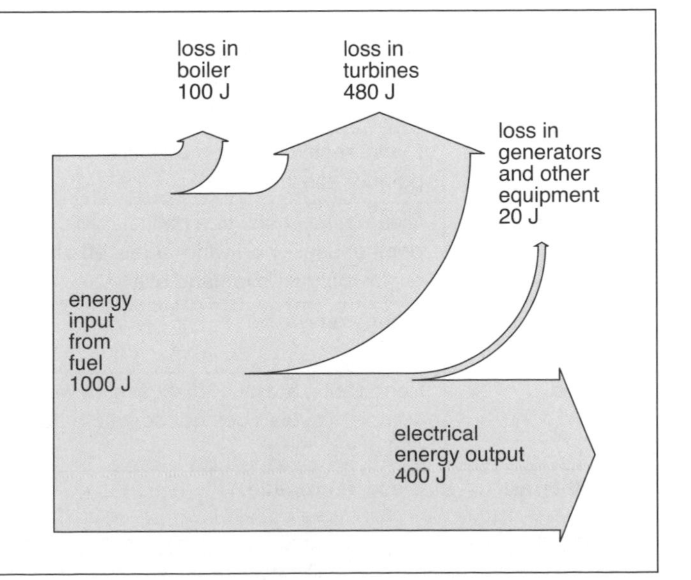

The main alternatives to fossil fuels

Scheme	Details
hydroelectric	Rainwater fills lake behind dam. Flow of water from lake drives generators.
tidal	Lake behind dam fills and empties with tides. Water flow drives generators.
nuclear	See J7.
solar	See panel on p 142.
wind turbine	See panel on right.
wind-wave	See panel on p. 142.
biofuel	Wood (for burning). Methane gas from plant and animal waste. Alcohol (fuel) produced from sugar cane.
geothermal	Using hot water from natural geysers and springs in volcanic areas. Hot rocks underground used to produce steam for turbines in power station. Pipes circulate water deep under the ground below buildings. The water is warmed by geothermal heat.

Scheme	Renewable energy	Fuel costs	Greenhouse gas emissions
hydroelectric	YES	NO	NO
tidal	YES	NO	NO
nuclear	NO	YES	NO*
solar	YES	NO	NO
wind turbine	YES	NO	NO
wind-wave	YES	NO	NO
biofuel	YES	YES	YES**
geothermal	YES	NO	NO

* In the reactor; extracting and preparing fuel, and managing the radioactive waste may contribute to greenhouse gas emissions.

** With managed crops, there is no overall addition to global warming because of the carbon dioxide absorbed.

Wind power

Generators driven by wind turbines ('windmills') are called ***aerogenerators***. The largest ones have power outputs of about 7 MW – compared with over 3000 MW for a large fuel-burning station. For increased power from one site, aerogenerators are grouped together in ***wind farms***.

Calculating power output The following calculation assumes that the wind loses all its velocity v when striking the turbine blades and transfers all its KE to them.

In the diagram on the right, all the air in the shaded region will transfer its KE to the blades in time t. As this air has a volume of Avt, its mass is $A\rho vt$, so

$$\text{KE of air} = \tfrac{1}{2} \times \text{mass} \times v^2 = \tfrac{1}{2} \times A\rho vt \times v^2 = \tfrac{1}{2} A\rho v^3 t$$

If the energy conversion efficiency is 100%, this KE is also the electrical energy output. So dividing by t gives

$$\text{electric power output} = \tfrac{1}{2} A\rho v^3$$

when $A = \pi r^2 = \tfrac{1}{2}\pi r^2 \rho v^3$

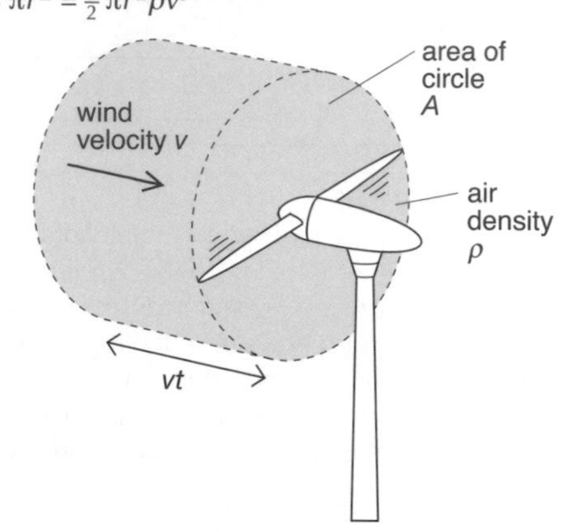

Note:

- In practice, a turbine cannot extract all the wind's KE, and there are other energy losses as well. Overall, the efficiency is reduced to about 40% or less.

Scheme	Advantage	Disadvantage
hydroelectric	Clean, efficient, renewable	Only for mountainous areas with high rain/snowfall; conflict over land use and water supplies
tidal	Clean, renewable, totally predictable	Few suitable sites; affects estuaries that are important for wildlife for miles upstream
nuclear	No carbon dioxide produced in reactor; lot of energy available from small amount of fuel	Long lifetime of radioactive waste; threat of terrorist or earthquake activity causing radioactive contamination
solar	Clean, renewable; can be installed near point of use	Limited sunshine/daylight in some areas; initial cost of photocells is high
wind turbine	Clean, renewable; can be installed close to point of use, or on wind farms; off-shore avoids conflict over land use	Very variable supply; cannot be used in calm or very high winds; some people object to noise and appearance; large land (or sea) area required
wind–wave	Clean, renewable	In early stages of development; must withstand severe weather conditions; very variable with weather
biofuel	Renewable; a way of disposing of waste; complete cycle does not contribute to global warming	Problem of ensuring a supply; competition with food crops can lead to shortages and higher food prices
geothermal	Clean, renewable	Expensive to install; in volcanic areas equipment may be damaged

Solar power

Solar power systems make *direct* use of solar radiation. This is done in two main ways:

Solar cells These use the ***photovoltaic effect***: a small voltage is generated when light falls on a slice of doped semiconductor. Such cells are expensive and their efficiency of energy conversion is normally less than 20%.

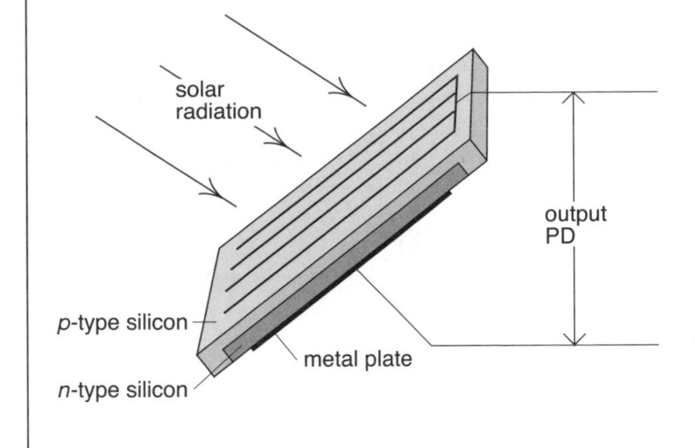

Solar panels These use the heating effect of solar radiation – for example, to pre-warm the water in a domestic hot water system.

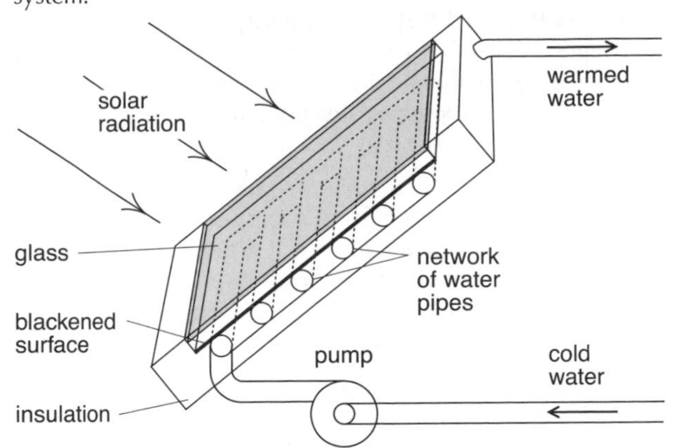

The maximum amount of available energy per second per square metre of panel is equal to the solar constant less corrections for atmospheric absorption, cloud cover, latitude, and angle of panel to the incoming radiation.

Intensity of power from the Sun $I = \frac{P}{A}$ (see D2)

Current research aims to improve efficiency to 30-60%.

Wind–wave power

There are several designs of wind–wave power plant. The oscillating water column design has been used in the LIMPET wave power plant in Scotland since 2000. In 2008 a 100 kW turbine started operating. There are plans for a 4 MW plant.

The waves force water to rise and fall inside the chamber. The air in the chamber is forced out, and sucked back, turning the turbine each time. The turbine is designed to turn in the same direction in both cases.

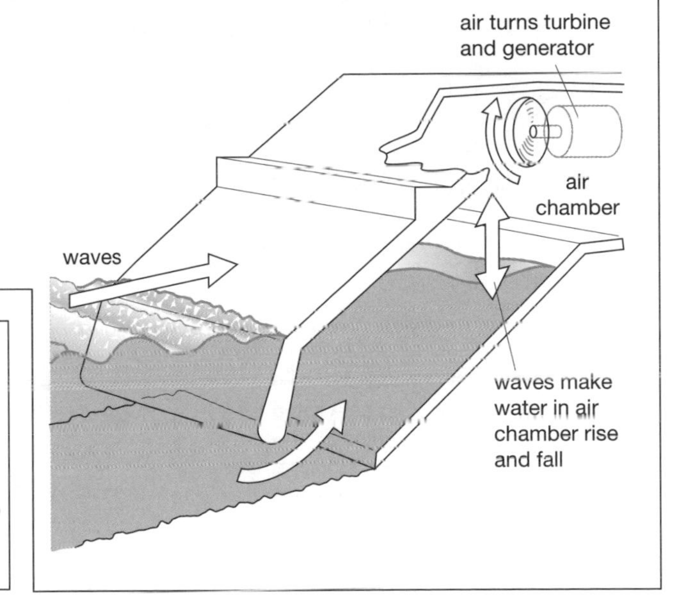

Energy (in kWh) and power

For practical reasons, energy is sometimes measured in units other than the joule (J). For example:

1 ***kilowatt hour*** (kWh) is the energy supplied when delivered at the rate of 1 kW (i.e. 1000 J s^{-1}) for 1 hour.

energy = power × time (see B7)

So 1 kWh = 1000 J s^{-1} × 3600 s = 3.6×10^6 J

Q1 Surveying

Geophones and seismic surveys

A geophone is similar to a microphone. When a seismic wave causes vibration of the geophone the magnet moves in the coil and induces an EMF in the coil. Geophones are very sensitive. The EMF can be used to trigger a timer.

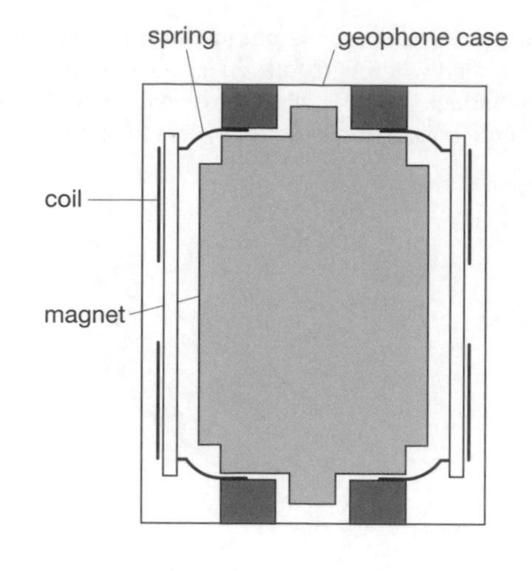

To conduct a survey, a controlled explosion is triggered and the time taken for the wave to reach each geophone is measured. The waves will not arrive at every point, because the paths will depend on the refraction, reflection, and transmission from the boundary between rock types. From the data logged, the length of the paths taken by the waves can be calculated and then used to deduce the underlying rock structure.

Hydrophones are similar instruments for use in the oceans. An array of hydrophones can be fixed, or towed behind a ship. They detect changes in water pressure caused by movements such as an underwater earthquake. They can be used in early warning systems for tsunamis, which can be caused by underwater earthquakes or volcanic eruptions.

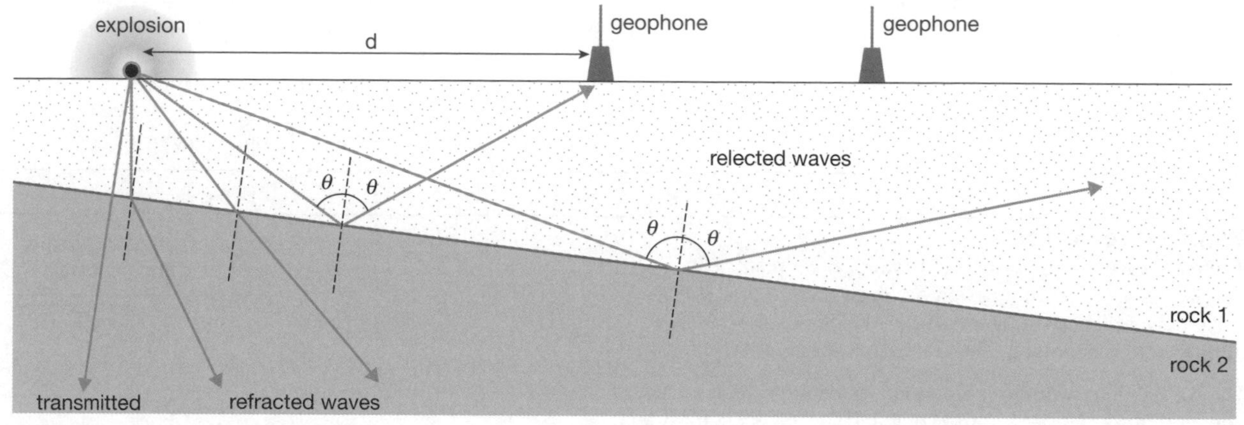

Resistivity surveys

In resistivity surveys the electrical resistance (see F2) between two probes in the ground is measured. This will depend on the composition of the ground between the probes, as well as the distance between the probes. Features such as a buried road, for example, will have a higher resistance due to the higher resistivity of the stones.

$R_2 > R_1$

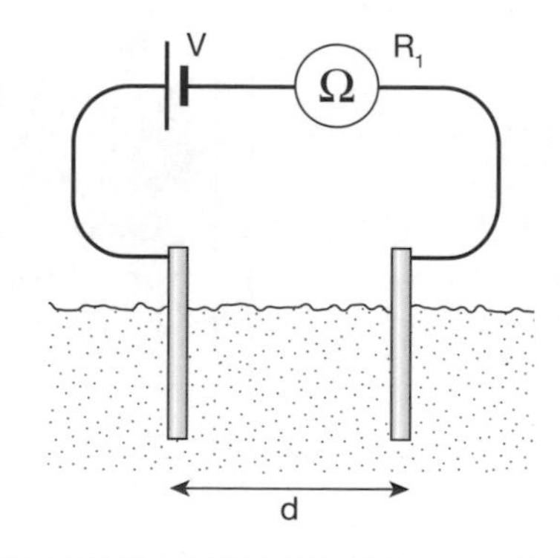

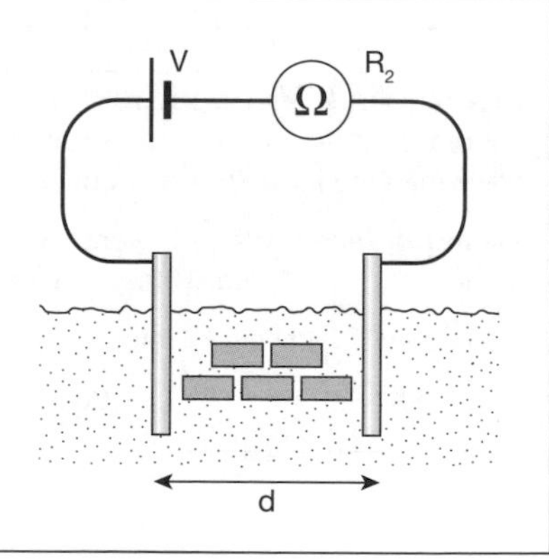

Variations in *g*

Read H4 first.

Across the Earth's surface, the measured value of g varies slightly, though by less than 1%. It is lowest at the equator
(1) mainly because of the effects of the Earth's rotation
(2) also because the Earth has a slightly greater radius towards the equator than towards the poles. g is also less where the crust is thick, because crust is less dense than mantle.

Variations in g can be detected using a ***gravimeter*** – a very sensitive spring balance with a mass attached. Local anomalies (unusual variations) in g can give clues about the presence of mineral deposits.

R1 Rockets and aeroplanes

Forces on rockets

Read B11 and B12 first

The rocket equation

Forces occur in interaction pairs. A rocket in free space has nothing to push against to propel itself forward.

A rocket engine burns fuel such as liquid hydrogen, in liquid oxygen in the combustion chamber to produce hot expanding gases. The hot exhaust gases are ejected from the back of the rocket. By conservation of momentum, the forward momentum of the rocket is equal and opposite to the momentum of the exhaust gases.

As the fuel is burned the mass of the rocket is reduced. The final velocity of the rocket, vf, (assuming there is no drag and no gravity) is

$$v_f = v_e \ln (m_o/m_f)$$

where v_e is the velocity of the exhaust gases, m_0 is the initial mass, and m_f is the final mass of the rocket.

The payload ratio $\pi = \dfrac{m_p}{m_r + m_p + m_f}$

where m_p = mass of the payload, m_r = mass of the rocket, and m_f = mass of the fuel.

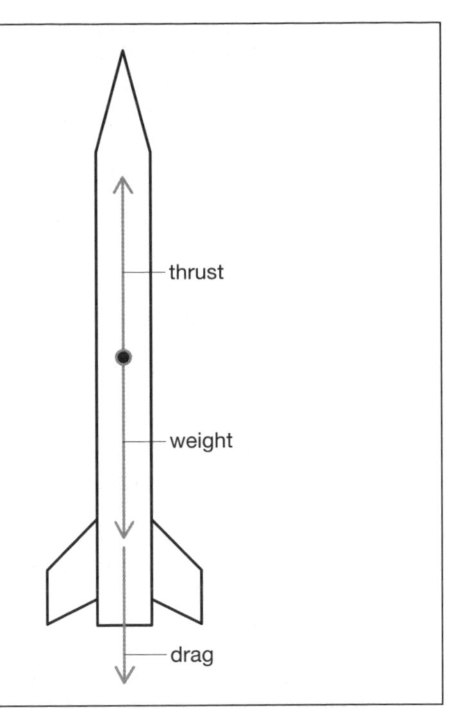

Lift and drag

Read B8 first.

A wing is an ***aerofoil*** – a shape which produces more lift than drag. For a wing of horizontal area S moving at velocity v through air of density ρ, the lift F_L is given by

$$F_L = \tfrac{1}{2} S C_L \rho v^2 \qquad (1)$$

where C_L is the lift *coefficient* of the aerofoil section.

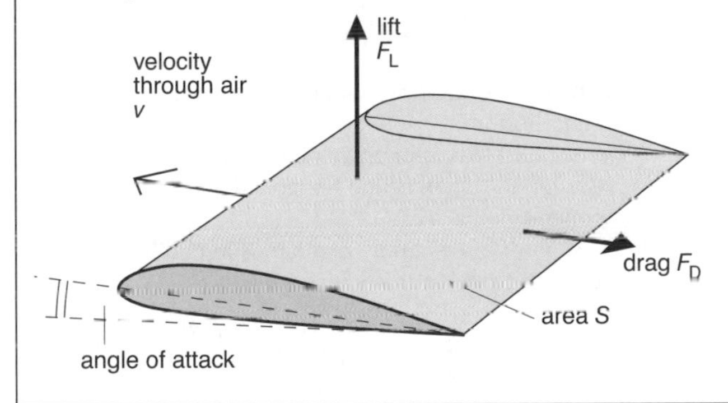

Angle of attack The value of C_L depends on this angle (shown above). Up to a certain limit, increasing the angle of attack increases C_L and, therefore, increases the lift.

For level flight, the lift must balance the aircraft's weight (see B10). If the speed decreases, then according to the above equation, the lift would also decrease if there were no change in C_L. To maintain lift, the pilot must pull the nose of the aircraft up slightly to increase the angle of attack.

Stalling If the angle of attack becomes too high, the airflow behind the wing becomes very turbulent and there is a sudden loss of lift. The wing is ***stalled***:

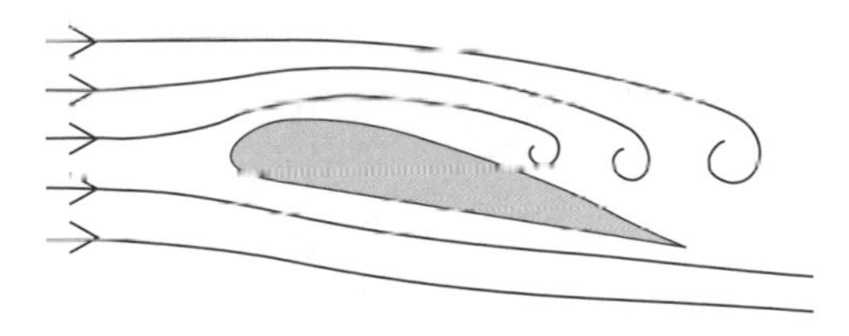

Forces on aeroplanes

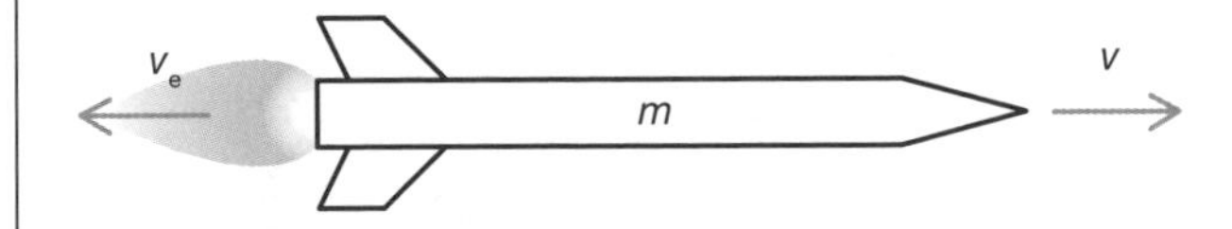

Aeroengines can use propellers or compressors to force air backwards so that the aeroplane is pushed forwards.

Lift is an important force used to keep an aeroplane in the air.

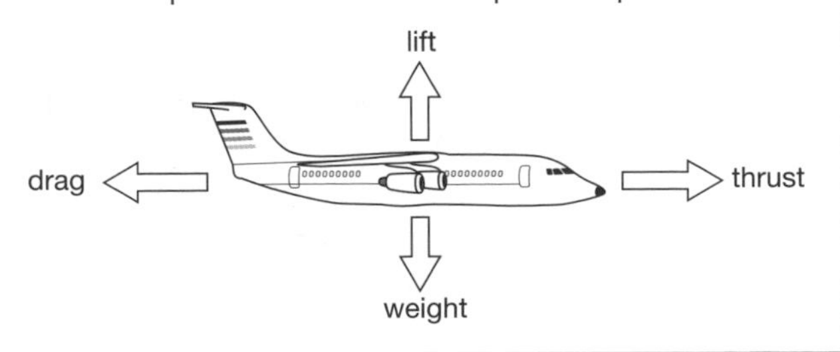

Self-assessment questions

After revising a section you should try these questions.
Questions are only given for those sections which relate to compulsory material in the specifications.
Answers, including references to sections where you can find more detail, begin on page 150.
You should review the work relating to any questions that you were unable to do or when you obtain incorrect numerical answers.
Where necessary assume that the acceleration of free fall $g = 9.8$ m s^{-2}.

Sections A1–A2

1. Write down the following quantities in standard form:
   (a) 3.5 MΩ **(b)** 220 pF **(c)** 15 mm s^{-1} **(d)** 25 mm^2
2. Convert the following quantities to include a number with a suitable prefix:
   (a) 5.0×10^7m **(b)** 3.2×10^{-3} A **(c)** 39×10^{-9} s
3. **(a)** Write down the defining equation for pressure.
   (b) Use the defining equation to arrive at a unit for pressure in terms of base units.
4. State what is meant by a dimensionless quantity.
5. State the two types of uncertainty that may occur in scientific experiments.
6. A quantity is quoted as $(3.7 \pm 0.2) \times 10^{-3}$ m. Calculate the percentage uncertainty in this quantity.
7. The diameter of a wire is measured as 1.2 ± 0.1 mm. The length of the wire is 73.0 ± 0.5 cm.
   Determine
   (a) the volume of the wire
   (b) the uncertainty in the volume.

Sections B1–B2

1. State the difference between a vector quantity and a scalar quantity. Give one example of each.
2. A student runs round a circular track of radius 40 m in 30 s. Calculate the average speed of the student.
3. The graph shows how the speed of a car varies with time from the instant when a driver sees a dog running into the road:

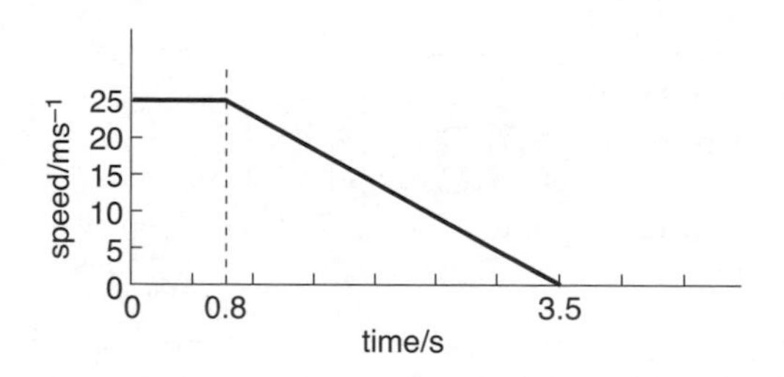

   (a) Determine the distance travelled before the driver applies the brakes.
   (b) Calculate the deceleration produced when the brakes are applied.
4. The graph shows how the speed of an athlete varies with time during the run-up to a long jump.

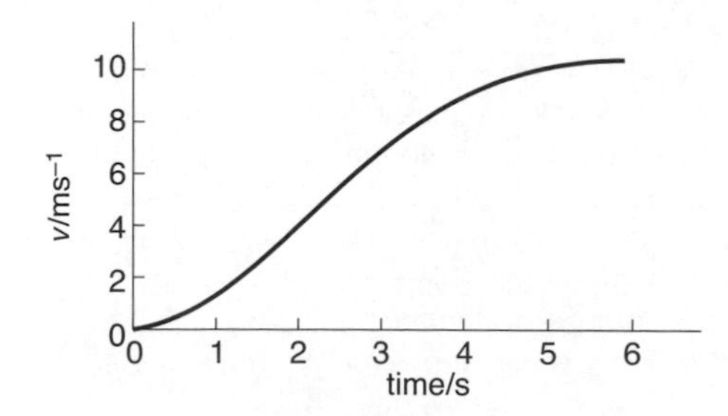

   Determine
   (a) the maximum acceleration of the athlete
   (b) the length of the run-up.
5. **(a)** State the difference between *mass* and *weight*.
   (b) State the unit in which each is measured.
6. Assuming no air resistance, how long would it take a ball to fall a distance of 100 m from rest?
7. A car travelling at 10 m s^{-1} accelerates uniformly at 2.5 s^{-2} for 3.5 s.
   (a) How far does it travel while accelerating?
   (b) Calculate its final speed.

Sections B3–B5

1. Determine the resultant force in each of the following:

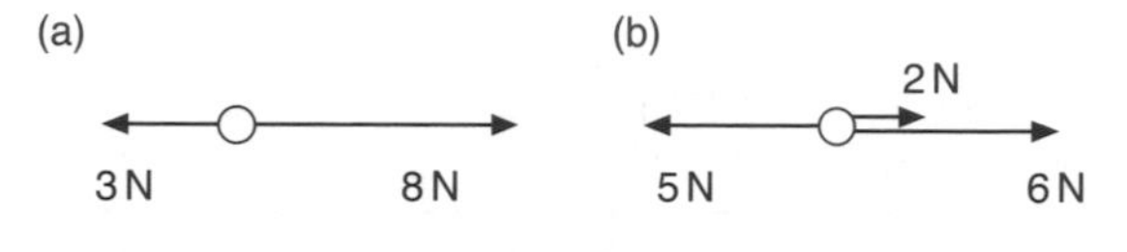

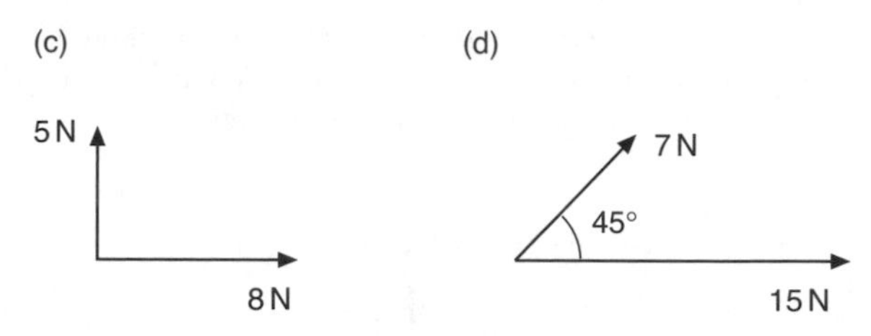

2. State the two conditions that are necessary for a system to be in equilibrium.
3. A force of 250 N is applied to a box at an angle of 35° to the horizontal. Calculate the horizontal and vertical components of this force.
4. The system of forces in the diagram is in equilibrium. Calculate the tension in each of the strings.

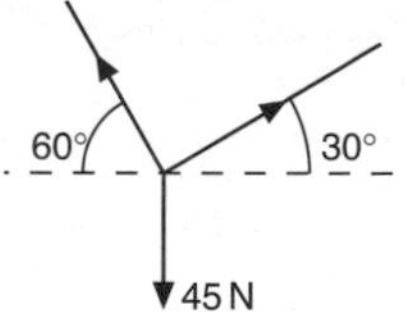

5. The diagram shows one arm of a 'mobile'. Calculate the weight of the ship.

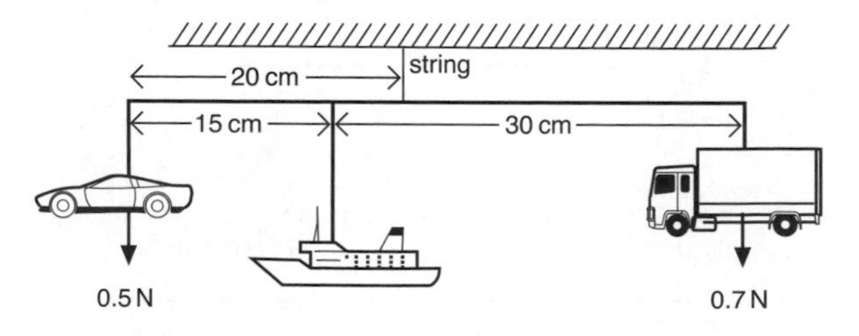

6. State the principle of moments.

7. A person of weight 550 N stands on a plank of weight 150 N in the position shown in the diagram:

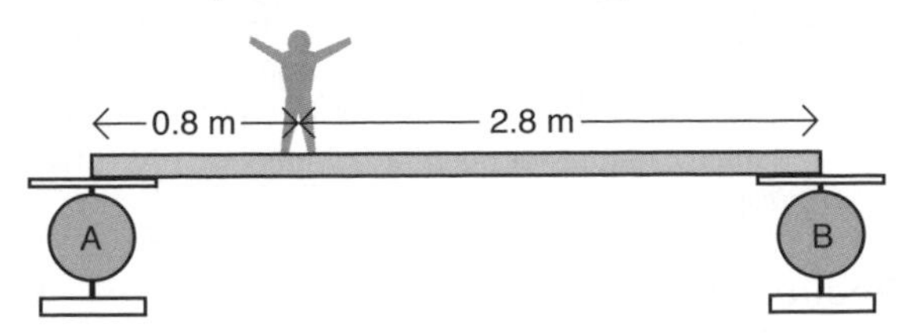

Calculate the reading in N of each of the scales A and B.

8. Explain what is meant by *stable equilibrium*. Explain how you would design a table lamp so that it will be in stable equilibrium.

9. A dart player throws a dart that hits the dartboard at the same height as that from which it was thrown.
The dart was thrown from a distance of 3.0 m from the board and takes 0.25 s to reach the board.
Calculate
(a) the maximum height reached by the dart
(b) the horizontal speed of the dart
(c) the vertical speed of the dart when it reaches the dartboard.

Sections B6–B10

1. **(a)** List 6 different forms of energy.
(b) Explain what is meant by *internal energy*.
(c) State the principle of conservation of energy.

2. Calculate
(a) the kinetic energy of a car of mass 850 kg travelling at 110 km h^{-1}
(b) the change in potential energy when a skier of mass 70 kg skis from a height of 1500 m above sea level to a height of 1210 m above sea level.

3. A weight lifter lifts a mass of 110 kg through a height of 0.5 m in 0.40 s. Calculate the useful power developed during the lift.

4. A van that develops an output power of 3.0 kW moves at a steady velocity of 30 m s^{-1}.
(a) Calculate the total forward force developed between the tyres and the road.
(b) Explain why the air resistance must be exactly equal to the forward force in this case.

5. State what is meant by *viscosity*.

6. A ball bearing of mass 3.1×10^{-5} kg and diameter 2.0 mm falls through glycerol. Glycerol is a liquid of viscosity 1.5 N s m^{-2}. Calculate
(a) the weight of the ball bearing
(b) the terminal velocity of the ball bearing.

Sections B11–B12

1. The kinetic energy of a body changes from 120 J to 40 J. It travels 3.5 m as this change occurs. Calculate the force acting on the body.

2. State the similarities and differences between an *elastic* and an *inelastic* collision.

3. A truck of mass 5000 kg travelling at 30 m s^{-1} collides with the rear of a car of mass 800 kg travelling in the same direction at 20 m s^{-1}.The two vehicles move together after the collision. Calculate the final velocity.

4. A nucleus of mass 4.00×10^{25} kg splits up into two parts. One of mass 0.07 kg moves off at a speed of 1.0×10^{7} m s^{-1}. Calculate the speed of the other particle.

Sections B14–B16

1. **(a)** State what is meant by *angular velocity*.
(b) State the unit in which it is measured.

2. A disc of radius 0.080 m rotates at a rate of 1500 revolutions per minute. Calculate
(a) the period of rotation of the disc
(b) the speed of the rim of the disc.

3. A proton of mass 1.7×10^{-27} kg and speed 2.0×10^{6} m s^{-1} moves in a circular path of radius 0.40 m. Determine the magnitude of the force acting on the proton. State the direction of this force at any instant.

4. Explain why a body moving at constant speed in a circular path is accelerating.

5. How is the centripetal force provided in the following situations?
(a) A cyclist moving round a bend.
(b) A car going round a bend on a very slippery banked track.
(c) An orbiting satellite.

6. Draw two cycles of a displacement–time graph for SHM. Show on your diagram the amplitude and period of the oscillation.

7. A particle is released and moves with SHM of period 5.0 Hz and amplitude 0.040 m.
Calculate
(a) the period of the motion
(b) the maximum acceleration of the particle
(c) the maximum velocity of the particle.

8. Show that the period of a simple pendulum is simple harmonic.

9. A car and its suspension system behaves as a mass–spring system. A car has a mass of 1250 kg. The period of oscillation after going over a bump in the road is 1.5 Hz. Calculate
(a) the total stiffness of the suspension
(b) the stiffness of each of the 4 'springs'.

10. The period of a simple pendulum is 1.2 s. It swings with a maximum displacement of 0.018 m. The mass of the bob is 0.050 kg. Calculate
(a) the total energy of a pendulum
(b) the kinetic energy when the bob is at half its maximum displacement
(c) the potential energy in this position.

11. **(a)** Explain what is meant by a forced oscillation.
(b) Under what conditions does resonance occur?

Sections C1–C5

1. A piano wire made of steel (Young modulus = 2.0×10^{11} Pa) has a length of 1.30 m and a diameter of 2.0 mm. It stretches 2.5 m when it is tightened. Calculate the tension in the wire.

2. Into what classes of materials do the following fall? copper, rubber, nylon, diamond, glass

3. A tendon stretches 3.5 mm when the tension is 14 N. It obeys Hooke's law up to this tension. Calculate the energy stored in the tendon.

4. Draw a diagram showing how the strain varies with stress for rubber. Explain what is happening during the different stages at a molecular level.

5. What makes brittle materials brittle?

6. Marble is strong in compression and has a Young modulus of 5.0×10^{10} Pa. A pillar of length 5.0 m and cross-sectional area 1.5 m^2 supports a load of 20 000 kg. Calculate
(a) the stress in the pillar
(b) the strain in the pillar
(c) the compression of the pillar.

7. Explain what is meant by a *composite material*. Give two examples of such a material.

8. Explain using band theory why some materials are electrical conductors while others are electrical insulators.

Sections D1–D2

1. Draw a sketch to show how the displacement of the medium varies with distance from the source for a sinusoidal wave. Indicate on your sketch the wavelength and the amplitude of the wave.
2. Distinguish between
   (a) refraction and diffraction
   (b) a longitudinal and a transverse mechanical wave
   (c) a progressive and a stationary wave
   (d) a node and an antinode in a stationary wave.
3. Magenta (red and blue) light is incident on a parallel-sided glass block. Draw a diagram to show the passage of the light through the block.
4. Place the following parts of the electromagnetic spectrum in order of increasing frequency.
   ultraviolet, UHF radio, gamma rays, visible light, microwaves
5. Explain what is meant by total internal reflection and state the conditions under which it takes place.
6. A ray of light is incident on an air–glass interface at an angle of incidence of 25.0°. The refractive index of the glass is 1.55. Calculate the angle of refraction.
7. Calculate the critical angle for water of refractive index 1.33.
8. The core of a fibre optic cable has a refractive index of 1.52 and the cladding a refractive index of 1.51. Calculate the critical angle for a ray of light travelling in the core.
9. In an experiment, it is found that it is possible to polarize the radiation used. State and explain the conclusions that can be drawn from this observation.
10. The intensity of the radiation reaching the outer atmosphere of the Earth is 1400 W m^{-1}. Calculate the intensity on Pluto.
    Distance from Earth to Sun = 1.5×10^{11} m.
    Distance from Pluto to Sun = 5.9×10^{12} m.

Section D3

1. **(a)** Explain what is meant by superposition of waves.
   (b) State the conditions necessary for observation of an interference pattern from two sources.
2. Explain the role of diffraction in setting up an experiment to observe interference using light.
3. Calculate the fringe spacing produced using laser light of wavelength 650 nm, at a distance of 2.5 m from two sources that are separated by 1.5 mm.
4. An observer notices that if she walks 1.8 m in the direction shown in the diagram, the sound intensity falls from a maximum to a minimum. The speakers are emitting notes of the same frequency.

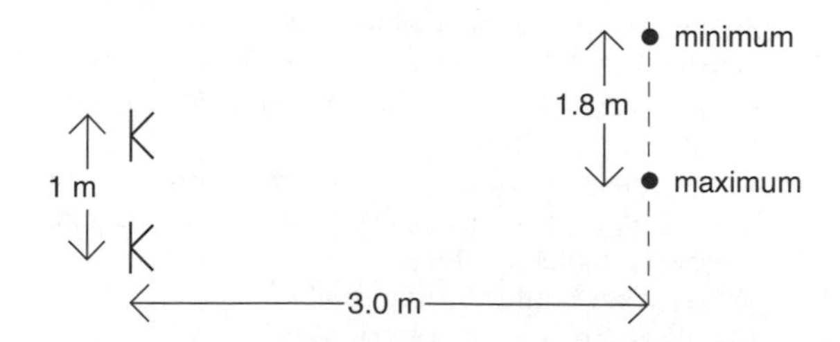

   (a) Explain this observation.
   (b) By means of a scale diagram determine the wavelength of the sound.
5. A diffraction grating has 300 slits per mm. Calculate the angular separation of the second-order maxima for red light (wavelength = 650 nm) and blue light (wavelength = 450 nm).
6. A string resonates at a fundamental frequency of 200 Hz. The length of the string is 0.65 m.
   (a) Sketch a diagram to show this mode of vibration.
   (b) Calculate the frequency and wavelength of the third harmonic vibration for this string.
   (c) By how much would you need to increase the tension to double the fundamental frequency?
7. A pipe that is closed at one end has a length of 0.45 m. It resonates at a fundamental frequency of 190 Hz. Estimate the speed of sound in air.

Section D4

1. Draw a diagram showing the passage of a parallel beam of light through a converging lens. Indicate the position of the principal focus of the lens and the focal length.
2. An object is 0.35 m from a converging lens of focal length 0.12 m. Determine the position of the image formed by the lens.
3. A lens of focal length 0.045 m is used as a magnifying glass. The virtual image is formed 0.25 m from the lens. Determine
   (a) the position of the object
   (b) the magnification produced.

Sections E1–E2

1. The work function of zinc is 4.31 eV.
   (i) Calculate the energy in J of an electron emitted from a zinc target by UV radiation of wavelength 200 nm.
   (ii) Calculate the threshold frequency for zinc.
2. Calculate the frequency of the radiation emitted when an electron in a hydrogen atom undergoes a transition from the level at –1.5 eV to the ground state at –13.6 eV.
3. Why do the energy levels have negative values?
4. A carbon dioxide laser emits infrared radiation of wavelength 10.6 μm. Calculate the difference between the energy levels that give rise to this emission.
5. Calculate the wavelength of a photon of energy 2.2 eV.
6. Calculate the photon energy for radiation of wavelength 1.5×10^{-10} m.
7. An atomic nucleus has a diameter of about 1.0×10^{-14} m. Calculate the minimum energy of an electron that could exist in the nucleus.

Sections F1–F5

1. Calculate the number of electrons passing each point in a circuit when the current is 30 mA.
2. **(a)** Explain what is meant by a supply having an *EMF* of 12 V and an *internal resistance* of 0.5 Ω.
   (b) Calculate the current that flows when a 4.0 Ω resistor is connected between the terminals of the supply.
   (c) Determine the current that would flow if the supply terminals were short circuited.
3. Calculate the total resistance of each of the following combinations of resistors:

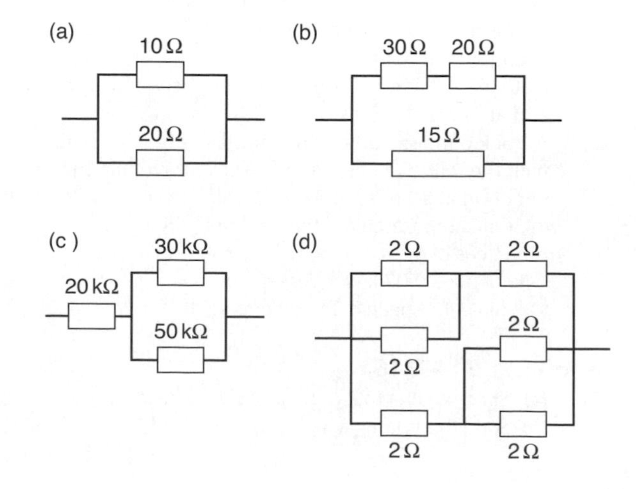

4. A car headlamp is connected to a 12.0 V supply of negligible internal resistance. The lamp then works at its rated power of 24 W. Calculate
   (a) the current in the circuit
   (b) the resistance of the filament when the lamp is working normally.
5. **(a)** Two components of resistance 4.0 Ω are connected in series to a 5.0 V supply of negligible internal resistance. Calculate
   (i) the current in the circuit
   (ii) the power dissipated by each resistor.
   (b) The components are now connected in parallel to the same supply. Calculate the new current and power dissipated for each resistor.
6. Explain in microscopic terms why as temperature increases
   (a) the resistance of a filament lamp increases
   (b) the resistance of a thermistor decreases.
7. A copper wire of diameter 0.80 mm carries a current of 1.5 A. The copper contains 8×10^{28} electrons per m^3. Determine the drift speed of the electrons down the wire.
8. When a potential difference of 9.0 V is placed across a wire there is a current of 1.6 A in the wire. The wire has a length of 1.8 m and an area of cross-section of 0.15 mm^2. Calculate the resistivity of the material from which the wire is made.
9. **(a)** Determine the readings of the meters in the following circuit:

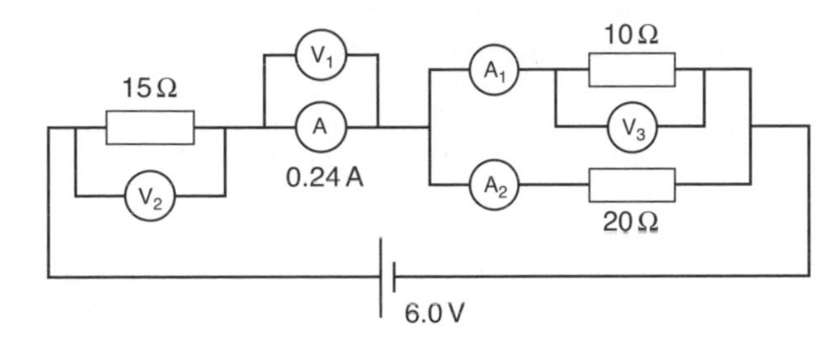

   (b) State the terminal potential difference of the supply.
   (c) Explain why this is lower than the EMF of the supply.
10. **(a)** Determine the magnitude of the output voltage in the circuit below.

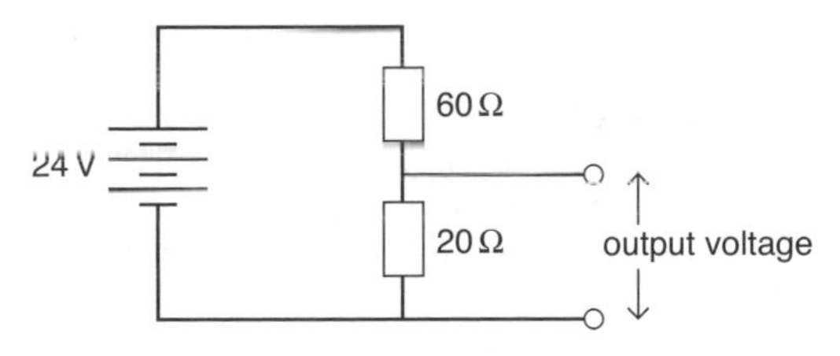

   (b) A 4.0 Ω resistor is connected across the output terminals. Determine the new value of the output voltage.
   (c) Determine the current drawn from the supply in **(a)** and **(b)**.
11. **(a)** Determine the peak value of the voltage in a supply that is producing 12 V RMS.
   (b) What will be the average power dissipated in a 8.0 Ω resistor connected between the terminals of the supply?
   (c) Calculate the maximum power dissipated by the resistor.
12. A 7.0 V RMS supply is produced from 230 V mains.
   (a) Determine the turns ratio of the transformer needed.
   (b) The current in the circuit connected to the 7.0 V supply is 30 mA. Calculate the current drawn from the mains assuming that the transformer
   (i) is 100% efficient
   (ii) has an efficiency of 85%.

Sections G1–G2

1. A 'square wave' is made up from a sinusoidal frequency equal to that of the square wave together with odd harmonics of this frequency. The amplitude of the harmonics decreases as frequency increases. A square wave of frequency 400 Hz is transmitted using a base bandwidth of 50 Hz to 4 kHz. Draw a frequency spectrum for the transmitted wave.
2. Determine the total bandwidth required for a radio station that is transmitting a base bandwidth of 50 to 4000 Hz.
3. A sinusoidal frequency of 1500 Hz is to be transmitted by amplitude-modulation on a carrier of frequency 200 kHz.
   (a) State the frequencies that will be received for this transmission.
   (b) Draw the frequency spectrum of the received signal.
   (c) Sketch the waveform of the signal received.
4. What is meant by *demodulation*?
5. **(a)** State the frequency band that is used for TV transmission.
   (b) Explain why this band is chosen rather than that used for medium-wave radio.
6. How many colour TV channels could be transmitted using the VHF waveband (30–300 MHz)?
7. The strength of a signal in a cable falls from 500 mW to 100 mW. Calculate the loss in dB.
8. **(a)** How often should a signal be sampled to transmit digitally music that has a base bandwidth from 15 Hz to 15 kHz?
   (b) If each sample uses 8 bits determine the minimum bit rate needed to transmit the information.

Sections H1–H5

1. State two similarities and one difference between electric and gravitational fields.
2. Calculate the force between two point charges of magnitude 3.2×10^{-19} C and 1.6×10^{-19} C separated by a distance of 2.5×10^{-10} m.
3. How much energy is required to separate completely the two particles in question **2**?
4. An electron of charge -1.6×10^{-19} C is placed in a uniform field of strength 20 000 V m^{-1}. Calculate the force on the electron.
5. State what is meant by a capacitance of 220 μF.
6. A 100 μF capacitor is charged to a potential difference of 6.0 V. It then discharges through a 2.2 kΩ resistor. How long will it take for the voltage to **(a)** halve **(b)** fall to 2.0 V?
7. A 100 μF capacitor is in series with a 200 μF capacitor. The PD across the combination is 12 V. Calculate
   (a) the total capacitance of the combination
   (b) the energy stored.
8. Mars has a mass of 6.4×10^{23} kg and a radius of 3.4×10^6 m. Determine the acceleration of free fall at its surface.
9. **(a)** Explain why gravitational potentials have negative values.
   (b) Sketch a graph to show how the gravitational potential varies with distance from the centre of the Earth over a range of R to $4R$.
10. The acceleration of free fall at the surface of the Moon is 1/6 that at the surface of the Earth. Calculate the speed and period of a satellite in orbit close to the surface of the Moon. Radius of the Moon is 1.7×10^6 m.

Section H6–H12

1. Sketch the magnetic field of a bar magnet.
2. Explain what is meant by a *neutral point* in a magnetic field.

3. (a) Define *magnetic flux density.*
(b) Calculate the force on a straight wire of length 0.20 m when it carries a current of 2.5 A in a magnetic field of flux density 50 mT.
4. Explain the operation of a brake that depends on electromagnetic induction.
5. The flux density through a 200 turn coil of area 8.5×10^{-4} m^2 changes from 0.03 T to 0.12 T in 15 ms. Calculate the induced EMF in the coil.
6. A 50 mH inductor and a 300 Ω resistor are connected in series. They are connected to a battery of EMF 12 V and negligible internal resistance. Calculate
(i) the initial rate of rise of current
(ii) the maximum current in the circuit.
7. Calculate the magnitude of the force on an electron moving at a speed of 1.5×10^6 m s^{-1} in a magnetic field of flux density 0.20 T.
8. Explain why the path of an electron in a magnetic field is circular while that in an electric field is parabolic.
9. Calculate the speed of an electron when it is accelerated through a potential difference of 2.5 kV.
10. In a mass spectrometer a designer wants to select ions with a velocity of 1.2×10^7 m s^{-1}. The magnetic field available has a flux density of 0.8 mT. Determine the strength of the electric field required to select this velocity.

Sections I1–I4

1. A cook forgets about a saucepan on a cooker. The pan initially contains 0.250 kg of boiling water. The power supplied to the saucepan from the cooker was 0.600 kW. How long will it be before the saucepan boils dry? (Specific latent heat of water = 2.3×10^6 J kg^{-1}.)
2. State the difference between evaporation and boiling.
3. (a) State Boyle's law.
(b) Explain what is meant by an *ideal gas.*
4. The temperature of a fixed mass of an ideal gas changes from 27 °C to 87 °C at constant volume. The original pressure was 0.9×10^5 Pa. Calculate the final pressure.
5. Calculate the final pressure for the same temperature change and initial pressure as in question **2** but allowing the volume of the gas to increase by 50%.
6. Calculate the number of moles of gas in a container of volume 2.5×10^{-3} m^3 containing gas at a pressure of 5×10^4 Pa at a temperature of 300 K.
7. Describe briefly an experiment to estimate the size of a molecule.
8. Calculate the kinetic energy of a molecule at a temperature of 300 K. The Boltzmann constant is 1.38×10^{-23} J K^{-1}.
9. Determine the root mean square speed of the molecules of a gas of density 0.12 kg m^{-3} when the gas pressure is $5 0 \times 10^5$ Pa.

Sections I5–I7

1 (a) Calculate the maximum efficiency of an engine that is taking energy from a source at a temperature of 500 K when the temperature of the sink is 300 K.
(b) When working at maximum efficiency the power from the source is 20 kW. Calculate the power that goes to the sink.
2. (a) A gas expands from 2.5×10^{-3} m^3 to 3.8×10^{-3} m^3 at a constant pressure of 1.2×10^5 Pa. Calculate the work done by the gas.
(b) The change is adiabatic. State and explain what happens to the gas as a result of the change.
3. Calculate the rate at which energy is conducted through the base of a copper saucepan of area 0.063 m^2 and thickness 0.0080 m when the bottom surface is at a temperature of 100.5 °C and the liquid in contact with the base of the saucepan is at 100 °C.

Sections J1–J7

1. Write down the symbols for the following particles including their mass and charge numbers.
Proton, electron, alpha particle, neutron, gamma ray photon
2. Describe the atomic structure of $^{206}_{82}$Pb.
3. State what is meant when we say that two nuclides are isotopes.
4. $^{206}_{82}$Pb is formed when a radioactive atom decays by alpha emission. Determine the proton number and mass number of the atom that has decayed.
5. Complete the nuclear equation for the fusion reaction below and identify the particle X.

$$4^1_1\text{p} \rightarrow \alpha + 2\text{X} + 2\nu_e + \gamma$$

6. A small source emits gamma rays. The count rate is 280 s^{-1} when the GM tube is 10 cm from the source. What would you expect it to be when the tube is
(i) 20 cm from the source
(ii) 15 cm from the source.
7. Thorium-234 has a half-life of 24 days. A sample initially contains 6.0×10^{12} atoms. Calculate
(a) the decay constant of thorium in s^{-1}
(b) the initial activity of the thorium-234
(c) the activity after 12 days.
8. Describe briefly one industrial and one medical use of radioactive materials. State the emission and the half-life of the source that is most appropriate for the application you choose.
9. In the fusion reaction in question **5** the proton has a mass 1.007 276 u, the alpha particle has a mass 4.001 506 u, and X has a mass 0.000 548 580 u.
(i) Determine the mass defect in u.
(ii) Calculate the energy in J liberated by the reaction. (1 u = 931 MeV; 1 MeV = 1.6×10^{-13} J.)
10. State the purpose in a nuclear reactor of **(a)** the moderator and **(b)** the coolant.

Sections K1–K2

1. The rest energy of an electron is 0.511 MeV. Calculate its rest mass. (1 eV = 1.6×10^{-19} J.)
2. Calculate the mass of an electron that is moving at a speed of 0.7*c*.
3. How much energy is released when an electron and a positron (each of mass 9.1×10^{-31} kg) annihilate each other? What form will the energy take?
4. An electron collides inelastically with an atom. Explain what is meant by an inelastic collision and why a collision with an atom may be inelastic.
5. Calculate the radius of a uranium-238 nucleus given that $R_0 = 1.2 \times 10^{-15}$ m.
6. To what class of particles do the following particles belong?
muon; omega-minus; electron-neutrino; neutron; kaon
7. Write down the quark structure of an antiproton.
8. Use conservation laws to determine which of the following are possible reactions:
(a) $\pi^+ \rightarrow \mu^+ + \nu_\mu$
(b) $p^+ + \pi^- \rightarrow \pi^+ + K^-$
9. Draw a Feynman diagram that shows the exchange of a π^+ between a neutron and a proton.

Section L1

1. State and explain two reasons why lenses may not produce a sharp image.
2. **(a)** State the position of the image formed by the objective of a refracting telescope.
   (b) What is the position of the final image formed by the eyepiece when the telescope is in normal adjustment?
3. A refracting telescope is made from two converging lenses of focal length 0.90 m and 0.030 m.
   (a) Determine the angular magnification of the telescope.
   (b) Calculate the distance between the objective and the eyepiece when the telescope is in normal adjustment.
4. A telescope has an aperture of diameter 0.25 m. Two stars viewed through a red filter that transmits light of wavelength 650 nm are just resolved. Determine the angular separation of the stars.

Sections L2–L6

1. A galaxy is travelling at 6 × 106 ms^{-1} and is 4 × 10^{24} m from Earth.
   (a) Calculate a value for the Hubble constant.
   (b) What value does this suggest for the age of the Universe?
2. The wavelength of the Sodium D line (λ = 589 nm) from a distant galaxy is shifted to 591 nm.
   (a) Is the galaxy moving towards us or away from us?
   (b) How fast is it moving relative to the Earth?
3. How far away, in parsecs (pc), is a star with an angle of parallax of:
   (a) 1 arcsec?
   (b) 0.5 arcsecs?
4. Observation of the spectrum from a distant star shows that the light on one side is shifted to longer wavelengths and on the other side to higher wavelengths. What does this tell you about the star?
5. The Sun is a main sequence star. Explain what this means and what will happen to the Sun when it moves off the main sequence.
6. Calculate the radius of a black hole with a mass 4 times the mass of the Sun. (mass of Sun = 1.99 × 10^{30} kg)

Sections M1–M5

1. Ultrasound is used to examine an eyeball. The speed of ultrasound in the eye tissue is 1.5 × 10^3 ms^{-1}.
   (a) Explain why a gel is used between the probe and the eyeball.
   (b) The piezoelectric crystal transmits and receives the ultrasound. It emits short pulses of waves. Explain why there is a time interval between the pulses.
   (c) A pulse from the back of the eyeball takes 3.2 × 10^{-5} s to be detected. Calculate the depth of the eyeball.
2. An X-ray source has an attenuation coefficient of 50 m^{-1} for bone and 6.7 m^{-1} for soft tissue. The initial intensity of the X-rays is 5 × 10^3 Wm^{-2}. Calculate the intensity after passing through
   (a) 2.5 cm of bone
   (b) 9.0 cm of soft tissue.
3. The magnetic field in an MRI scanner is 1.20 T.
   (a) Calculate the Larmor frequency.
   (b) The magnetic field is varied linearly along the length of the patient. What effect does this have on the Larmor frequency? Explain the advantage of this variation in field.

Sections N1

1. **(a)** State the factors that affect the moment of inertia of a body.
   (b) A torque of 3.0 N m is applied to a disc of moment of inertia 0.5 kg m^2. The disc is free to rotate and starts from rest. Calculate
   (i) the angular acceleration
   (ii) the angular speed after 5 seconds
   (iii) the number of revolutions during the first 5 s.

Self-assessment answers

Sections A1–A2

1. **(a)** 3.5 × 10^6 Ω **(b)** 2.20 × 10^{-10} F
   (c) 1.5 × 10^{-3} m s^{-1} **(d)** 2.5 × 10^{-5} m^2
2. **(a)** 50 Mm **(b)** 3.2 mA **(c)** 39 ns
3. **(a)** $P = F/A$ **(b)** kg m^{-1} s^{-2}
4. A quantity that has no units, e.g. a ratio of two lengths.
5. Random and systematic errors or uncertainties.
6. 5.4%
7. **(a)** 8.3 × 10^7 m^3
   (b) 2.3% or absolute uncertainty = ±0.2 × 10^3 m^3

Sections B1–B2

1. Vector quantity has magnitude (size) and direction. Scalar quantity has magnitude only.
2. 6.3 m s^{-1}
3. **(a)** 20 m **(b)** 9.3 m s^{-2}
4. **(a)** 5.2 m s^{-2} **(b)** ~26 m
5. Mass is the quantity of matter in a body (scalar). Weight is the force of attraction of the Earth on the mass (vector).
6. 4.52 s
7. **(a)** 50.3 m **(b)** 18.8 m s^{-1}

Sections B3–B5

1. **(a)** 5 N to the right **(b)** 3 N to the right
   (c) 9.4 N at 32° to horizontal
   (d) 20.5 N at 15° to horizontal (by scale drawing)
2. No resultant moment (torque) and no resultant force.
3. Horizontal force = 205 N; vertical force = 143 N
4. T_1 = 39 N, T_2 = 22 N, 45 N
5. 2.2 N
6. See B4.
7. A reads 503 N; B reads 197 N.
8. If given a displacement it returns to its original position. Wide base; heavy base (load it with metal) so that centre of gravity is low.
9. **(a)** 0.31 m **(b)** 12 ms^{-1}
   (c) 2.5 ms^{-1}

Sections B6–B10

1. **(a)** See B2. **(b)** Total PE and KE of all the particles in the object. **(c)** See B2.
2. **(a)** 4.0 × 10^5 J **(b)** 2 × 10^5 J
3. 1350 W
4. **(a)** 100 N
   (b) When velocity is constant there is no acceleration. This is the case only when there is no resultant force.
5. (See B9.) Viscosity is a frictional force that opposes relative motion between layers of fluid or between a solid and a fluid.

6. (a) 3.8×10^{-4} N (b) 6.7×10^{-3} m s^{-1}

Sections B11–B12

1. 22.9 N
2. Total energy and momentum is conserved in both types. In an elastic collision kinetic energy is conserved.
3. 28.6 m s^{-1} in original direction of motion.
4. 1.7×10^5 m s^{-1}

Sections B14–B16

1. (a) The angle swept out by a radial line per second.
   (b) rad s^{-1} (radian per second)
2. (a) 0.040 s
   (b) 12.6 m s^{-1}
3. 1.7×10^{-14} N toward the centre of the circular orbit.
4. Velocity is a vector, so when the velocity changes, as it does in circular motion, the velocity changes. There is acceleration because this is rate of change of velocity.
5. (a) This is equal to the friction between the cyclist and the road.
   (b) The horizontal component of the normal reaction to the track.
   (c) For an Earth satellite, the gravitational force between the Earth and the satellite.
6. 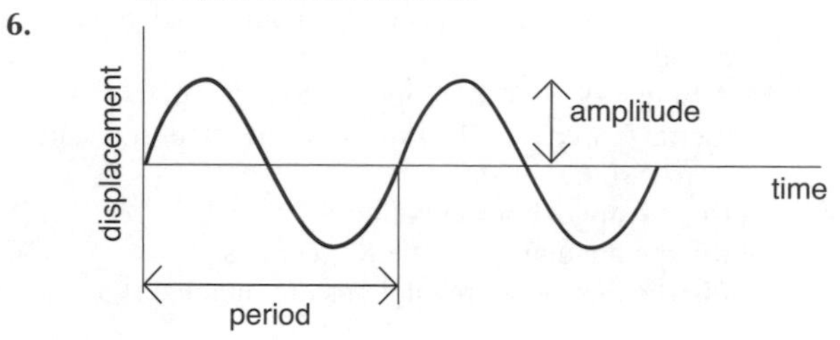

7. (a) 0.20 s (b) 39.5 m s^{-2} (c) 1.25 m s^{-1}
8. See B12.
9. (a) 111 kN m^{-1}
   (b) 28 kN m^{-1}
10. (a) 2.22×10^{-4} J ($A\omega$ = maximum v; $\omega = 2\pi/T$; $KE = \frac{1}{2}mv^2$)
    (b) 1.67×10^{-4} J (c) 0.055×10^{-4} J
    (At half maximum displacement PE is $\frac{1}{4}$ maximum and KE is $\frac{3}{4}$ maximum.)
11. See B13.

Sections C1–C5

1. 1210 N
2. Polycrystalline; polymer; polymer; crystalline; amorphous
3. 0.025 J
4. See H4.
5. See H4 (stretching glass).
6. (a) 1.3×10^5 Pa
   (b) 2.6×10^{-6}
   (c) 1.3×10^{-5} m
7. See H6.
8. See H6.

Sections D1–D2

1. 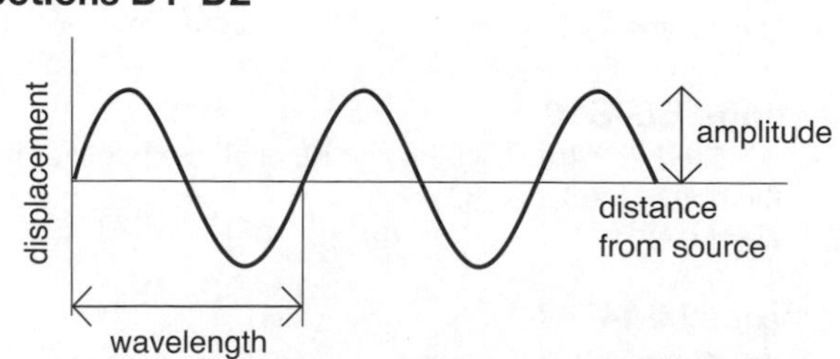

2 (a) Refraction: deviation of path of a wave at an interface between two media.
   Diffraction: deviation when wave meets an obstruction such as a gap.
   (b) Longitudinal: particles of medium oscillate in direction of energy transfer.
   Transverse: particles of medium oscillate perpendicular to direction of energy transfer.
   (c) Progressive: point of maximum displacement moves in direction of transfer of energy.
   Stationary: points of maximum and minimum amplitude are fixed.
   (d) Node: point of zero amplitude in a stationary wave.
   Antinode: point of maximum amplitude in a stationary wave.
3. 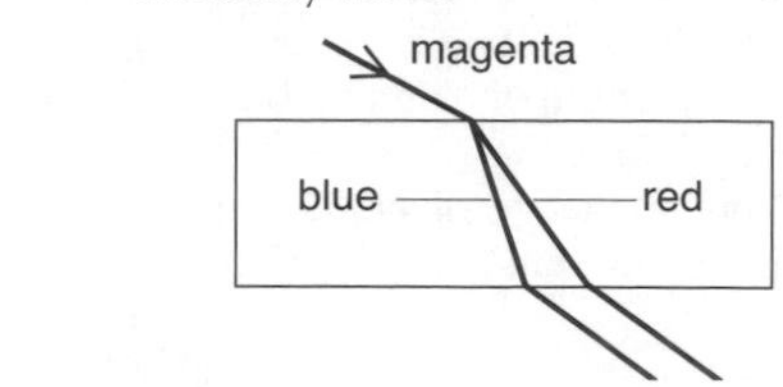

4. UHF radio, microwaves, visible light, ultraviolet, gamma rays.
5. See D2.
6. 15.8°
7. 40°
8. 83°
9. The radiation is in the form of a transverse wave. Unpolarized transverse waves contain vibrations in all planes perpendicular to the direction of propagation of energy. The process of polarization selects one of these planes. Longitudinal waves contain vibrations in the direction of energy so no further selection of oscillations is possible.
10. 9.0 W m^{-2} (using inverse square law)

Section D3

1. (a) See D3.
   (b) Two sources must be coherent (same frequency and constant phase difference) and have similar amplitudes to produce good contrast between interference maxima and minima.
2. Diffraction spreads light form a single slit so that it illuminates two slits to produce coherent sources. Diffraction spreads out the light from two slits so that the beams overlap and produce interference patterns.
3. 1.1 mm
4. (a) At the maximum position the waves are arriving in phase and interfering constructively. By moving 0.8 m the wave from the lower speaker travels half a wavelength further than the other so that the waves are antiphase and interfere destructively.
   (b) 0.95 m
5. 7.3°
6. (a)
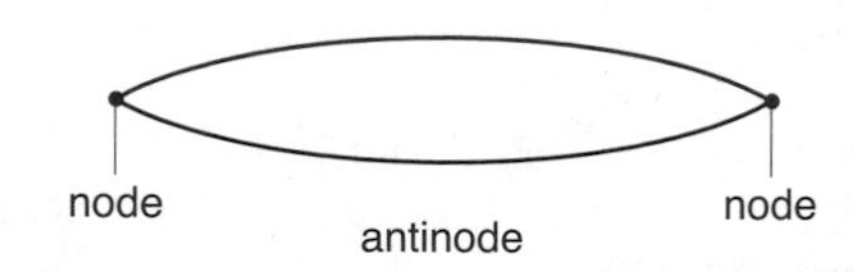

   (b) 600 Hz; 0.22 m
   (c) four times
7. 342 m s^{-1}

Section D4

1. 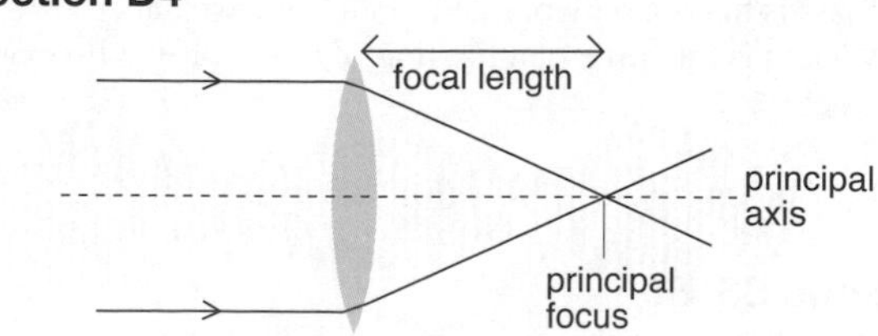

2. 0.18 m from the lens on opposite side to the object; a real image
3. **(a)** 0.038 m from the lens (on the same side as the image)
   (b) 6.6 ×

Sections E1–E2

1. **(i)** 3.0×10^{-19} J
   (ii) 1.0×10^{15} Hz
2. 2.9×10^{15} Hz
3. A free electron has zero energy. Energy has to be put in to raise an electron to zero energy so the electron energies inside atoms are negative.
4. 1.87×10^{-20} J
5. 5.3×10^{14} Hz
6. 1.32×10^{-15} J
7. 10×10^{-12} J
   (This is much greater than beta particle energies and suggests that there are no electrons in the nucleus.)

Sections F1–F5

1. 5.6×10^{18} electrons
2. **(a)** An EMF of 12 V means that 12 J of work is done when 1 coulomb of charge passes around the complete circuit. The resistance of the components inside the supply (chemicals, wires, etc.) is 0.5 Ω.
   (b) 2.7 A
   (c) 24 A
3. **(a)** 6.7 Ω **(b)** 11.5 Ω **(c)** 38.8 kΩ
   (d) 1.5 Ω
4. **(a)** 2.0 A **(b)** 6.0 Ω
5. **(a) (i)** 0.63 A **(ii)** 1.56 W
   (b) (i) 2.5A **(ii)** 6.3 W
6. **(a)** More collisions of electrons with the lattice ions
   (b) More electrons liberated; has a greater effect than the increased collisions of electrons with lattice ions
7. 2.3×10^{-4} m s^{-1}
8. 4.7×10^{-7} Ω m
9. **(a)** $V_1 = 0$; $V_2 = 3.6$ V; $V_3 = 16$ V; $A_1 = 0.16$ A; $A_2 = 0.08$ A
   (b) 5.2 V
   (c) Some EMF is 'lost' in producing current in the internal resistance of the supply.
10. **(a)** 3.0 V **(b)** 1.26 V
   (c) 0.3 A in **(a)**; 0.38 A in **(b)**
11. **(a)** 17 V **(b)** 18 W **(c)** 36 W
12. **(a)** 32.9:1 step down (more turns on coil connected to 240 V supply)
   (b) (i) 0.88 mA **(ii)** 1.04 mA

Sections G1–G2

1.

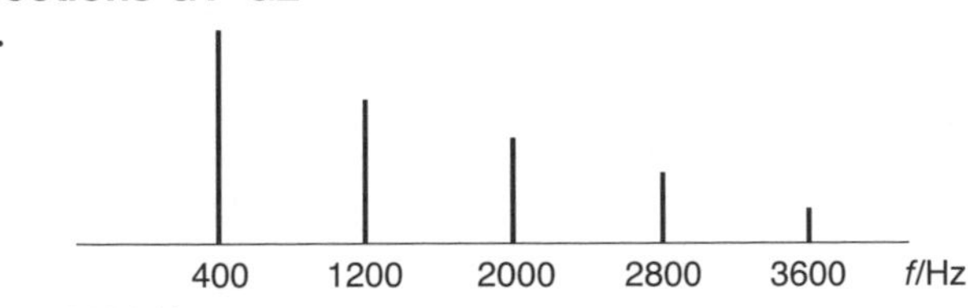

2. 8000 Hz
3. **(a)** 198.5 kHz, 200 kHz, 201.5 kHz

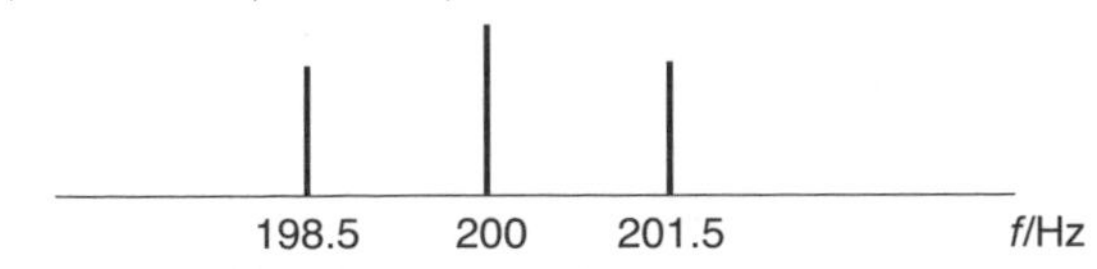

(b)

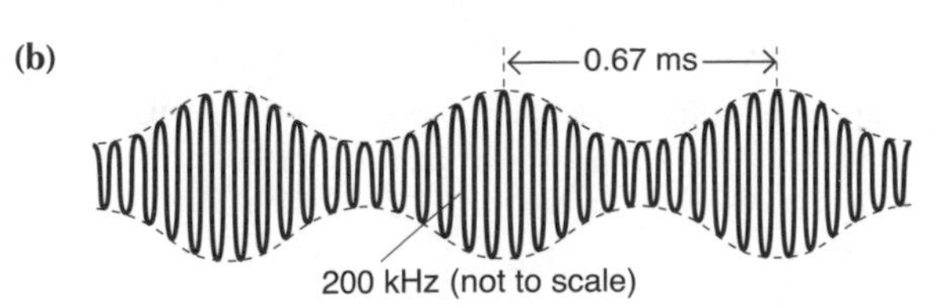

4. See H10.
5. **(a)** UHF
   (b) Large bandwidth (24 MHz for colour TV) required Medium-wave bandwidth has frequency range of less than 3 MHz
6. 10
7. –7 dB
8. **(a)** 30 kHz
   (b) 240 kHz

Sections H1–H5

1. Similarities: for point charges or masses
   Both obey inverse square law for variation of force with distance: both obey inverse *r* law for variation of potential with distance; both act at a distance with no requirement for a material medium.
   Difference: electric fields may produce attraction or repulsion; gravitational fields only produce attraction.
2. 7.3×10^{-9} N
3. 1.8×10^{-18} J
4. 3.2×10^{-19} N in opposite direction to the direction of the field.
5. 200 μC of charge is 'stored' for each volt of potential difference between the terminals of the capacitor.
6. **(a)** 0.15 s **(b)** 0.24 s
7. **(a)** 67 μF **(b)** 4.8×10^{-3} J
8. 3.7 m s^{-2}
9. **(a)** Potential at infinity (a very long way from the mass) is zero. Energy has to be supplied to a mass to move it to infinity. Since energy is added to get to zero potential the potential is negative.
   (b)

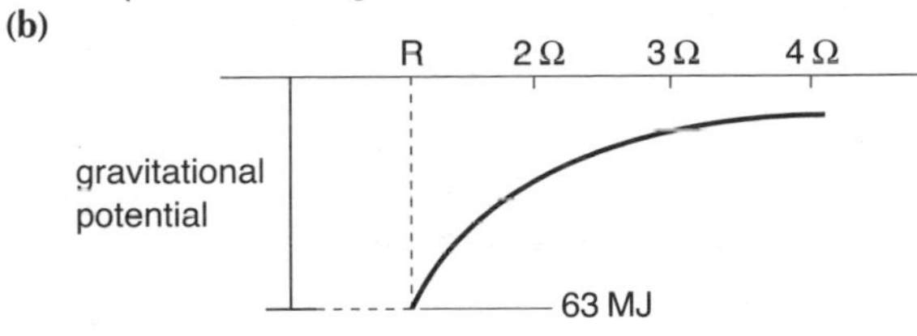

10. Speed = 1670 m s^{-1}; period = 106 min

Sections H6–H12

1. See H6
2. A point where there is no magnetic effect. The resultant magnetic flux density is zero.
3. **(a)** The strength of the magnetic field when force of 1 N is exerted on a wire of length 1 m when it caries a current of 1 A.
   (b) 25 mN
4. See E7
5. 1.0 V
6. (a) 240 A s^{-1} **(b)** 40 mA
7. 4.8×10^{-14} N
8. When an electron moves at right angles to a magnetic field the magnetic force is always perpendicular to the direction of motion of the electron and has constant magnitude. This is the condition for circular motion. An electron that starts at right angles to the field is accelerated in the direction of the field but has constant velocity perpendicular to the field. This leads to a parabolic path.
9. 3.0×10^{7} m s^{-1}
10. 9600 V m^{-1}

Sections I1–I4

1. 958 s
2. See I2
3. See I3
4. 1.08×10^{5} Pa

5. 0.72 × 10^5 Pa
6. 0.050 mol
7. See I4
8. 6.2 × 10^{-21}
9. 350 m s^{-1}

Sections I5–I7

1. **(a)** 40% **(b)** 12 kW
2. **(a)** 156 J
   (b) Work done in expansion so temperature falls
3. **10.** 1560 W

Sections J1–J7

1. **(a)** $^{1}_{1}p$ $^{0}_{-1}e$ $^{4}_{2}\alpha$ $^{0}_{1}n$ $^{0}_{0}\gamma$
2. 82 protons and 124 neutrons in the nucleus with 82 electrons 'orbiting' the nucleus
3. The nuclides have the same number of protons in the nucleus but a different number of neutrons.
4. Proton number 84; mass (nucleon) number 210
5. $4^{1}_{1}p \rightarrow {}^{4}_{2}\alpha + 2^{0}_{1}X + 2^{0}_{0}\nu_e + {}^{0}_{0}\gamma$; X is a positron.
6. **(i)** 70 **(ii)** 124
7. **(a)** 8.0 × 10^{-6} s
   (b) 4.8 × 10^8 Bq
   (c) 3.4 × 10^8 Bq
8. See G2
9. **(i)** 0.026 501u
   (ii) 3.95 × 10^{-12} J
10. See G3

Sections K1–K2

1. 9.1 × 10^{-31} kg
2. 1.27 × 10^{-30} kg
3. 1.6 × 10^{-13} J; gamma radiation
4. Inelastic collision – no loss of KE in the collision
   Ionization and excitation (see G7)
5. 7.4 × 10^{-15} m
6. lepton; baryon; lepton; baryon; meson
   (baryons and mesons are hadrons)
7. $\overline{u}\overline{u}\overline{d}$
8. **(a)** This is possible. Lepton and charge numbers are conserved.
   (b) This is not possible. Charge is conserved but baryon number is not. p is a baryon. All the others are mesons.
9. 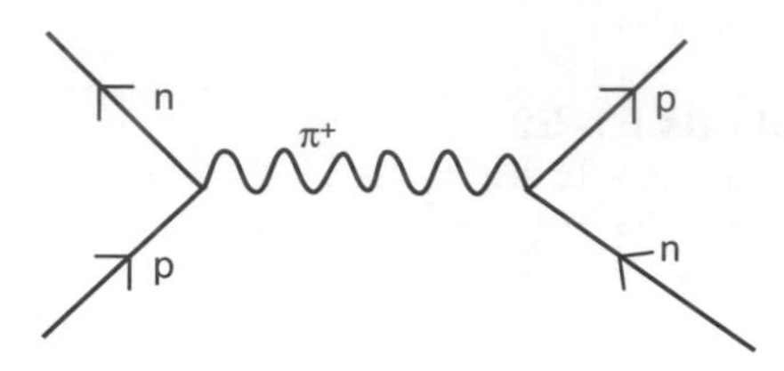

Section L1

1. Chromatic aberration and spherical aberration (see C4).
2. **(a)** At the principal focus of the objective lens
   (b) At 'infinity'
3. **(a)** 30 × **(b)** 0.93 m
4. 2.6 × 10^{-6} rad or 1.5 × $10^{-4\circ}$

Sections L2–L6

1. **(a)** 1.5 × 10^{-18} s^{-1}
   (b) 6.67 × 10^{17} s (2.1 × 10^{10} years)
2. **(a)** away **(b)** 1.02 × 10^6 ms^{-1}
3. **(a)** 1 pc **(b)** 2 pc
4. It is rotating (one side towards the Earth, the other side away from it).
5. See L4
6. 1.2 × 10^4 m

Sections M1–M5

1. **(a)** Reduces the amount reflected and maximizes transmitted energy.
   (b) To allow reflected waves to return and be detected without interference with transmitted waves.
   (c) 2.4 cm
2. **(a)** 1.43 × 10^3 Wm^{-2} **(b)** 2.74 × 10^3 Wm^{-2}
3. **(a)** 51.0 MHz
   (b) It will change the Larmor frequency. The location of the pulse along the body is known from the value of the Larmor frequency.

Section N1

1. **(a)** mass and distribution of mass
   (b) **(i)** 6 rad s^{-2} **(ii)** 30 rad s^{-1} **(iii)** 11.9 revs (75 rad)

Physical data

Physical quantity	*Symbol*	*Value*
speed of light in a vacuum	c	2.998 × 10^8 m s^{-1}
permittivity of free space	ε_0	8.854 × 10^{-12} F m^{-1}
permeability of free space	μ_0	4π × 10^{-7} H m^{-1}
proton rest mass	m_p	1.673 × 10^{-27} kg
neutron rest mass	m_n	1.675 × 10^{-27} kg
electron rest mass	m_e	9.110 × 10^{-31} kg
proton charge	e	1.602 × 10^{-19} C
electron charge	$-e$	−1.602 × 10^{-19} C
	(minus sign often omitted)	
specific charge: electron	e/m_e	1.759 × 10^{11} C kg^{-1}
Planck constant	h	6.626 × 10^{-34} J s
gravitational constant	G	6.672 × 10^{-11} N m^2 kg^{-2}
Avogadro constant	N_A	6.022 × 10^{23} mol^{-1}
universal molar gas constant	R	8.314 J K^{-1} mol^{-1}
Boltzmann constant	k	1.381 × 10^{-23} J K^{-1}
absolute zero		0 K, −273.15 °C
standard atmospheric pressure		1.013 × 10^5 Pa

Physical quantity	*Symbol*	*Value*
kilowatt hour	kW h	3.600 × 10^6 J
electronvolt	eV	1.602 × 10^{-19} J
unified atomic mass unit	u	1.661 × 10^{-27} kg
	(energy equivalent: 931.5 MeV)	
acceleration of free fall (mean, at Earth's surface)	g, g_0	9.807 m s^{-2}
mass of Earth		5.976 × 10^{24} kg
mass of Sun		1.989 × 10^{30} kg
mass of Moon		7.350 × 10^{22} kg
equatorial radius of Earth		6.378 × 10^6 m
mean distance of Earth from Sun		1.496 × 10^{11} m
mean distance of Moon from Earth		3.844 × 10^8 m
solar constant		1.352 × 10^3 W m^{-2}
astronomical unit	AU	1.496 × 10^{11} m
parsec	pc	3.086 × 10^{16} m
light year	ly	9.461 × 10^{15} m

Equations

You will be given a data sheet in the exam. The equations you need will be provided in the question or on the data sheet. You should print a copy of the actual data sheet (it is in the specification) to familiarize yourself with all the equations and the symbols used. Make sure you know what each equation means and how to use it.

This is a list of some of the equations that you will need to understand and use at AS and/or A level.

speed $= \frac{\text{distance}}{\text{time}}$ $\quad v = \frac{\Delta s}{\Delta t}$

acceleration $= \frac{\text{change in velocity}}{\text{time}}$ $\quad a = \frac{\Delta v}{\Delta t}$

force = mass × acceleration $\quad F = ma$

density $= \frac{\text{mass}}{\text{volume}}$ $\quad \rho = \frac{m}{V}$

work done = force × distance moved in direction of force $\quad \Delta W = F\Delta s$

power $= \frac{\text{energy transferred}}{\text{time taken}} = \frac{\text{work done}}{\text{time taken}}$ $\quad P = \frac{E}{t}$

weight = mass × acceleration of free fall
= mass × gravitational strength $\quad$ weight $= mg$

kinetic energy $= \frac{1}{2}$ mass × velocity^2 $\quad E_k = \frac{1}{2}mv^2$

change in gravitational potential energy = mass × gravitational field strength × change in height $\quad \Delta E_p = mg\Delta h$

current $= \frac{\text{charge}}{\text{time}}$ $\quad I = \frac{\Delta Q}{\Delta t}$

potential difference = current × resistance $\quad V = IR$
electrical power = potential difference × current $\quad P = VI$

energy = potential difference × current × time $\quad E = VIt$

potential difference $= \frac{\text{energy transferred}}{\text{time}}$ $\quad V = \frac{E}{Q}$

resistance $= \frac{\text{resistivity × length}}{\text{cross-sectional area}}$ $\quad R = \frac{\rho l}{A}$

% efficiency $= \frac{\text{energy (or power) output}}{\text{total energy (or power) input}}$ × 100%

pressure $= \frac{\text{force}}{\text{area}}$ $\quad P = \frac{F}{A}$

momentum = mass × velocity $\quad p = mv$

pressure × volume = number of moles × molar gas constant × absolute temperature $\quad pV = nRT$

pressure × volume = number of molecules or atoms × Boltzmann constant × absolute temperature $\quad pV = NkT$

wave speed = frequency × wavelength $\quad v = f\lambda$

Period $= \frac{1}{\text{frequency}}$ $\quad T = \frac{1}{f}$

centripetal force $= \frac{\text{mass × velocity}^2}{\text{radius}}$ $\quad F = \frac{mv^2}{r}$

force between point charges $= \frac{\text{charge (1) × charge (2)}}{4\pi \text{ × permittivity × (distance between charges)}^2}$ $\quad F = \frac{Q_1Q_2}{4\pi\varepsilon r^2}$

force between point masses $= \frac{\text{universal gravitational constant × mass (1) × mass (2)}}{\text{(distance between masses)}^2}$ $\quad F = \frac{Gm_1m_2}{r^2}$

capacitance $= \frac{\text{charge stored}}{\text{potential difference}}$ $\quad C = \frac{Q}{V}$

For a transformer:

$\frac{\text{potential difference across coil (1)}}{\text{potential difference across coil (2)}} = \frac{\text{number of turns on coil (1)}}{\text{number of turns on coil (2)}}$ $\quad \frac{V_1}{V_2} = \frac{N_1}{N_2}$

moment of a force	$T = Fx$
torque of a couple	$T = Fx$
power	$P = Fv$
stress	$\sigma = \frac{F}{A}$
strain	$\varepsilon = \frac{\Delta l}{l}$
Hooke's Law	$F = K\Delta x$
the Young modulus	$E = \frac{\sigma}{\varepsilon}$
electric current	$I = \frac{\Delta Q}{\Delta t}$
Elastic energy	$E = \frac{1}{2}Fx$
power	$P = I^2R$
	$P = \frac{V^2}{R}$
current	$I = nqvA$
resistors in series	$R = R_1 + R_2 + \ldots$
resistors in parallel	$\frac{1}{R} = \frac{1}{R_1} + \frac{1}{R_2} + \ldots$
force on a current-carrying conductor	$F = BIl\sin\theta$
photon energy	$E = hf$
photoelectric effect	$hf = \Phi + \frac{1}{2}mv_{max}{}^2$
de Broglie equation	$\lambda = \frac{h}{mv} = \frac{h}{p}$
refractive index	$n = \frac{c_i}{c_r}$
	$n = \frac{\sin i}{\sin r}$
critical angle	$\sin\theta_c = \frac{n_2}{n_1}$
diffraction grating	$d\sin\theta = n\lambda$
double-slit interference	$\lambda = \frac{ax}{D}$
force	$F = \frac{\Delta p}{\Delta t}$
kinetic energy	$E = \frac{p^2}{2m}$
Stoke's law	$F = 6\pi\eta rv$
kinematic equations	$v - u + at$
	$s = ut + \frac{1}{2}at^2$
	$v^2 = u^2 + 2as$
angular velocity	$w = \frac{v}{r} = \frac{2p}{T}$
centripetal acceleration	$a = \frac{v^2}{r} = r\omega^2$
simple harmonic motion	$a = -(2\pi f)^2x$
	$x = A\sin 2\pi ft$
	$x = A\cos 2\pi ft$
mass-spring system	$T = 2\pi\sqrt{\frac{m}{K}}$
Simple pendulum	$T = 2\pi\sqrt{\frac{l}{g}}$
gravitational field of strength	$g = \frac{F}{m}$
	$g = \frac{GM}{r^2}$
electric field strength	$E = \frac{F}{Q}$
	$E = \frac{Q}{4\pi\varepsilon_0 r^2}$
	$E = \frac{V}{d}$
capacitance	$C = \frac{Q}{V}$
energy of charged capacitor	$W = \frac{1}{2}QV$
time constant of CR circuit	$\tau = CR$
capacitor discharge	$Q = Q_o e^{\frac{-t}{RC}}$
force on moving charged particle	$F = Bqv\sin\theta$
radius of path of charged particle	$r = \frac{p}{Bq}$
magnetic flux	$\Phi = BA$
induced EMF	$\xi = {}^-N\frac{\Delta\Phi}{\Delta t}$
thermal energy change	$\Delta Q = mc\Delta\theta$
mass–energy	$\Delta E = \Delta mc^2$
radioactivity	$A = \lambda N$
radioactive decay	$N = N_0e^{-\lambda t}$
half life	$\lambda = \ln 2/t_{\frac{1}{2}}$
apparent magnitude	$m = -2.5\log I + \text{constant}$
apparent/absolute magnitude	$m - M = 5\log\left(\frac{r}{10}\right)$
Hubble's law	$v = H_0d$
age of Universe	$t \approx \frac{1}{H_0}$
Doppler formula	$\frac{\Delta\lambda}{\lambda} = \frac{v}{c}$
lens formula	$\text{power} = \frac{1}{f} = \frac{1}{u} + \frac{1}{v}$
attenuation	$I = I_0e^{-\mu x}$
radiant flux	$F = \frac{L}{4\pi d^2}$
Stefan-Boltzmann Law	$L = \delta T^4\eta = 4\pi r^2\delta T^4$
Wein's Law	$\lambda_{max}T = 2.898 \times 10^{-3}\ mK$

Index